FISH, FISHING AND FISHERIES

TOWARDS SUSTAINABLE FISHERIES MANAGEMENT

A PERSPECTIVE OF FISHING TECHNOLOGY WEAKNESSES AND OPPORTUNITIES WITH A FOCUS ON THE MEDITERRANEAN FISHERIES

Fish, Fishing and Fisheries

Additional books in this series can be found on Nova's website under the Series tab.

Additional e-books in this series can be found on Nova's website under the e-book tab.

FISH, FISHING AND FISHERIES

TOWARDS SUSTAINABLE FISHERIES MANAGEMENT

A PERSPECTIVE OF FISHING TECHNOLOGY WEAKNESSES AND OPPORTUNITIES WITH A FOCUS ON THE MEDITERRANEAN FISHERIES

ALESSANDRO LUCCHETTI
ANTONELLO SALA
SUZAN KHOLEIF
AND
EMILIO NOTTI

nova publishers
New York

Library of Congress Cataloging-in-Publication Data

Towards sustainable fisheries management : a perspective of fishing technology weaknesses and opportunities with a focus on the Mediterranean fisheries / editors: Alessandro Lucchetti, Antonello Sala, Suzan Kholeif, Emilio Notti (Consiglio Nazionale delle Ricerche (CNR) – Istituto di Scienze Marine (ISMAR), Ancona, Italy. National Institute of Oceanography and Fisheries (NIOF), CAIRO, Egypt.
pages cm. -- (Fish, fishing and fisheries)
Includes index.
ISBN 978-1-63463-698-8 (hardcover)
1. Fishery technology--Mediterranean Sea. 2. Sustainable fisheries--Mediterranean Sea. 3. Fishery management--Mediterranean Sea. I. Lucchetti, Alessandro, 1973-
SH213.4.T69 2015
639.209163'8--dc23

2014047571

Published by Nova Science Publishers, Inc. † New York

CONTENTS

PREFACE

In the last fifty years the human harvesting of marine resources became more efficient as the understanding of the habits and behaviours of the various species available in the sea gradually increased. Thus technologies naturally developed over time and fishing systems became more sophisticated and increasingly adapted to catch the most desirable species. This evolution is still in progress, since technological development goes on. These technical advances in fishing gears have generally led to more efficient economic fishing operations and improved access to resources. On the other hand, general awareness of environmental problems due to the exploitation of fisheries resources has increased also. Therefore, in order to find ways to preserve fishery resources and marine environment, fishing gears and their impact should be further investigated and less impacting techniques should be developed. In this regard, fishing technologists are responsible to provide fishery managers as well as fishermen and stakeholders with useful advice and technical solutions for mitigating adverse impact of fishing gears.

The current book is mainly conceived to transfer the expertise on the fishing technology to Mediterranean fisheries scientists and end-users such as fisheries managers, administrators, controllers etc. The Code of Conduct for Responsible Fisheries of the FAO (1995) clearly states that "*States and sub-regional or regional fisheries management organizations and arrangements should promote, to the extent practicable, the development and use of selective, environmentally safe and cost effective gear and techniques*". Taking this statement into account, the book aims at providing information on sustainable fishing technology. This involves three main issues that will be tackled in the book, namely selectivity, impact and cost efficiency.

The book contents are based on the requirements of the ecosystem-approach to fisheries management. It is structured to provide information on fishing gear technology, selectivity, bycatch-impact reduction and fishing vessel performance and optimization. The information included might be used to learn how to assess the impacts of different fishing gears and find out technical solutions to mitigate these impacts and to address scientific studies.

The contents of this book might be considered as a useful tool when preparing management measures as well as for the establishment, enforcement, or enhance of fishing rules. Furthermore, technical advice on the improvement of energy efficiency in fishing activity has been also considered.

The first section implies an overview of the fishing gears used in the Mediterranean professional fishing, with a focus on the technological features. The impacts of different fishing gears, alternative solutions as well as methods and strategies that mitigate adverse environmental impacts have been also described; in this section, the problem of bycatch and available solutions (Bycatch Reducer Devices) are also tackled.

The second section deals with the technical measures of fishing gears affecting size and species selectivity in order to highlight eventual problems and suggest new technical details that can be inserted in the Regulation and/or Legislation. The selectivity theory, the methods for the study of trawl selectivity, the setup of a selectivity study have been presented. In addition to these, a section on trawl net design, how to understand and to produce a trawl net drawing, is included. In this section, an overview of existing trawl net designs and riggings has been carried out in order to give a complete picture of the situation in the Mediterranean. The methods to reduce catch of juveniles or unwanted sizes, the factors affecting the selectivity of a fishing gear as well as the influence of gear design on the selection properties have been described. Finally, some simple and practical rules on the measurement and/or estimation of net performances have been provided.

The final part is mainly focused on the fishing vessel technology for non-specialists. Vessels may deploy a variable amount of fishing gear (number of towed gears, different net dimension etc.). Thus, the use of different gears often requires different tactics as well as a different use of energy. The fishing vessels technology section addresses the need to reduce energy consumption and the associated running costs. Increasing engine efficiency while reducing polluting emissions and fuel consumption is also an expectation shared by all scientists and managers involved in the marine engines. The main aim of this section is to cover the basic relevant concepts of naval architecture and ship design through a general overview of naval architecture principles, definitions and principal dimensions of a vessel, types of vessels, hydrostatics, resistance and powering, layout of fishing vessels.

This book was conceived in the framework of Ritmare flagship project (National Research Programme funded by the Italian Ministry of University and Research) and FORCE Project – Fisheries & Aquaculture-Oriented Research Capacity in Egypt – (INCO. 2011- 6.2, Collaborative project - Contract no. 295004), financed by the Commission of the European Communities. It does not necessarily reflect the European Commission's views and in no way anticipates its future policy

Chapter 1

RATIONALE

Millions of people in the world depend upon fish and fishing activities for their livelihoods. Fish represents an important source of food and fisheries (catching, processing, marketing of fish stocks) are a fundamental source of employment and income, especially in some regions of the Mediterranean Sea, where small scale fisheries are generally community and family based, in a society rooted in traditions, local knowledge, culture and both men and women play an important role. As stated in the "Code of Conduct for Responsible Fisheries" (FAO, 1995), that provides a collection of principles, goals and elements for the reasonable management of fishing activities, "*Fisheries, including aquaculture, provide a vital source of food, employment, recreation, trade and economic wellbeing for people throughout the world, both for present and future generations [...]*".

However, fishing implies the harvesting of marine organisms directly from their natural environment, therefore it requires a clear pattern of principles followed by rules to conserve and manage fish resources and their habitats (*fisheries [...] should therefore be conducted in a responsible manner;* FAO, 1995). In particular, States should prevent overfishing and implement management measures to ensure that fishing effort is commensurate with the productive capacity of the fishery resources and their sustainable utilization. Presently, over 70% of World's fish species are either fully exploited or depleted and fishing fleets are generally oversized in relation to the resources available.

Growing concern on the environmental consequences of fishing impact (fishing mortality, catch of juveniles, habitat damages etc.) and more lately on the ecosystem effects of fishing, has stimulated intense research activity in the last 30 years (Lindeboom & de Groot, 1998; Jennings and Kaiser, 1998; Collie et al., 2000). It has been recognized that it is not just fish populations that need protection but also the environment that supports them. Some fishing techniques may cause habitat destruction or may disrupt food webs by targeting only certain species. In this sense, trawl fishery is the most problematic among fishing techniques causing serious impacts on benthic habitats and ecosystems, vulnerable species, fish populations and marine communities structure as a whole. A number of recent studies have established that intensive fishing with poorly selective gears can have dramatic immediate effects on the benthos (Watling and Norse, 1998) and particularly where complex biogenic structures are found at the seabed surface (Hall,1999). Moreover, certain fishing gears are usually not selective enough so that the catch of undersized specimens as well as the catch of unwanted species is generally high.

Marine environment can be affected by fishing activities in several ways (ICES, 2003), including:

1. Direct and indirect effects on the abundance and composition of the target fish populations (e.g., number of fish in a particular stock, their size distribution and their genetic makeup);
2. Direct and indirect effects on the abundance and composition of the non-target populations (bycatch) including under-sized individuals of commercial species, non-commercial species, benthic invertebrates and species which deserve conservation interest such as seabirds, marine mammals, turtles, elasmobranches;
3. Alterations to the seabed and associated habitats of bottom living fish and invertebrates;
4. Changes in the structure, function and integrity of ecosystems, including effects on the food webs and multispecies predators-preys relationships (Kaiser and Spencer, 1994; Jennings and Kaiser, 1998; Ramsay et al., 1998; Coll et al., 2007).

Therefore, the development and diffusion of technical innovations in fishing gears, together with prohibition of certain gears and limitations to the technical properties and size of some others, are considered as having a paramount importance in order to increase selectivity and reduce fishing impact on the benthic ecosystems. The main goal of a reasonable management of fisheries is that fishing and fish processing are conducted in such a way that the environmental impact is minimized, the waste reduced, the biodiversity of fish caught and aquatic ecosystems preserved, the abundance of stock maintained within safe limits, the population structure preserved, the fish quality guaranteed.

1.1. Fisheries Management in the Mediterranean

The Mediterranean Sea is a semi-enclosed sea with a total surface area of approximately 2.5 million square kilometres, representing 0.8% of the total World marine surface. Overall, Mediterranean Sea is poor in nutrients and the primary production is generally low, with the supply of nutrients gradually declining from the Straits of Gibraltar towards the Eastern basin. Due to its geographical location being placed at a relatively narrow range of latitudes (from 30°N to 46°N), the Mediterranean shows a marked seasonal cycle. Water masses are stratified in summer, but the deep-water (below 400 m) temperature is 13±0.3°C throughout the year.

However, the Mediterranean is considered as one of the most important marine regions in the world for its particularity and high biological diversity which supports a locally significant and diversified fishing sector (Nike Bianchi and Morri, 2000; Coll et al., 2010). Mediterranean fisheries is characterized by a wide range of fishing gears and methods some of which are unique, and by a great variety of target species resulting in a mixed composition of catches and a highly diversified fleet structure.

Fishing is a major economic activity in the Mediterranean in terms of employment and jobs, revenues and food supply. Besides, fishing products are an important ingredient of the Mediterranean diet and fisheries have always been one of the main food providers of the Mediterranean. In this respect, fisheries are extremely important from a socioeconomic point of view, especially in certain coastal communities where fishing is the only available work opportunity to earn living.

The Mediterranean fishing fleet is characterized by a large number of small and relatively old vessels (80% are less than 12 m in length) that are widely distributed along the coastline with fragmented landing sites and sales and a high degree of polyvalent techniques (Lleonart and Maynou, 2003). These vessels are active in the small-scale sector, fishing from the coastal area to 100 meters depth, or in the lagoons where biodiversity is really high. This sector has a very high economic value targeting high unit-value species such as Sea breams, Sea bass, Sole, Bluefin tuna, Albacore, Swordfish, Cephalopods, Crustaceans and many others. With technical progress, some small-scale vessels are now fishing beyond 100 meters depth. Most living marine resources exploited by fishing activities live within continental shelves at shallow depths (less than 400 m); however in recent years deep-water fisheries have strongly improved the operational depth, targeting shrimps, Hakes, sharks and other marine organisms at a depth greater than 400 m.

In the Mediterranean, artisanal fisheries are important and complex in view of the mixed-species catches, the diversity and seasonality of gears used, and the nature of many stocks that are shared between countries. Artisanal fishery involves a consistent fleet made up of small vessels with low tonnage. It is characterized by the use of several different and diversified fishing gears and by its targeting of a large variety of species (Quignard and Farrugio, 1982; Fredj et al., 1992). Artisanal fishing activities exhibit great variations from one area to another not only in terms of different biological and environmental conditions, but also of social, economic and historical contexts in which fishermen live as well as tradition (Farrugio et al., 1993). Furthermore, artisanal fisheries play a key role in providing an important source of food and contribute to poverty reduction and sustainable development in several areas of the world (FAO, 2005). In spite of its importance, knowledge of artisanal fishery's catches, fleet structure and their characteristics is scarce (Vitale et al., 2011).

Larger vessels using towed gears or purse seines, account for a large proportion of the total catches, targeting economically very important demersal (Red mullets, Hakes, Soles, Sparids, Red shrimps, Norway lobsters, Cephalopods) as well as pelagic species (Bluefin tunas, Swordfish, Albacore, Anchovies, Sardines, Mackerel etc.). Moreover the Mediterranean region is subjected to a very high level of anthropogenic pressure, since certain fishing grounds are exploited by vessels coming from more than 20 countries, including also non-Mediterranean countries, sharing the same pool of resources (Lucchetti and Sala, 2010).

Managing the Mediterranean fisheries is a complicated process in view of the large number of fishing fleets using a wide array of fishing gears in the same shared fishing areas. Therefore, it is necessary to establish general rules to ensure the conservation of fisheries resources.

Moreover, the traditional consumption of small fish in many countries often encourages the commercialization and targeting of the juvenile fraction of some species, sometimes in violation of laws regarding Minimum Landing Sizes (MLS). Consequently, demersal populations are generally overfished especially by bottom trawls, shallow coastal areas (generally within the three-mile coastal limit or on bottom depths less than 50 m, depending on the country) are illegally trawled, and small illegal mesh sizes are used. Due to this situation, both pelagic and demersal stocks living in the open sea as well as inshore waters have shown a clear decline in the last 20 years. Several studies show that most demersal stocks are fully exploited or overexploited, and even some pelagic stocks show overexploitation trends in the northern areas (Farrugio et al., 1993; Aldebert and Recasens, 1996; Sardà et al., 1998; Papaconstantinou and Farrugio, 2000; Lleonart and Maynou, 2003; Bas et al., 2003; Palomera et al., 2007). Overall, the Scientific, Technical and Economic Committee for Fisheries (STECF) highlighted the recent status of resources in the Mediterranean: 32 out of 36 stocks were assessed as overfished (89%), while only 4 stocks were considered sustainably exploited consistent with high long term yields. All demersal fish stocks (100% of 18 stocks) were found overexploited (Cardinale et al., 2010; Colloca et al., 2013).

The Mediterranean, with the exception of the Adriatic Sea, has a narrow continental shelf such that about 90% of fishing in the region is considered as coastal. For these reasons, fisheries management in the Mediterranean tends to focus on coastal fisheries.

From a technical point of view there are three main methods to manage a fishery; limiting the catch (TACs and quotas), limiting the effort (limiting licenses, capacity, fishing time, fishing areas, engine power, boat dimension etc.), and limiting the technical characteristics of the gear or their quantity. At present, the management of fishing stocks in the Mediterranean is mainly based on the definition of input controls: limiting fishing effort (Lleonart and Maynou, 2003), defining closed areas and seasons, Minimum Landing Sizes and Minimum Mesh Sizes (MMS). When considering the particular conditions of fisheries in the Mediterranean, it is clear why the traditional conservation policy has been differently applied than in other areas. The biological, social and economic characteristics of the Mediterranean fisheries implied, for several years, the need to establish a specific management framework for this basin.

The characteristic features of Mediterranean fisheries can be briefly summarized as:

a) High diversity in terms of catch composition: the commercial catches are composed of even more than 100 species (multispecies fishery);
b) High diversity of the fisheries sector (vessels dimensions, gears, technology etc.);
c) A short average life span, permitting rapid renewal of resources;
d) A narrow continental shelf (excluding the Adriatic Sea);
e) The absence of large single stocks especially in the demersal regime compared to those which inhabit other seas (Farrugio et al. 1993);
f) The high marketability of small fish in many countries encouraging the targeting of the juvenile fraction of some species often in violation of laws regarding minimum sizes and meshes.

1.2. Bycatch Issue

Most fishing operations, whether they employ towed or fixed gears, catch organisms that are not the primary target of the fisher. Thus, there is growing evidence on the negative impacts of commercial fisheries in catching unwanted marine organisms and for the habitat damage that occurs in commercial fisheries (Watling and Norse, 1998; AMCC, 2001; POC, 2003).

Unwanted organisms or non-target species are commonly referred to as "bycatch". The bycatch is that part of the catch consisting of living creatures that are caught unintentionally by fishing gear and are retained (commercial part of the bycatch) or discarded (discards).

Over the last four decades, a growing concern has been expressed by fishery managers and conservation/environmental working groups on the consequences of bycatch and discards; in particular the biological overfishing and the alteration of marine ecosystems structure deserve a paramount attention. Thus, the bycatch and its management have become an important issue in global fisheries management and the search for solutions to bycatch problems has intensified.

Bycatch is a problem of primary importance mainly in bottom trawling, where most of the rejected catch might not survive when dumped to the sea. Moreover, whether this fishing mortality affects large quantities of juveniles, bottom trawling may seriously affect the recruitment and the future biomass of stocks. Furthermore, large quantities of discarded species can affect the habitat at ecosystem level, influencing the prey-predators relationships and the marine food webs.

In this regard, the FAO Code of Conduct for Responsible Fishery (1995) highlights that the aim of a reasonable fisheries management is the long-term conservation of quality, diversity and availability of fishery resources; in particular the management measures should ensure not only the conservation of target species but also of non-target species belonging to the same ecosystem. This is what is actually known as the Ecosystem Approach to Fisheries (EAF). EAF is one of the most important concepts recently introduced for the comprehensive management of fisheries sector within the goal of sustainable use of marine resources. Prudent responsible fisheries management should take account of the profound interactions between fisheries and their supporting ecosystem. This ecosystem-based approach to fisheries management is necessary to understand the overall impacts of fishing (Dayton et al., 1995; Pitcher and Pauly, 1998). The purpose of EAF is to setup, conduct, develop and manage fisheries in such a way that addresses the multiple needs and desires of societies without threatening the options for future generations to benefit from the full ranges of goods and services provided by marine ecosystems.

The definition of an EAF can consequently be ascertained as "*An ecosystem approach to fisheries that strives to balance diverse societal objectives by taking into account the knowledge and uncertainties about biotic, abiotic and human components of ecosystems and their interactions and applying an integrated approach to fisheries within ecologically meaningful boundaries* (FAO, 2003)".

The ecological impacts of discards remain widely unquantified in the Mediterranean. In terms of catching and then discarding unwanted marine organisms, this can be separated into two issues: the effect on the marine ecosystem and the loss of potential income (and human food source). The loss of potential income is incurred through the loss of potential growth and contribution to stock replacement when small/juvenile commercial fish are caught and killed (growth overfishing). The extensive discarding of commercial species results in substantial forgone potential yield, and for depleted stocks, discards are considered a serious impediment to rebuilding. Every year, around one-quarter of global fisheries catches are discarded at sea, dead or dying. Bycatch and discards are therefore amongst the main concerns in fisheries management. In the EU, data from discard monitoring programs have indicated that discarding is most severe in demersal and beam trawl fisheries; in some fisheries, up to 90% of catches can be discarded, including non-commercial species, benthic organisms etc.

1.2.1. Target Catch Definitions

This term refers to catch of a species, a particular size or sex, or an assemblage of species that is primarily sought in a fishery such as shrimp in a shrimp fishery or a sole in a sole fishery etc. The definition of target catch within a fishery is not static as the mix of species targeted and caught in a multispecies fishery may change over time. Thus, the target species can change with fishing gears, area, season and daytime in a mixed fishery.

The Mediterranean fisheries are considered quite peculiar in this issue since the bycatch is relevant in multispecies fishing grounds where several species coexist together at the same time and place. Most of fishes and invertebrates caught in the Mediterranean attain different sizes when fully grown, have different shapes and behaviours, and finally have different Minimum Landing Size (MLS) making it hard to target only one of them in their shared habitat.

Therefore, in the Mediterranean, it is very difficult to define a single "target species". Probably it is possible to identify some key species such as the Norway lobster (*Nephrops norvegicus*), the European hake (*Merluccius merluccius*), the cuttlefish (*Sepia officinalis*), the common sole (*Solea solea*), the pink shrimp (*Parapenaeus longirostris*), the Anchovy (*Engraulis encrasicolus*), the Bluefin tuna (*Thunnus thynnus*) etc which vary with fishery, areas and seasons.

1.2.2. Bycatch Definitions

On the basis of terminology proposed in the "Bycatch workshop" held in Newport (USA) (McCaughran, 1992), the bycatch is "*the part of the catch that is not targeted and it consists of discarded catch plus incidental catch*" (Figure 1). A portion of the bycatch may have commercial value including non-target species that are retained (incidental catch) (Marasco and Terry 1982). In other cases, the bycatch could be unused and rejected at sea as a result of economic, legal or personal consideration.

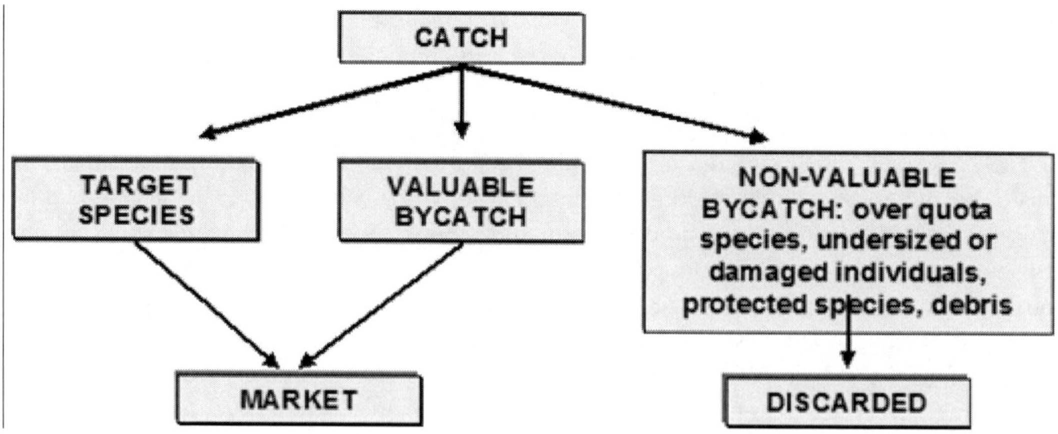

Figure 1. Catch composition of a fishing gear.

Several different definitions of bycatch have been stated in the last 20 years. Murawski (1992) noted that "*the use of the term bycatch adds considerable confusion to a topic that is already complex to both scientists and managers*". The term is relatively imprecise because it constitutes a value judgment and may be inaccurate when used over any extended time to describe an element within a multi-species catch. Moreover, many species and types of fish that were previously considered to be bycatch are now included in a broader range of target species; i.e., *yesterday's bycatch may be today's target species*. Some national bycatch definitions have been put but there is inconsistency between them. The United States of America Magnuson–Stevens Act (MSA-2007) defines bycatch as: "*fish which are harvested in a fishery, but which are not sold or kept for personal use, and includes economic discards and regulatory discards. Such term does not include fish released alive under a recreational catch and release fishery management program*". This effectively means that bycatch is equivalent to discards under the Act. This definition has been reinterpreted (NMFS, 1998) in the United States within the context of specific fishery management plans and publications: "*Bycatch: discarded catch of any living marine resource plus retained incidental catch and unobserved mortality due to a direct encounter with fishing gear*". In Australia's bycatch management, the term "bycatch" refers to "*all non-targeted catch including by-product, discards and the biomass that does not reach the deck of the fishing vessel but is affected by interaction with the fishing gear*".

Bycatch is sometimes defined as "*discarded catch plus incidental catch*" where incidental catch is considered to be retained non-target species. However, if target species (e.g., juveniles) are discarded this may cause some confusion as target species are not usually considered to be "bycatch" (Kelleher, 2005).

On the basis of a recent FAO paper (Kelleher, 2005), "*bycatch is the total catch of non-target animals. Discards are not a subset of bycatch since the target species is often discarded*". Thus, older definitions of bycatch include retained incidental species and discards while in more recent definitions the retained incidental catch is excluded. Some of the changes to bycatch definitions (e.g., Australia and USA) reflect the change from single species to multispecies management and the recognition that ghost fishing mortality and encounter mortality may be high for some gear types in some situations.

The term bycatch is confusing. While there is a broad-based agreement that all species retained for sale should be part of an effective fisheries management plan and that discards should be minimized, the use of the term bycatch in this context is extremely problematic. But bycatch can also include protected species because fishing gear can accidentally capture species such as marine mammals, sea turtles, sea bird etc.

Finally, it is possible to define the bycatch as living creatures that are caught unintentionally by a fishing gear. Sometimes bycatch may be kept or sold (for example, a trawler who catches a legal-sized hakes while fishing for shrimps may keep the catch to eat at home or to sell, if he has a proper license. In other cases bycatch cannot be used (for example, if it is undersized or a protected species, or not marketable) and must be thrown back. This returned bycatch is called discard. Thus, the catch obtained by a fishing gear can be simply divided as follows in Figure 2.

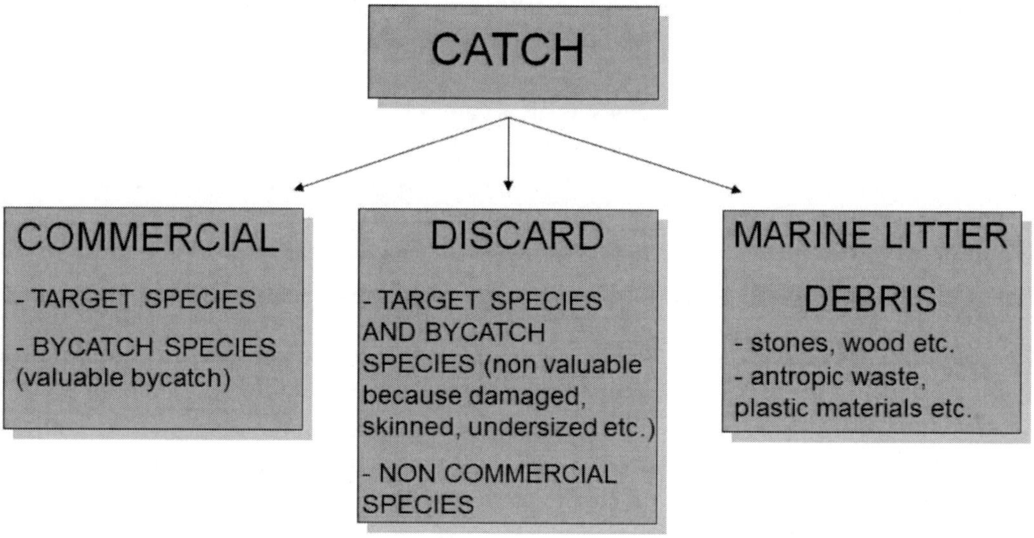

Figure 2. Definition of different catch components.

1.2.3. Discard Definitions

The term "discard" has distinctly different meanings in different jurisdictions, resulting in frequent confusion between "discard" and "bycatch". European Commission (EC) paper (European Commission, 2002) defines discards as commercial species retained by a fishing gear that have been brought on board a fishing vessel and are thrown back into the sea, effectively ignoring non-commercial species. The Nordic workshop (Nordic Council of Ministers, 2003) defined "discard" as "*The proportion of the catch which is taken on board, or brought to the surface by the vessel and which is subsequently thrown back to sea, dead or dying, or likely to die*".

On the basis of Workshop on Discard Sampling Methodology and Raising Procedures (ICES 2004), the "*Discard is the proportion of the catch, which has been taken on board the catching vessel and which is subsequently thrown back to sea dying or dead. In addition, fish that has been taken to the surface during the capture process and handled in such a way that they are likely to die after release, including 'slipping' is also considered as discards*".

The definition of discards is reported in the FAO Fisheries Report No. 547 (FAO, 1996) as "*Discards, or discarded catch is that portion of the total organic material of animal origin in the catch, which is thrown away, or dumped at sea for whatever reason. It does not include plant materials and post-harvest waste such as offal. The discards may be dead or alive*". Discarding is considered to be an act of volition requiring a decision by fishers to reject or dump the fish. Discards include slipped fish; i.e., fish caught in a net and subsequently released into the sea without being brought on board the vessel. Discards do not include dead corals or empty shells.

Recently the new Regulation EU 1380/2013 has defined the discard as "*catches that are returned to the sea*".

There are number of reasons for discarding. They are mainly legislative and/or economic or just belong to fishing practices. In many instances, the individual reasons operate simultaneously. In a number of cases, legislation makes discarding mandatory affecting both juvenile and adult specimens. For example, several legislations in the Mediterranean require discarding of fish, molluscs and crustaceans for specimens smaller than defined minimum landing sizes, catches in excess of defined percentage compositions of catches taken with nets of a given mesh size, and catches in excess of quotas. National legislation may also imply or mandate discarding. In some countries, national quotas of fish are allocated at the national level, into sectorial quotas or ITQs. In practice, when a vessel or sector of the fleet has taken its quota and if no further quota becomes available, catches in excess of a sectorial quota or ITQ may not be landed and should be discarded even if a national quota is not exhausted. In other cases, international commissions may fix quota for some stock as in the case of the International Commission for the Conservation of Atlantic Tunas (ICCAT) that annually fixes the quota for the Bluefin tuna in the Mediterranean. The reasons for discarding may be listed as follows:

- Fish of the wrong size;
- Fish economically too small (no markets);
- Undersized fish (landing legally prohibited);
- Fish of wrong species, Low-value species (e.g., in coastal mixed-species gillnet fishery);
- Species quota reached or vessel not licensed to land that species. This can cause problems in mixed species fishery regulated by quotas;
- Fish damaged or spoiled;
- Damages caused by gear or fishing operation;
- Fish spoiled by waste substances and by other marine pollution (debris);
- Fish spoiled by too long soaking time (e.g., in gillnet fishery);
- Fish spoiled too fast because of too high temperature conditions;
- Damages caused by depredation in gear. For example, the cetaceans depredation into pelagic trawl nets (this is really rare in the Mediterranean Sea) or passive nets;
- Discard of less valuable fish to save space for more valuable catch due to lack of space on board (this is not the case of Mediterranean);
- Retention of large fish in preference to smaller size groups even though the latter may be greater than the minimum landing size;
- Preservation of individual quotas for later use;
- Capture of large numbers of small fish due to inadequate gear selectivity; and
- A fish specimen caught in a wrong season; e.g., a closed season may or must be discarded.

Information on discards in Mediterranean trawl fisheries confirm the magnitude of the problem though they vary considerably in amount and composition depending on region, boat size, season, bottom type and depth of the exploited ground. High discard levels have been reported for several Mediterranean trawl fisheries. One of the main parts of the catch is sometimes represented by the *debris* or marine litter. At its broadest definition, marine litter includes any manufactured or processed solid materials resulting from human activities that are discarded at sea or reach the sea through waterways of domestic and industrial outfalls (National Academy of Sciences, 1975). It also comprises plastic garbage washed or blown from land into the sea, fishing gear abandoned by recreational and commercial fishers, and solid non-biodegradable materials (such as plastics) disposed of by ships at sea (bags, bottles, strapping bands, sheeting, synthetic ropes, synthetic fishing nets, floats, fiberglass, piping, insulation, paints and adhesives). It also includes leachates from landfill sites, and spoils from dredging and trawling operations (Kiessling, 2003). However, this definition does not describe in depth the debris. The term debris also includes non-human materials but also rocks, sand, mud, organic materials such as dead shells, dead coral and plant materials (seaweed).

Marine litter currently poses a growing threat to the marine and coastal environment conservation. Number of studies carried out in different regions clearly indicate that the situation with regard to marine litter is continuously getting worse, mainly because most marine litter consists of material that degrades slowly.

Recently several research projects and surveys have been funded in order to assess the magnitude of this problem. Most of these projects usually categorize marine litter according to the material (wood, plastic, glass etc.) an type of items (box and containers, clothes, fishing nets etc.). The last frontier is the study of environmental impact of micro-plastics present in marine litter.

Chapter 2

FISHING GEAR TECHNOLOGY

More than 20 main fishing techniques are commonly used in the Mediterranean fisheries. Fishing gears are commonly classified in two main categories: passive and active gears. Active or mobile fishing gears are towed by the propulsion system of the fishing vessel or by means of winches with the fishing vessel, either anchored or slowly under way, including in particular towed nets, dredges and seines. These gears can be towed by one or two vessels on the bottom or at any depth in mid-water or on the surface to catch demersal as well as pelagic species.

Passive gears include fishing gears such as fixed nets (gillnets, entangling nets, trammel nets), traps, pots and drifting nets the use of which does not require an active movement by the fishing vessel. Passive gears may be anchored to the bottom or drifting on the surface or in the water column. Demersal and pelagic marine organisms can be caught or entrapped by a passive gear because they are unable to see the gear (passive nets) or because voluntary enter the gear (traps) or are attracted towards a bait (longline). The catch is harvested just once the gear is hauled back after a period of setting (12-48 hours depending on the gear type). Passive gears are commonly used by small vessels of the artisanal fisheries.

The definition of passive and active gears is based on the relative behaviour of the target species and the fishing gear. With passive gears, the capture of fish is generally based on the movement of target species towards the gear, while capture by active gears is based on a aimed chase of the target species and combined with different ways of catching it. Fishing gears are worldwide classified on the basis of the International Standard Statistical Classification of Fishing Gears (ISSCFG, Nédélec and Prado, 1990):

2.1. BOTTOM TRAWLING

Trawling is the operation of towing a net to catch marine organisms. Trawling encompasses a wide range of gear designs and methods of operation.

Three categories of bottom trawls can be distinguished based on how their horizontal opening is maintained: bottom otter trawls, beam trawls and bottom pair trawls (not used in the Mediterranean Sea). In bottom otter trawl, the vertical opening is maintained by weights on the bottom and floats on the top of the net, in beam trawl the vertical and horizontal opening is defined by a rigid frame.

Table 1. International Standard Statistical Classification of Fishing Gears with a selection of gears used in the Mediterranean

Gear Categories	Standard abbreviations	ISSCFG
SURROUNDING NETS		01.0.0
With purse lines (purse seines)	PS	01.1.0
Without purse lines	LA	01.2.0
SEINE NETS		02.0.0
Beach seines	SB	02.1.0
Boat or vessel seines	SV	02.2.0
TRAWL NETS		03.0.0
Bottom trawls	TB	03.1.0
Beam trawls	TBB	03.1.1
Bottom otter trawls	OTB	03.1.2
Bottom pair trawls	PTB	03.1.3
Midwater trawls	TM	03.2.0
Midwater otter trawls	OTM	03.2.1
Midwater pair trawls	PTM	03.2.2
Otter twin trawls	OTT	03.3.0
DREDGES		04.0.0
Boat dredges	DRB	04.1.0
Hand dredges	DRH	04.2.0
LIFT NETS		05.0.0
Portable lift nets	LNP	05.1.0
Boat-operated lift nets	LNB	05.2.0
Shore-operated stationary lift nets	LNS	05.3.0
FALLING GEAR		06.0.0
Cast nets	FCN	06.1.0
Falling gear (not specified)	FG	06.9.0
GILLNETS AND ENTANGLING NETS		07.0.0
Set gillnets (anchored)	GNS	07.1.0
Driftnets	GND	07.2.0
Encircling gillnets	GNC	07.3.0
Fixed gillnets (on stakes)	GNF	07.4.0
Trammel nets	GTR	07.5.0
Combined gillnets-trammel nets	GTN	07.6.0
TRAPS		08.0.0
Stationary uncovered pound nets	FPN	08.1.0
Pots	FPO	08.2.0
Fyke nets	FYK	08.3.0
Stow nets	FSN	08.4.0
Barrier, fences, weirs, etc.	FWR	08.5.0
HOOKS AND LINES		09.0.0
Handlines and pole-lines (hand operated)	LHP	09.1.0
Handlines and pole-lines (mechanised)	LHM	09.2.0
Set longlines	LLS	09.3.0
Drifting longlines	LLD	09.4.0
Trolling lines	LTL	09.6.0
GRAPPLING AND WOUNDING		10.0.0
Harpoons	HAR	10.1.0
HARVESTING MACHINES		11.0.0
Pumps	HMP	11.1.0
Mechanised dredges	HMD	11.2.0
MISCELLANEOUS	MIS	20.0.0

In general trawl nets, large or small, used on the bottom or in mid-water, are basically funnel shaped, with side netting panels extended forward to form wings which guide the fish harvested into the center of the net entrance (net mouth). The net may consist of two panels (upper and lower panels), or four panels (upper, bottom and side panels). The basic requirements for operating the trawl net are sufficient power to tow the net, a system to keep the net mouth of open while towing, a system of wires to connect the net and gear to the source of towing power and the ability to haul and lift the gear.

Harvesting of marine resources in the proximity of the seabed requires fishing techniques that often interact with the bottom habitat. It is a common practice to refer to the towed nets on the bottom as “bottom or demersal trawling”. Towed gears may interact with the bottom substrate and organisms living close to the seabed or in the bottom substrate.

Towing speed ranges from 2 knots (1 $m \cdot s^{-1}$) to 7 knots (3.6 $m \cdot s^{-1}$) and fishing in the Mediterranean Sea might be conducted at depths from 10 to 800-900 metres. Two main fishing practices belonging to the demersal trawling have a very high contact with the seabed in the Mediterranean: the bottom otter trawl and the beam trawl. The degree of contact between the fishing gear and the bottom habitat mostly depends on the gear properties:

dimension, weight, presence of chains etc. influence the area swept by the particular gear and the pressure exerted on the bottom by the various gear components. The consequences of trawling impact are also affected by the bottom substrate properties (type of bottom, grain size, presence of seagrass etc.).

2.1.1. Otter Trawl

Bottom otter trawl is a type of mobile or "active" fishing gear consisting of very large cone-shaped net that is towed on the seabed to catch benthic or demersal fish. Otter trawls take their name from the rectangular trawl "doors" or "otterboards" used to spread the net horizontally (Figure 3). The net is designed and rigged to catch species living on, near or within the bottom. Actually bottom trawl is towed by a single vessel even if in the past certain types of nets were towed by two vessels (two vessels one net). A more sophisticated rigging allows the towing of two (or more) nets by the same vessel (one vessel two nets). This rigging, recently re-introduced in the Mediterranean Sea is known as "twin trawl". In the twin trawl rigging towing two nets side by side enhance the effective swept area, and hence the catch as well as the bottom impact.

The bottom otter trawl nets are usually towed at a speed between 2.5 and 4.5 knots. Duration of a tow depends mainly on the bottom depth, expected density of fish (whether fish is aggregated or not), the bottom properties (mud or sand), the slope in the fishing area. It ranges from 30 minutes up to 5 hours commonly 1-2 hours.

The basic bottom trawl rigging implies the towing vessel, two wires connecting the vessel to the otterboard, two otterboards (trawl doors) that are used to spread the connecting wires and hence hold the net open horizontally, two bridles connecting the otterboards to the net, and finally the trawl net.

The basic bottom trawl design generally implies two wings, a netting body ending in a codend, which retains the catch (Figure 3). The front part of a trawl is framed with a headline on the top, equipped with a series of floats and a groundrope on the bottom, equipped with leads and/or chain, which is designed to provide a good contact with the bottom and increase catch efficiency. The under-trawl escapement for many species is significant (Ingolfsson and Jørgensen, 2006) and therefore the challenge for efficient capture is often related to the ground gear design. The groundrope can be equipped with various types of ground gears (rubber discs, bobbins, or spacers etc.) the purpose of which is to protect the netting of the trawl from damage and to allow continuous towing without hook-ups. Bottom contact with the gear is needed for successful operations.

The body of the trawl is funnel-shaped, i.e., wide at its "mouth" and narrowing towards the codend, and is usually fitted with wings of netting at the both sides of the mouth. It is long enough to ensure the adequate flow of water and prevent fish escaping from the mouth after being caught. It is usually made of diamond-meshed netting with the size of the meshes decreasing from the front of the net towards the codend. The size of mesh in the codend is a determinant of the size of fish which the net catches. Consequently, regulation of mesh size is a common way of managing mortality of juvenile fishes in trawl nets.

The otterboards have three main functions:

- to spread the trawl horizontally during towing;
- to provide the front bottom contact points of the trawl gear;
- to herd and stimulate fish to swim towards the trawl path.

The otterboards are connected to the trawl wings by sweeps or bridles (generally ropes). These connections vary in length from a few metres up to 300 metres. The bridle or sweep has normally bottom contact during towing. Two warps or wires, long and flexible steel ropes, connect the winch of the boat with the otterboards (Figure 3). The length of the warps strongly depends on the bottom depth. At depths less than 70 meters, the length of warps is 7-10 times the bottom depth, while it is much reduced to be 2-2.5 times the bottom depth in fishing grounds 70-100 meters deep. In trawl gears, the winches installed on deck control the trawling warps and store them when not in use.

The sweeps/bridles and otterboards herd the fish and stimulate them to swim towards the trawl path. The contact of the otterboards and bridles with the bottom as well as the water turbulence behind the boards generate a cloud of mud and sand that, together with the noise created, leads to a herding effect of the fish into the net (Lindeboom and de Groot, 1998). Therefore, the capture success depends on how many fish are present in the swept area and are overtaken by the trawl and enter the net and how many are able to escape over or under the trawl. The escapement of fish entered the net is a matter of selectivity.

In brief, the traditional rigging of a otter trawl comprises (Figure 3):

- An on board winch having one or more horizontally mounted drums and/or warping ends on which a rope may be wound; auxiliary drums may be used to hold the net after hauling;
- Two warps or wires, long flexible steel ropes connecting the vessel to the otterboard;
- Two otterboards or trawl doors to hold open horizontally the wings and mouth of a trawl;
- Two bridles or sweeps, the ropes between the otterboards and the net; bridles substantially increased the area of seabed swept by the gear at very little cost in terms of additional towing power and improved productivity; certain types of trawl design imply the use of a really short bridle so that the otterdoad is joint to the net wing with a short rope ("Americana net");
- The net.

The traditional design of a bottom trawl includes (Figure 4):

- Mouth is the open end through which fish enter; otterboard guarantees the horizontal opening, while headline and groundrope guarantee the vertical opening (Figure 4);
- Groundrope or foot rope is a connected section of rope usually made of polyamide (PA) or polyethylene (PE) and maintaining ground contact through the use of several weights. Very often a chain may be joined to the groundrope to increase the bottom contact.

- Headline or headrope is the principal upper frame rope of a net to which the net is attached. Several floats are attached to the headline to lift the upper part of the net (Figure 4);
- Wings are the tapered sections extending forward from two sides of the main body of the net with each wing being divided into lower wing and top wing (Figure 4);
- Body is usually the central main part or section of the net or trawl between the wings and the extension piece. It is a tapered section and consists of a lower and upper panel between wings and extension piece;
- Extension piece is the untapered or cylindrical section of netting either inserted between body and codend or attached to end of codend to increase its length. Sometimes the extension is missing and the codend is directly joint to the body;
- Codend is the rearmost part of the trawl net where the catch accumulates having either a cylindrical shape or a tapered shape. Traditionally a diamond mesh codend with a mesh opening of 40-50 mm is used in the Mediterranean. The use of this type of mesh results in a gear which is not selective enough for protecting juveniles of most of species. Moreover, the high marketability of fry fish in some areas strongly incentivizes the use of small meshes at the codend.

There are numerous different types of otterboards in use (Figure 5). In general, they are large and heavy steel, or wood, plastic and iron structures weighted on their base by a protective iron shoe. Otter trawls are most frequently towed over relatively flat soft-bottoms, but can also be towed over relatively rough bottoms (e.g., boulders) when fitted with "rock-hopper" gear.

The bottom trawl is considered the less selective gear in the Mediterranean given that it allows the catch of juveniles as well as of huge amounts of non-target species.

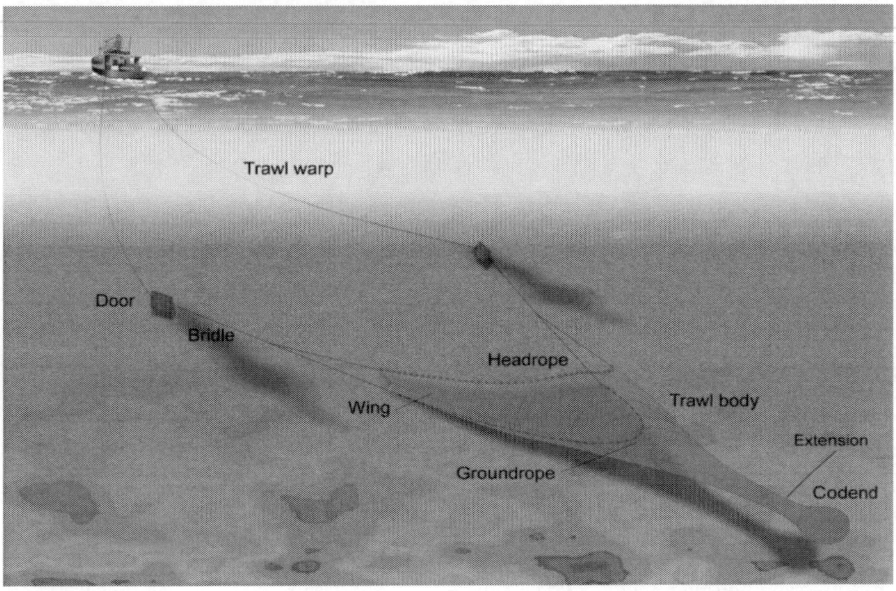

Figure 3. An otter trawl showing some of the main components in its design. Rigging and characteristics of a bottom trawl.

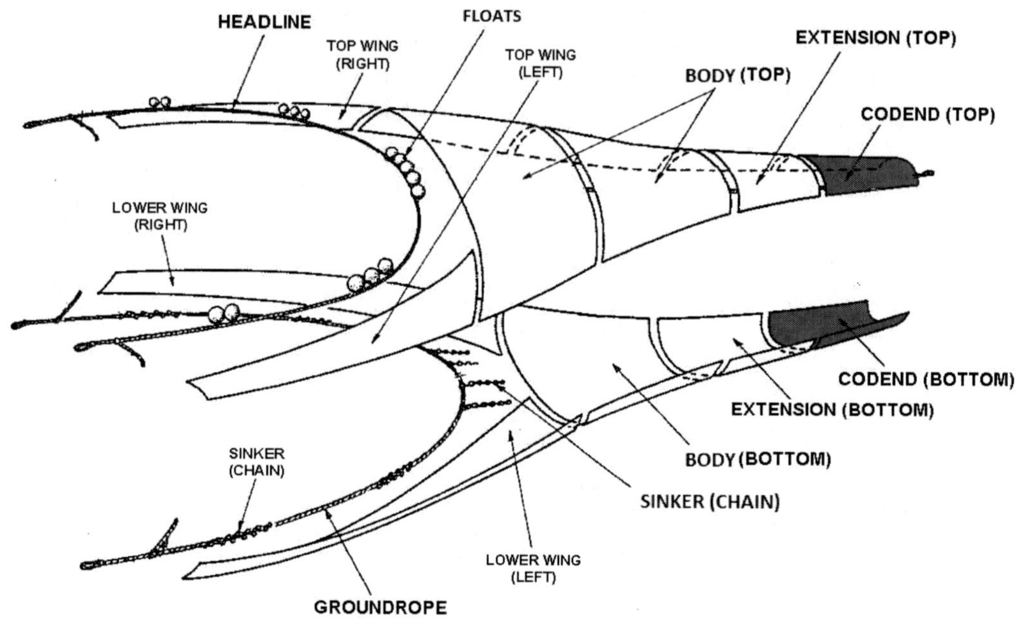

Figure 4. Bottom trawl netting panels and details of wings and mouth of a bottom trawl net.

Figure 5. (Continued).

Figure 5. Different types of otter trawl or doors used in the Mediterranean bottom trawl.

Regarding the codend meshes, the REC. CM-GFCM/33/2009/2 on the minimum mesh size in the codend of demersal trawl nets encourages that "*The Members and Cooperating entities of GFCM shall adopt and implement, at latest by 31 January 2012, a minimum 40 mm square mesh codend or a diamond mesh size of at least 50 mm of acknowledged equivalent or higher size selectivity, for all trawling activities exploiting demersal stocks when operating in the GFCM Area*".

2.1.2. Beam and Rapido Trawl

Beam trawl consists of a cone-shaped body ending in a bag or codend that retains the catch. In these trawls, the horizontal opening of the net is provided by a rigid frame that is made of wood or metal and is up to 12 m long (Figure 6). The vertical opening is provided by two hoop-like trawl heads/shoes mostly made of steel. No hydrodynamic forces are needed to keep a beam trawl open.

Different types of beam trawl are used in Mediterranean. They are generally used in shallow waters by small units within some small-scale fisheries. Provençal "gangui" (from the Southeast of France) and Catalan (NW Spain) "ganguils", Greek "kankava" for sponges, Italian "gangamo" for prawns and sea urchins (used in Sicily) are the most common examples. In the Provençal French fishery using "gangui", the target species are scorpaenidae, red mullets and other high value species used in traditional dishes. In Italy, large bottom trawlers of the Adriatic coast can use a kind of beam trawl, the so-called "Rapido trawl", to catch scallops (*Pecten jacobeus*) and queen scallops (*Aequipecten opercularis*) on sandy on detritic bottoms ("maerl") at 40-50 m depth at some distance from the coast, and common sole (*Solea solea*) in muddy inshore areas. The gear consists of a box dredge 3-4 m wide and about 170 kg weight rigged with 5-7 cm long teeth and a lower leading edge, and net bag to collect the catch (Giovanardi et al., 1998).

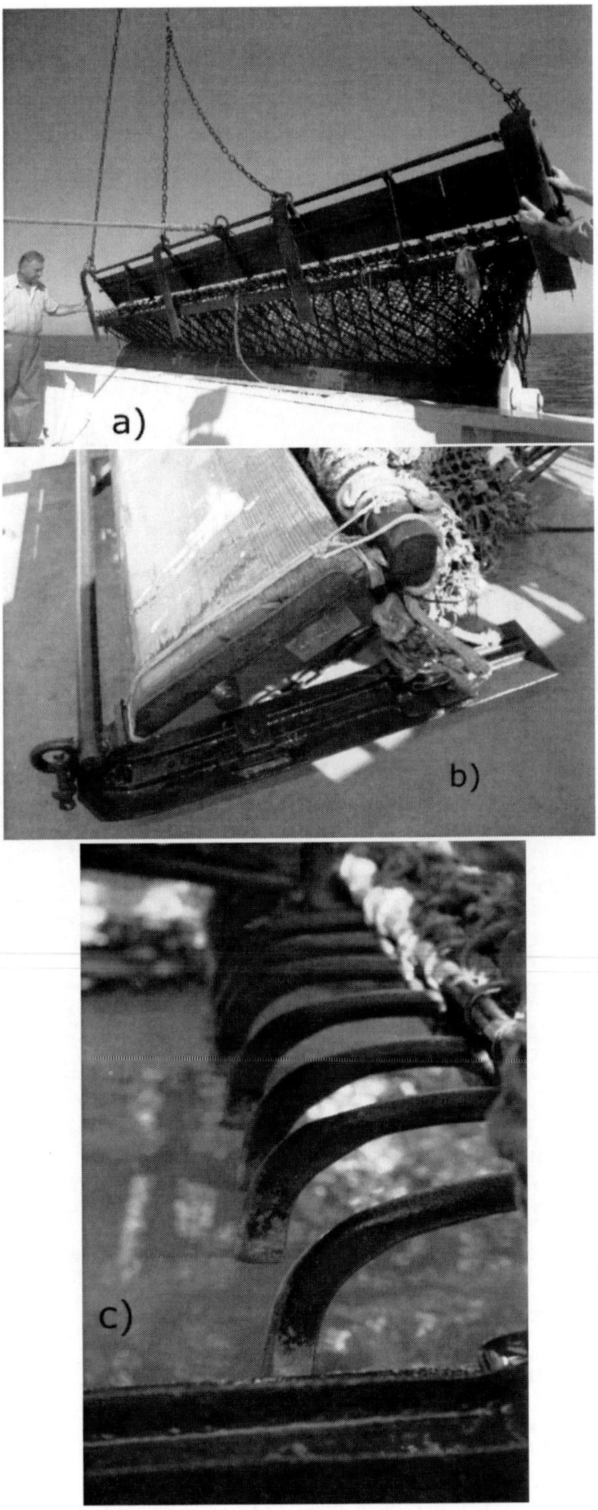

Figure 6. (Continued).

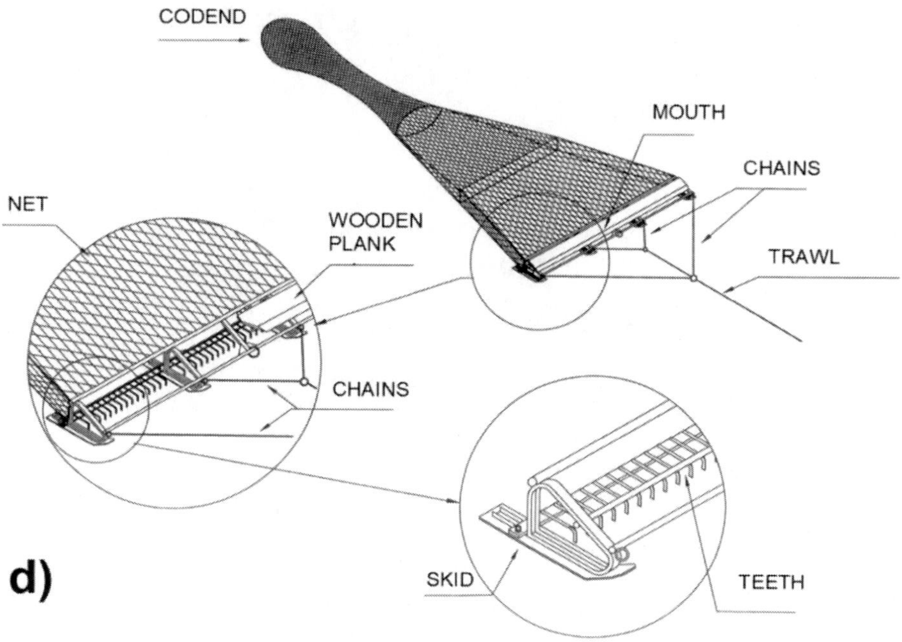

Figure 6. a) Commercial Rapido trawl used in GSA 17; b) particular of the inclined wooden board fitted in front of the metallic frame act as depressor; c) teeth; d) scheme of Rapido trawl.

Figure 7. Towing methodology with rapido trawling.

An inclined wooden board is fitted to the front of the metallic frame to act as depressor that keeps the gear in contact with seabed and, even more, presses it on the bottom to facilitate the penetration of the teeth in the sediment (Figure 6). A single vessel may tow four or even six gears simultaneously (Figure 7). The towing speed is about 5-7 knots.

2.1.3. Bottom Trawl Impact

The effects of fishing on ecosystem and on the resources are due both to the technical characteristics of the fishing gears and to the characteristics of the fishing practices. Demersal towed gears are considered the most important anthropogenic activities aimed at exploiting the living resources within the marine environment (Jennings and Kaiser, 1998; Lucchetti and Sala, 2012). The technological advances in fishing gears and the increased engine power of the Mediterranean vessels have recently lead to a situation where physically undisturbed areas of seabed have become rare in areas that have been substantially fished by bottom trawls. Thus, concerns about the impacts of towed fishing gears such as trawls and dredges on benthic habitats and organisms have increased over the last two decades. The reasons for these concern are that benthic habitats, among several positive effects, provide shelter and refuge for juveniles, and the associated fauna comprise important food sources for demersal fish.

Several studies have been conducted to evaluate the physical and biological effects of dragged gear operations on benthic habitats and communities. Recent studies in the Mediterranean have highlighted the massive impacts of towed gears on the marine benthos of shelf areas particularly scallop dredging (Franceschini et al., 2002; Morello et al., 2005); bottom- (Demestre et al., 2000; Smith et al., 2007; Sala et al., 2009; 2010) and Rapido trawling (Giovanardi et al., 1998; ICES, 2003; Pranovi et al., 2000).

Basically, two methods have been used to study and assess such impacts. One method is to compare physical and biological parameters on a site before and after disturbance by experimental fishing or alternatively compare an exposed site with an adjacent undisturbed control site (Freese et al., 1999; Hansson et al., 2000; Kenchington et al., 2001). The other method is to compare commercial fishing grounds that have been heavily fished with areas that are lightly fished or not fished at all (Engel and Kvitek, 1998; Thrush et al., 1998; McConnaughey et al., 2000).

Physical Disturbance

Widespread otter bottom trawling is potentially one of the most invasive forms of anthropogenic disturbance to benthic habitats in the Mediterranean.

The effects of otter trawling vary greatly depending on the amount of gear contact with the bottom. Gear type and rigging are the main factors affecting the level of disturbance caused by fishing activity. Trawl doors, weights, bridles/sweeps and the ground gear represent different components of otter trawl rigging, each with a specific impact on the bottom habitat.

Trawling basically alters the physical environment of the benthic communities by creating furrows or scars from trawl otterboards, by scouring and flattening the seabed thorough ground gear and weights, by removing and re-suspending sediments. The results of very recent research projects (Lucchetti and Sala; 2012) show that hydraulic dredges and Rapido trawls are basically similar in their seafloor impact by flattening and ploughing seabed features.

Hall (1994) in a review of physical disturbance of otter trawling noted that impacts of otter trawling were largely restricted to the disturbance caused by the doors. However, the effect of the ground gear should not be underestimated. The height of the groove depends on the weight of the otterboard, the angle of attack (the higher the attack angles, the higher the groove's height), and the nature of the substrate being deepest in soft mud.

In soft bottoms, the towed gears scrape or plough the seabed re-suspending sediment, changing grain size and sediment texture, and destroying bed forms. Moreover, the sediments might be whirled up into the water masses and suspended on flat sandy/muddy bottoms (Pranovi et al., 2000). Therefore, intense and continued trawling on continental shelves has a noticeable effect on water turbidity that has to be considered together with natural processes (Lucchetti and Sala, 2012; Palanques et al. 2001; Figure 8). This interaction may induce turnover of the sediment surface, oxygenation of deeper sediments, and release of organic matter and nutrients. These factors may stimulate local production even if the negative consequences may be more important: habitats destruction, benthos mortality and release of pollutants which could induce changes in biogeochemical cycles and primary production processes (Churcill, 1989; Pilskan et al., 1999; Cloern, 2001; Duplisea et al., 2001). These factors can lead to an overall reduction in production. Moreover, the eutrophic processes in closed basins and low depths (as in the northern Adriatic) may be enhanced by trawling leading to hypoxia in sensitive soft bottom areas and the quantity of hydrogen sulphide released from sediments may increase (Caddy, 2000).

The turbidity can also negatively affect the photosynthetic processes, particularly in coastal areas, by decreasing light penetration (Tudela, 2004). This is believed to have contributed to the disappearance of seagrass meadows in some areas of the Mediterranean.

Figure 8. Re-suspension of sediments caused by Rapido trawling. Evidence of the furrows left by Rapido trawls in earlier tows can be seen in the lower part. Images obtained by sidescan sonar investigation (from Lucchetti and Sala, 2012).

The furrows left by Rapido trawls are 4 m in width, equivalent to the dimension of the frames, and 5-15 cm in depth. Beam trawls cause direct mortality to non-target organisms in two ways. Firstly, the shoes, tickler chains or chain mat impact on animals on the seabed (Bergman and van Santbrink, 2000) and secondly, animals are caught in the net and die from sustained injuries during hauling or when the bycatch is processed and discarded (Lindeboom and De Groot, 1998).

Rapido trawling produces the most severe impacts on benthos because it captures epifaunal and infaunal components. Contact with this heavy gear is also responsible for high mortality (Giovanardi et al., 1998).

In the case of dredges, the furrows have an average height of 5-15 cm but are up to 20 cm where the cage is lowered into the water and the dredging starts. Dredge marks appearing as 3 m-wide long stripes equivalent to the dredge width, were relatively clear and numerable. Lateral sledge marks noted to be paired (e.g., two continuous parallel marks) have a width of about 50 cm.

The Bottom Trawling Ecosystem Impacts

The ecosystem effects of bottom trawling can be divided in two types of impacts: direct and indirect impacts. The direct impact includes changes in the habitat as well as in the populations of the targeted fish and invertebrates (Molluscs and Crustaceans) species. The direct impact also involve the mortality of non-target populations of fish and other animals or vulnerable species (via bycatch and discards), and physical impact caused by the gear on benthic organisms.

The indirect impact include the reduction of the biodiversity at higher levels such as community structure of benthos, changes in the food web, as well as the environmental effects of dumping discards and organic detritus (offal) and the mortality caused by lost gear (ghost fishing).

Bottom trawling disturbances on benthic communities may cause a change in species composition and species diversity. Physical interaction of bottom trawls with bottom sediment might result in the removal or damage of sedentary living organisms including seaweeds and corals, and in the displacement of stones or other larger objects of uneven bottom surface. The most serious effects of trawling have been demonstrated for hard-bottom habitats dominated by large sessile fauna such as sponges, anthozoans and corals which have been shown to decrease considerably in abundance in the path of the ground gear (Freese et al., 1999; Moran and Stephenson, 2000). Such habitats may thus be severely affected by fishing operations. These systems are characterized by high habitat complexity and, hence, high biodiversity indices.

According to comparative experiments carried out in different biocenoses of the Northern Adriatic Sea (Pranovi et al. 2000), a single rapido tow may induce changes in the structure of benthic infauna which last up to 9 months before being completely recovered.

Mediterranean studies on otter trawling impact (Smith et al., 2000) showed that trawling implied functional change on the megafaunal community structure with sessile or discretely mobile filter-feeding organisms being replaced by mobile scavengers and opportunists.

Scientific studies carried out in the Mediterranean showed that certain types of dragged gears (such as hydraulic dredging) imply the destabilization and partial modification of sediment conditions, decreasing the habitat complexity and leading to fluctuations within benthic communities (Brambati and Fontolan, 1990). Radical reductions in the abundances of infaunal organisms in favour of few dominant opportunistic species are widely demonstrated as a consequence of mechanical impact and/or suction by hydraulic dredging also in the Adriatic Sea (Hall, 1994; Pranovi and Giovanardi, 1994; Tuck et al., 1998).

Morello et al. (2005) observed that the macrobenthic community of a highly trawled area was in a permanently altered condition. This was confirmed by the high abundance of taxa typical of disturbed sediments (e.g., *Owenia fusiformis* and *Ampelisca* sp.) or moderately disturbed transitional zones (e.g., *Abra alba* and *Corbula gibba*). However, the authors found that the effects of fishing on community structure were still discernible above natural variation despite intensive fishing that has been going on for decades. This can be explained by the fact that the area investigated comprises a shallow dynamic habitat that's subjected to intense hydrodynamism and environmental fluctuations like storms making the benthic community there having an inherent resilience to natural physical disturbance (MacDonald et al., 1996; Kaiser, 1998). These habitats may be resistant to trawling because they are subjected to high degree of natural disturbances such as strong currents and large temperature fluctuations.

Effects of trawling on the biodiversity and habitats of non-target species are a key issue also. In some areas and fishing grounds, trawling can imply biodiversity changes of endangered or threatened marine species including elasmobranches, turtles and their habitats.

The trawling can also impact the seagrass beds by both suspending sediments and directly damaging the vegetal mass. Trawling has the most dramatic consequences on *Posidonia oceanica* beds (Sánchez Jerez and Ramos Esplá, 1996; Ardizzone et al., 2000). Sediment suspension affects macrophyte photosynthesis by decreasing light intensity. This is believed to have contributed to the disappearance of seagrass meadows, and to affect fish recruitment and the quality of juvenile feeding areas in the Mediterranean Spanish coast (Sánchez-Jerez and Ramos-Espla, 1996). Seagrass beds are spatially complex and biologically productive ecosystems that provide habitats, shelters and food resources for diversified fish fauna and act as an important nursery area for many species (Harmelin-Vivien, 1982). Meadows regress significantly for two main reasons; anthropic changes in sediment structure and composition, and the direct mechanical impact of fishing (Ardizzone et al., 2000). Posidonia beds also play an important role in stabilizing sediments. The loss of such plants can Lead to sediment erosion by waves and currents making it difficult for restoration.

The Impact of Bottom Trawling on Non-target Species and Sizes

Discarding and bycatch occur in the Mediterranean because most of fishing gears and fishing practices are not selective enough for targeted species and sizes, and because target species inhabit also areas which are occupied by a wide range of other species.

The discarding of targeted species and bycatch has long been recognized as a problem. Usually discarding constitutes a reduction of future harvesting opportunities and it might have negative consequences for the environment and ecosystem.

There are two main reasons for discards in bottom trawling. Firstly, the direct incidental killing or damaging effects exerted by the gear, and secondly the voluntary decision during sorting operations on board. On deck, fishermen sort out fish that can be sold and the remaining catch (discard) is returned to the sea with most of specimens being already dead or dying. Biocenoses composition, seasonality, and exposure to air during sorting operations seem to play a key role in determining the ultimate physical damage, physiological stress and mortality of species discarded by rapido fishing activities (Pranovi et al., 2001; Mattiello et al., 2004).

In certain areas of the Mediterranean, the amount of discard from bottom trawling can exceed that of fish landed. A species is discarded in bottom trawling mainly because it's either too small or below minimum landing size, damaged during capture (probably due to too long towing), or because it is not a marketable species.

Bottom trawling strongly affects the Mediterranean elasmobranch populations due to the low selectivity of fishing practices and direct exploitation. Analyses of historical data on experimental surveys and fishery landings have proved a clear decline of demersal elasmobranch species populations in some Mediterranean regions since the 1960's (Aldebert, 1997). Results indicate that the decline of elasmobranch started on the continental shelf, and recently extended to the slope. Mediterranean bottom trawling mainly impacts small sharks such as the smooth-hounds (*Mustelus mustelus* and *M. asterias*), the small spotted catshark (*Scyliorhinus stellaris and S. canicula*), the Spiny dogfish (*Squalus acanthias*), the blackmouth catshark (*Galeus melastomus*), and the longnose spurdog (*Squalus blainvillei*) as well as most of the rays (Figure 9).

The use of large meshes as well as a significant reduction of fishing in the nursery areas has been suggested as essential management measures for the conservation of Elasmobranch species in the Mediterranean.

Moreover the long duration of commercial trawl tows, together with exhaustion and damage to the fish as it passes along the trawl and is concentrated and compacted with other fish in the cod end, results in limited escape of younger undersized fish through the cod-end meshes, and reduced survival of escapees. Thus the majority of the under-sized fish in the cod end are typically dead when hauled on deck at the end of the tow, and the under-sized fish are sorted out and discarded at sea. The poor size selectivity of the trawl can result in the discarded bycatch of undersized specimens (Figure 10).

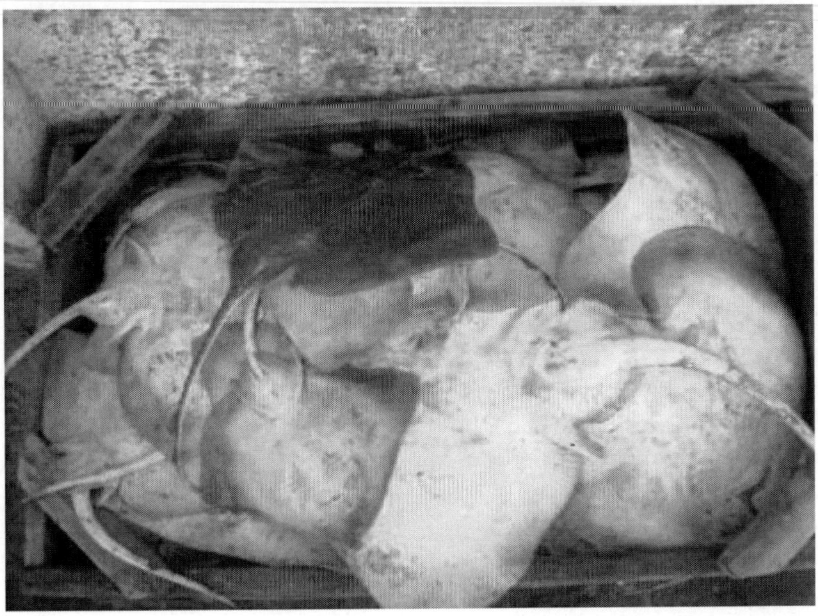

Figure 9. Rays caught by a bottom trawler in Egypt.

Figure 10. Undersized specimens of Zeus faber and Lopihus spp caught by bottom trawling.

2.2. Pelagic Trawling

In mid-water or pelagic trawling, the net can be towed by one or two vessels (pair trawling) and it is fit and rigged to operate in mid-water for the capture of small pelagic species such as Anchovy (*Engraulis encrasicolus*) and Sardine (*Sardina pilchardus*), and to a lesser extent other pelagic fish such as horse mackerel (*Trachurus* spp.) and mackerel (*Scomber* spp.).

In its simplest design, the net is usually a cone- or pyramid-shaped body closed at the end by a codend and it can extend at the opening by the wings. Pelagic trawling use higher vertical opening nets than for bottom trawling. The nets consist of four panels the front parts of which are generally made of large meshes or long rope to progressively lead fish schools toward the posterior part of the nets with smaller meshes. The codend mesh size is about 20 mm in stretched length. Pelagic trawl nets are mainly used in the Gulf of Lions (France), in the north of the Adriatic by the Italian fleet ("volante") and in Tunisia.

In pair trawls, the horizontal opening is maintained by the distance between the two boats during towing operation (Figure 11 A). In single boat pelagic trawl, the horizontal opening is maintained by spreading the two wings by means of 2 or 4 doors; 2 small pelagic- and 2 traditional bottom-doors (Figure 11 B).

Floats on the headline and weights on the groundline often maintain the vertical opening. Modern large mid-water trawls, however, are rigged in such a way that floats are not required, relying on downward forces from weights to keep the vertical opening during fishing. Two weights (about 300 kg each) are joined to the ends of lower wings to quickly keep the groundrope deep.

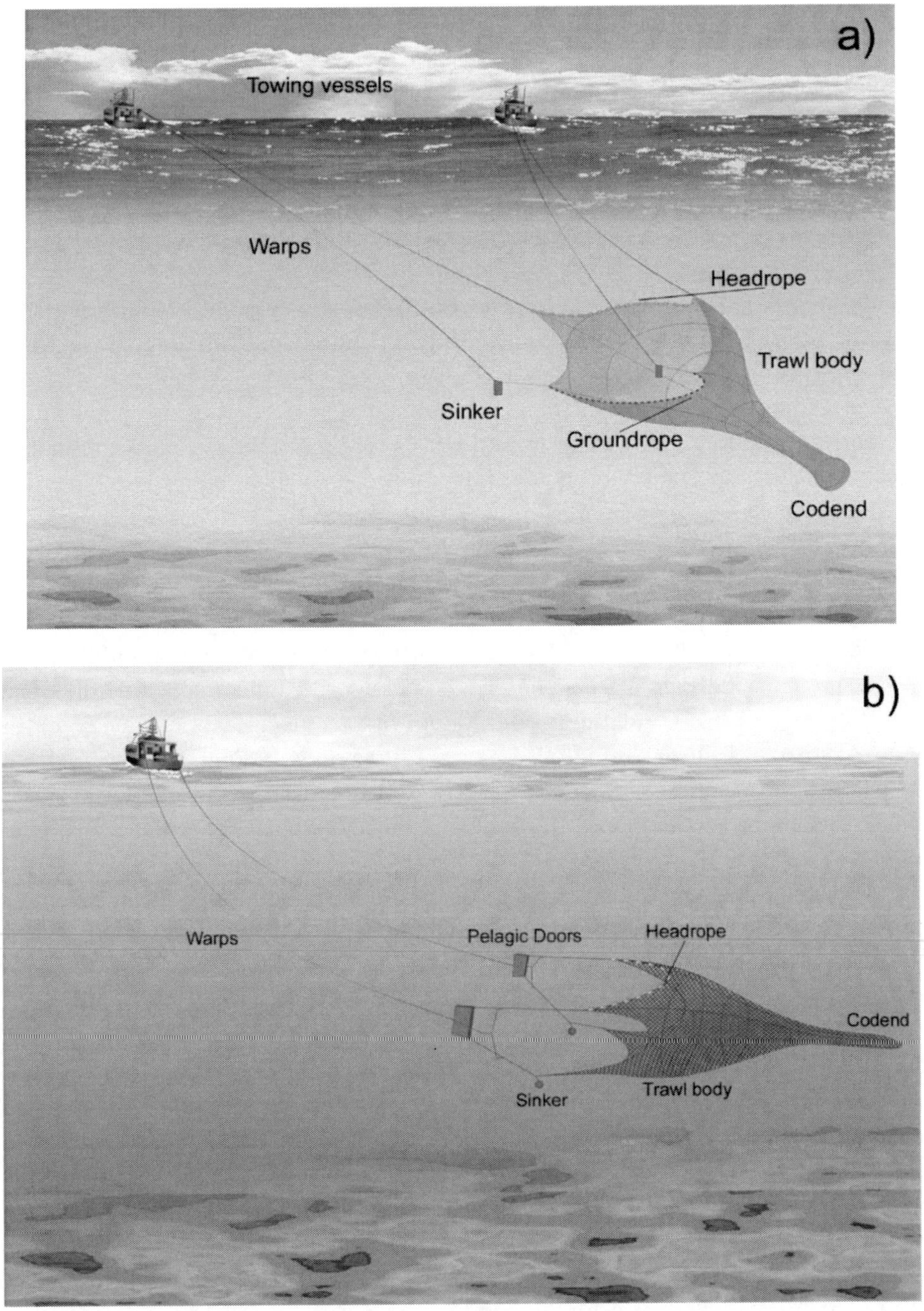

Figure 11. Mid-water pair trawl (a) and single boat pelagic trawl (b) rigging.

Pelagic trawl is towed at the appropriate level in water column to intercept target shoal with gear depth being controlled by altering towing speed and/or warp length. Electronic equipment such eco-sounder is actually common and is an essential tool to detect fish concentration ahead the net and, hence, the trawl path and depth can be adjusted accordingly. Sonar is sometimes used to identify small pelagic schools at great distance from the boat. Trawl winches installed on deck control the trawling wires and store them. Net drums are common tools to handle mid-water trawls on board vessels. Towing speed commonly used is about 4 knots or even more.

The dimension of boats used in trawling varies from small inshore boats up to large deep-sea trawlers, with the size of gear scaled on the basis of the boat dimension.

2.3. Surrounding Nets

Surrounding nets are large nets set framed with floatline on top and leadline at the bottom used for surrounding school fish both from sides and from underneath, thus preventing them from escaping by diving downwards.

The surrounding nets can be divided in two main types of nets; surrounding net with purse line and surrounding net without purse line. According to the type of surrounding nets gear, specific equipment may be required the main of which being some facility for manoeuvring large net.

Generally the net design implies three main netting compartments: a single net wing, the main netting body and the bunt, which is mounted on a side of the net. Sometimes the purse seines have a symmetrical net design: two lateral net wings having the same dimensions and characteristics, the main netting body and a bunt or bag, positioned in the middle of the net for the collection of the catch. The bunt has small meshes and thicker netting material compared with the main body netting. The bottom of purse seines is characterized by the presence of a purse line which runs into steel rings (purse rings) joint to the lower edge of the net. The purse line, which is normally a wire made of steel, once recovered by means of winches, enables the net to be closed from the bottom like a purse, and thus to retain all the fish encircled (Figure 12).

Purse seines are considered as the most important surrounding nets from a commercial point of view since they can be used in the Mediterranean for the catch of small (Anchovies and Sardines) and large (mainly Bluefin tuna, Albacore) pelagic species. Purse seines targeting small pelagic species are normally operated by means of skiffs (usually one-three small boats) that are used to carry out aggregation of the schooling fish by means of lights (Figure 13). Small pelagic species are caught during night since lamps mounted on small tenders are used to aggregate the school fish. Sometimes the lamps used to attract the school fish are mounted directly on the main vessel. Lights seem to attract the target species directly or indirectly by 44 attracting and illuminating prey organisms. Hydroacoustic instruments like echosounder are important tools for locating locate fish aggregations.

When a fish school has been located (with echosounder) and aggregated (by means of lights), the net shooting start by lowering the net wing into the sea (the net wing is kept on board the boat skiff); gradually even the main body of netting and the bunt are shot into the sea encircling the school of aggregated fish. Usually, all the purse seine is set and the circle is closed in less than 10 minutes. Once the net has completely encircled the school, the main boat comes back to the wing. When the setting is completed, the net is closed by hauling in the purse line and fish can no longer escape. From now on, the main vessel starts to haul back the net by using a mechanized power-block and the catch is slowly herded into the bunt.

In shallow water this operation is usually carried out completely by hand. When most of the purse seine has been retrieved, the fish have been grouped within a restricted area along the portside of the vessel. Then the fish are harvested from the purse seine using a large scoopnet. The fish caught by means of this gear remain alive within the net and the live fish caught are transferred into large tanks with water and ice. Small purse seines can be operated entirely by hand in small scale fisheries. The purse seine can have a symmetrical net design: two lateral net wings having the same dimensions and characteristics, the main netting body and a bunt or bag positioned in the middle of the net for the collection of the catch (Figure 14).

The bunt has small meshes and thicker netting material compared with the main body netting. This is the section of the purse seine where the schooled fish are gathered at the end of hauling process. The shooting of the net around a detected school of fish starts from one of the two wings hold by the skiff boat. Than the purse seiner set the net around the school and comes back to the skiff. This operation usually lasts less than 5 minutes. The last netting panel lowered into the water is the second wing. Once all the net is set out, the net hauling starts immediately by recovering the two net wings simultaneously before the net touches the bottom (see the section on the bottom impact). In case of small nets, the hauling operation except for the pursing line, is often carried out completely by hand, until the catch is aggregated in the middle of the net (the bunt). Furthermore, in order to facilitate the hauling operation, the boat is positioned downstream, and the current gradually help to drive the net toward the boat (Figure 15). The hauling procedures normally last less than one hour.

In case of purse seines having bigger dimensions the net design usually implies three main netting compartments: a single net wing, the main netting body and the bunt, which is mounted on a side of the net (Figure 16). The net shooting start by lowering the wing into the sea (the net wing is kept on board the boat skiff); gradually even the main body netting and the bunt are shot into the sea encircling the school of fish aggregated by means of lights. Once the net has completely encircled the school the main boat comes back to the wing already lowered into the sea at the haul beginning, from now on it starts to haul back the pursing wire by means of 47 winches and the net by using a powerblock. Thus the catch is slowly herded into the bunt. Summarizing the net hauling starts by recovering the net wing and ends with the bunt. The pursing phase normally lasts less than five minutes. The purse wire recovering starts within 5-10 minutes from the beginning, according with the fishing depth and fishermen experience. The complete purse wire recovering normally requires 5 minutes. On average the fishing operations from the net shooting until the purse wire closure last between 13-23 minutes.

With large purse seines some cases fishermen take advantage of the current flow to give to the net proper round shape. During the hauling operation the boat is positioned upstream and a small workboat can be adopted in order to keep the main boat in position, or to avoid that the current push the boat onto the net (Figure 17).

Purse seine fisheries targeting small pelagic species, namely Anchovy (*Engraulis encrasicolus*) and Sardine (*Sardina pilchardus*), represents a relevant fishing sector in the Mediterranean Sea from an economic as well as social point of view. Existing studies on the Mediterranean purse seine fisheries suggest that the discard ratio is rather low (Kelleher, 2005; Tsagarakis et al., 2012) because the gear is highly selective and vessels mainly target small pelagic fish with a low diversity of species and sizes. However, the discarded portion could be affected by several factors such as quantity and composition of the catch as well as market prices (Santojanni et al., 2005).

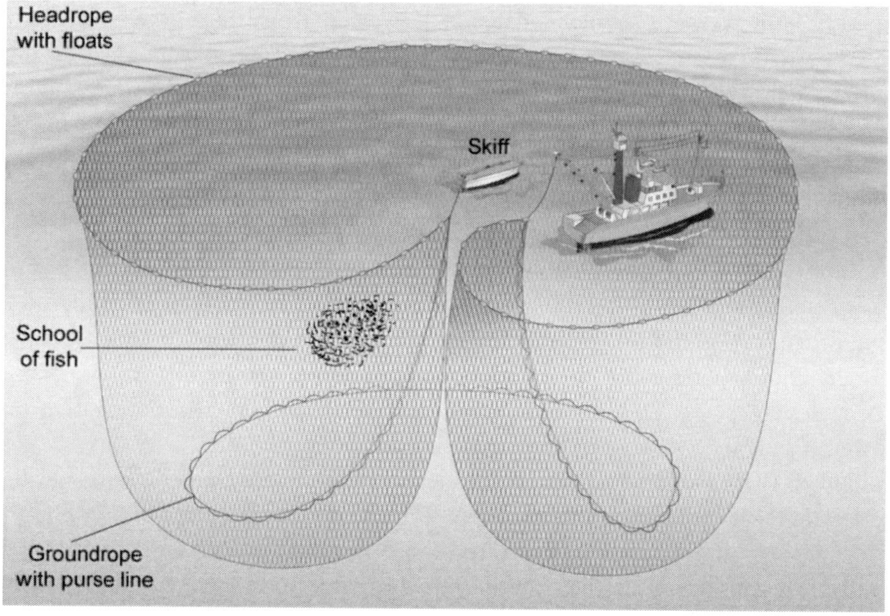

Figure 12. Scheme of the purse seine fishing procedures (adapted from www.eurocbc.org).

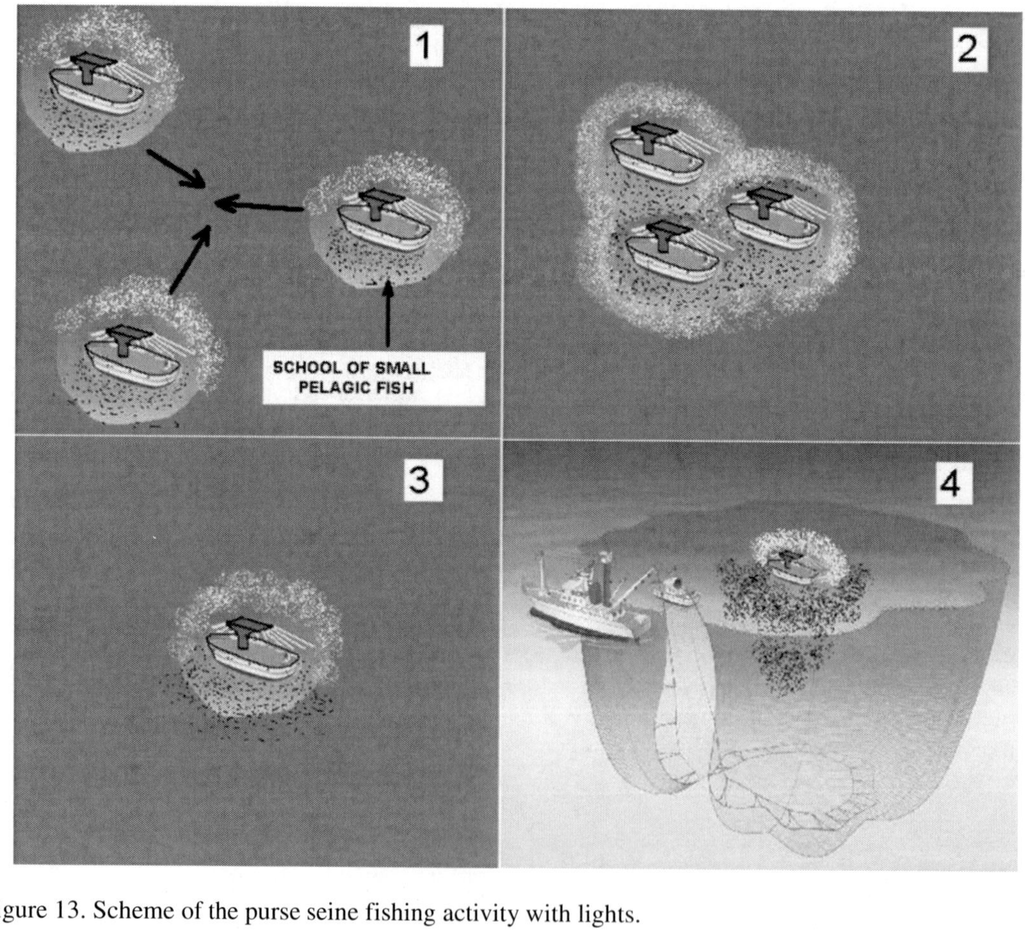

Figure 13. Scheme of the purse seine fishing activity with lights.

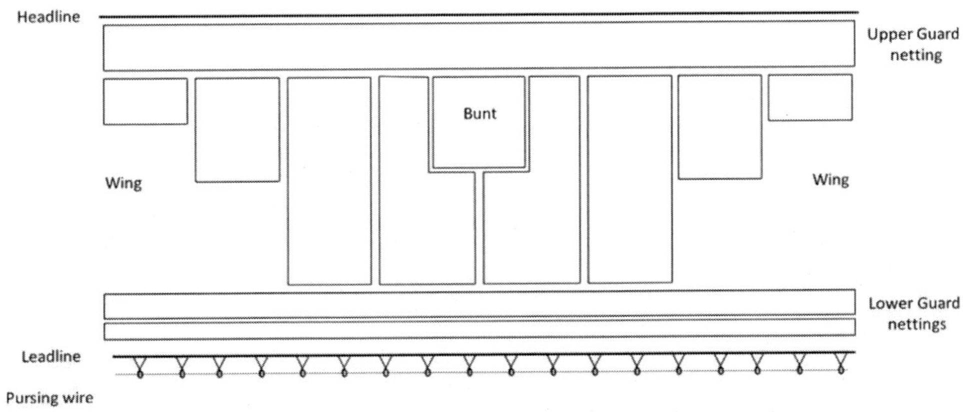

Figure 14. Technical drawing of a typical North Adriatic purse seine.

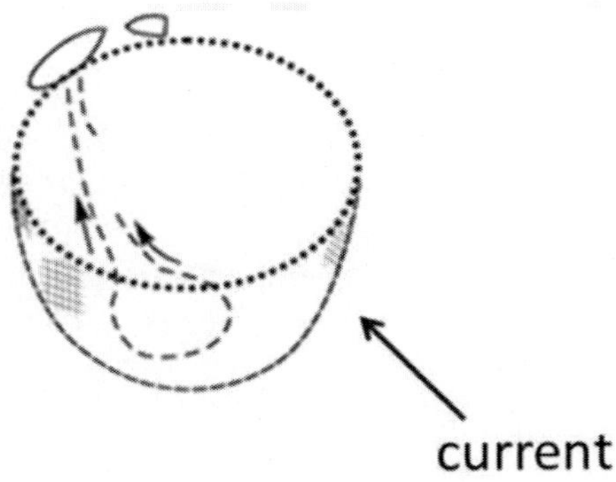

Figure 15. Hauling operation of small purse seine.

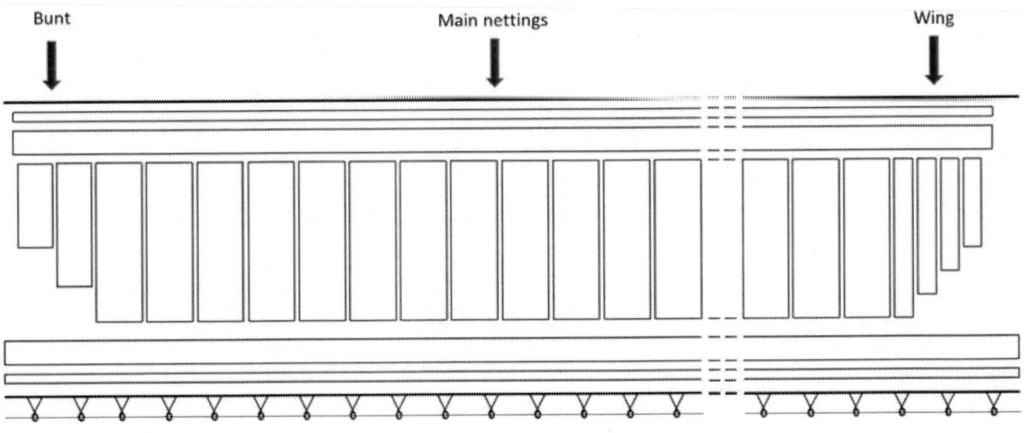

Figure 16. Design of a typical purse seine net used in deep-waters.

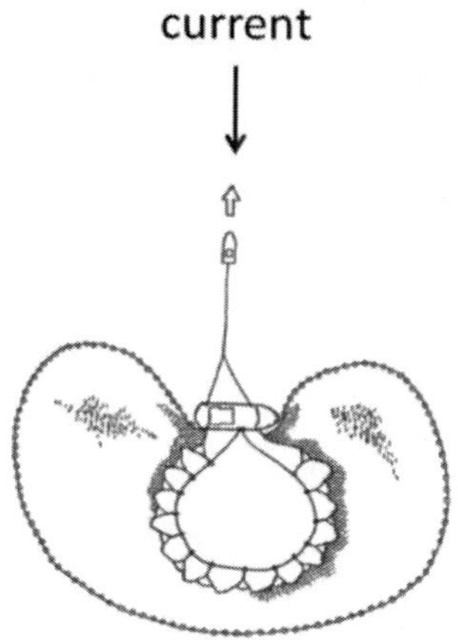

Figure 17. Hauling operation of big purse seine. A workboat can be used in order to keep the main vessel in position.

The fishing operations carried out by a purse seiner first of all imply the identification of the school. Large pelagic species can be located through sightings on the sea surface (like concentration of sea birds, ruffling of the water surface and presence of groups of dolphins). Small pelagic species are caught during night since lamps mounted on small tenders are used to aggregate the school fish. Hydroacoustic instruments like echo-sounder are important tools for locating locate fish aggregations. In the case of large vessels targeting the Bluefin tuna, the sonar can be used to locate the school.

When a school is located, the main vessel encircles the school at maximum speed. Usually, all the purse seine is set and the circle is closed within 4 to 8 minutes. In the Mediterranean, purse seiners fishing Bluefin tuna (*Thunnus thynnus*) often do not shoot all the net and close the circle with only a portion of the seine. This is mainly because tunas can change their direction and escape before the circle is closed. Once the school is surrounded, the end of the net that stays attached to the skiff is transferred aboard the purse seiner and the two extremities of the purse line cable are hauled with the winch (rarely by hand) as quickly as possible in order to close the net at its bottom. In fact, the fish can escape below the net as long as the purse seine is not closed,. The net is then pulled aboard the purse seiner with a hydraulic power block or rarely by hand. When most of the purse seine has been retrieved, the fish is grouped within a restricted area along the portside of the vessel. Then, the fish are harvested from the purse seine using a large scoopnet. Shooting lasts about 5 minutes, while pursing is longer taking between 15 to 25 minutes depending on the efficiency of the boat. Net hauling takes between 60 and 90 minutes. Besides deck equipment, 8-12 men are required for these operations.

Purse seines usually have very large dimensions measuring 1500 to 2000 m in length and 120 to 250 m in drop. Small purse seines can be operated entirely by hand in small scale fisheries.

Presently, purse seine is the most widely used fishing practice for the catch of Bluefin tuna. Fish caught by means of this gear remain alive within the net and most of the live fish caught are transferred to transport cage for tuna farming. In the last ten years, no-global organizations have denounced the serious depletion of Bluefin tuna in the Mediterranean.

Figure 18. Purse seine hauling.

Figure 19. Mediterranean purse seine on board a fishing vessel; on the left it can be possible to observe the purse rings, on right the floats of the headrope.

Figure 20. Tuna cage carrying the tuna transferred from the purse seine.

In tuna fishing outside the Mediterranean, purse seine fishing is well-known for the high levels of dolphin bycatch. Some technical modifications to the gear (MEDINA panels) or to fishing practices (backing down operation) can be applied to mitigate this adverse impact of purse seine. In the Mediterranean, however, the problem of incidental catch mainly involves non-target species of fish and rarely the catch of sea turtles. In this latter case, sea turtle are still alive when hauled on board and, therefore; the negative impact on turtle population can be reduced by applying the right procedures on board. However this is a minor issue in the Mediterranean.

The most important purse seine nets technical data can be summarized as follow:

- Headline length (*HLL*): Length of the headline;
- Stretched net length (*NL*): it is the length of the netting mounted on the headrope when fully extended;
- Horizontal number of meshes (*HNM*): it is the number of meshes measured along a horizontal line along a netting;
- Vertical number of meshes (*VNM*): it is the number of meshes measured along a vertical line along the netting;
- Mesh size (*MS*): the Mesh length can be selected as the most suitable parameter for the description of the mesh size;
- Fully extended or stretched net drop (*ND*): it is defined as the sum of the length of the meshes (including knots) when wet and stretched perpendicular to the float line. In order to compare the net drop of different purse seines the *ND* can be measured considering the main netting panel only, without taking into account the horizontal strengthening panels. The fully extended net drop is calculated by multiplying the

mesh length and the number of meshes from 51 the headrope to the leadrope (VNM) as follows:

vertical number of meshes x correspondent mesh length: VNM x MS

- Usually one, two or three horizontal netting panels made of larger meshes and thick twine are joint to the headrope and groundrope to strength the net during hauling.

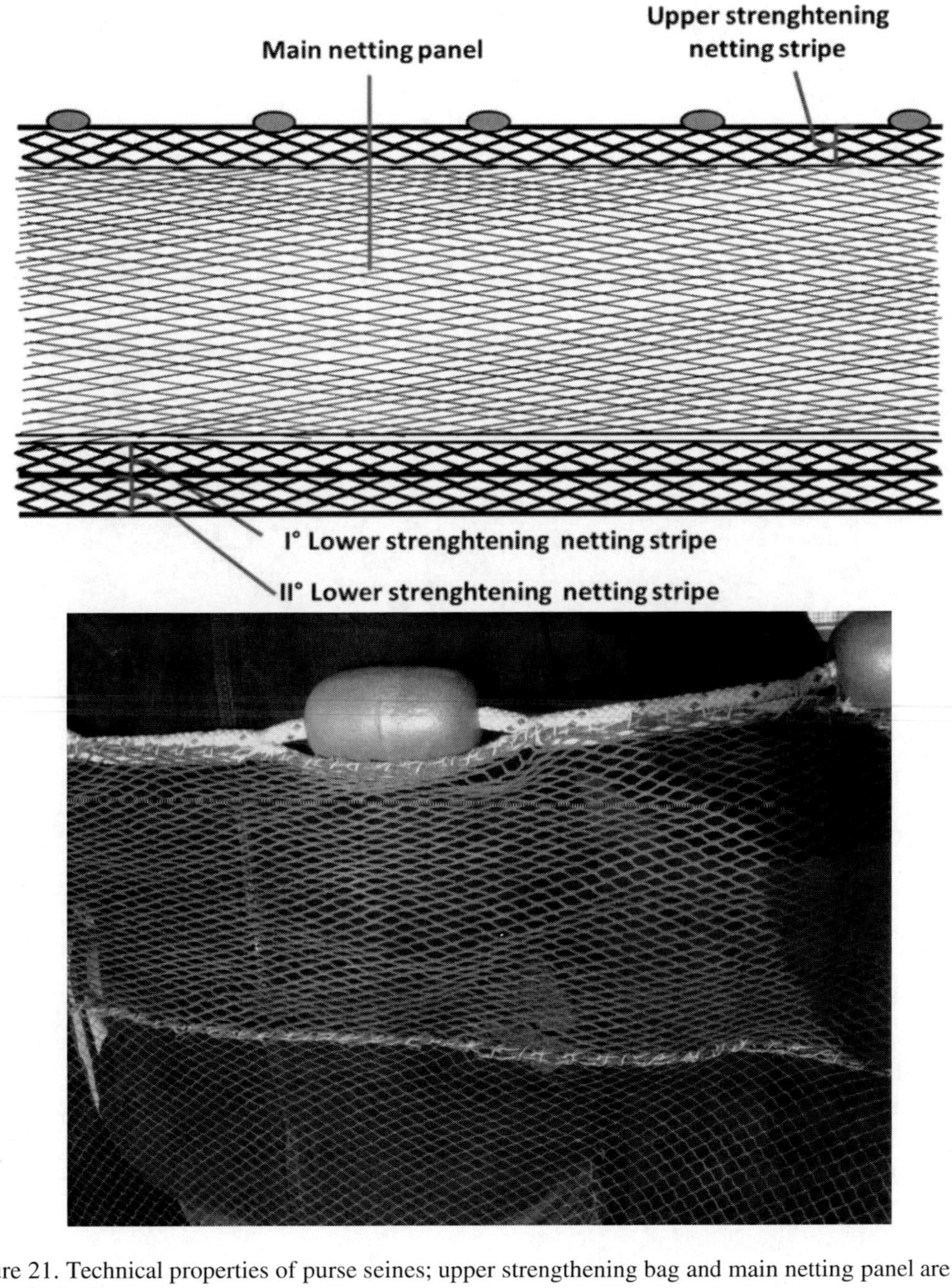

Figure 21. Technical properties of purse seines; upper strengthening bag and main netting panel are shown in the lower picture.

The fully extended net drop or fictitious drop is a useful but theoretic parameter. The fully extended or stretched net drop is also called "Fictitious net drop" because in real situation, when the net is operated, the net height is never completely stretched. Therefore, the real net drop during fishing operations is usually quite different from the fully extended net drop. The real net drop can be measured with high precision only by using sensors mounted on the leadrope. However, a possible way to estimate the real net drop is by calculating the vertical hanging ratio starting from the horizontal hanging ratio. The hanging ratio (*E*) is defined as the ratio between the length of the final rope (*HLL*) and the fully extended length of the side of the netting to which it is attached prior to being hung (British Standard 4440: 1974) and is expressed as:

$$E_1 = \frac{L}{L_0} = \frac{HLL}{MS \cdot HNM}$$

where *L* is the hung length of the netting or the mounted length of the main mounting rope and *Lo* is the length of the same netting when fully extended (Figure 22). The side of the netting is usually calculated by multiplying the mesh size (*MS*) and number of meshes wide (*HNM*). The numerator and denominator are both expressed in the same unit of length. This ratio may be given as a fraction or as a decimal or percentage. Sometimes this information may be given as the number of meshes per unit length of rope.

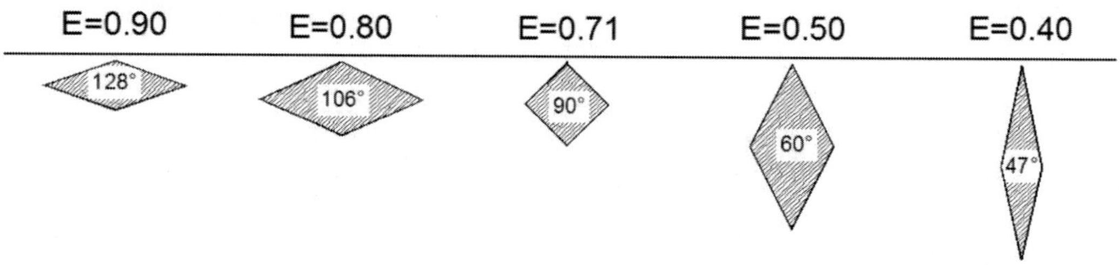

Figure 22. Example: effect of changing hanging ratio on mesh geometry (Prado, 1990).

The secondary hanging is defined as:

$$E_2 = \frac{H}{Ho}$$

where *H* is the hung height of the netting or the mounted height of the main mounting rope and H_o is the height of the same netting when fully extended. Therefore, the real net height during fishing operation could be defined as the hung height or mounted length of the side hanging line, that is *H*, with a certain margin for error. This is mainly because the weight of leads of the groundrope obviously could influence the real net drop, but there is not any other option to guess the real net drop. The horizontal and vertical hanging ratio are related by the formula:

$$E_1{}^2 + E_2{}^2 = 1$$

Therefore, the real net drop or Effective net drop or Hung height (*H*) can be calculated as:

$$H = H_0 \times E_2$$

$$H = H_0 \times \sqrt{1 - E_1{}^2}$$

In any case *it should be taken into consideration that this is a geometric and theoretic measure that could not fully reflect the real situation.*

Moreover if we assume that the ideal shape of purse seine during fishing operation is equal to a hemisphere, it is possible to calculate the minimum (*Min. ND*) and maximum (*Max. ND*) theoretical net drop.

We may think the floatline of a purse seine set on the surface as a circle, thus having a length equal to a circumference. In order to close the bottom of the purse seine completely (by hauling in the purse line), without hauling in any of the net or deforming it, the depth of the net should be at least equal to the diameter of the circle formed by the float line at the surface (Vold et al., 2012; Figure 23). Therefore, the minimum net drop (*Min. Drop*) will be equal to:

$$Min.Drop = Length\ of\ headline/\pi$$

This means that the minimum drop would be about 30% of the headline length.

In the same way, if we consider the extreme net configuration without any "distortion", we can assume that the net assume a shape of a perfect hemisphere once the purse line is completely hauled in (Figure 24). In this case the net drop, that is the theoretical maximum net drop without netting distortion (*Max. Drop*) will be equal to half circumference.

$$Max.ND = Length\ of\ headline/2$$

This means that the Net drop of a purse seine could range between two limits

$$Min.ND < Net\ Drop < Max.ND$$
$$\boldsymbol{Length\ of\ headline/\pi} < Net\ Drop < \ Length\ of\ headline/2$$

With these theoretical relationship firmly established it is possible to immediately verify the minimum net drop required for a purse seine having a headline length of 800 m (EC Reg. 1967/2006) is:

$$\boldsymbol{Net\ Drop = Length\ of\ headline/\pi\ = 800/\pi = 254.6\ m}$$

Therefore, a net drop of a least 254 m is essential in order to avoid any net distortion during hauling operations, by operating with a purse seine with a maximum headline length of 800 m.

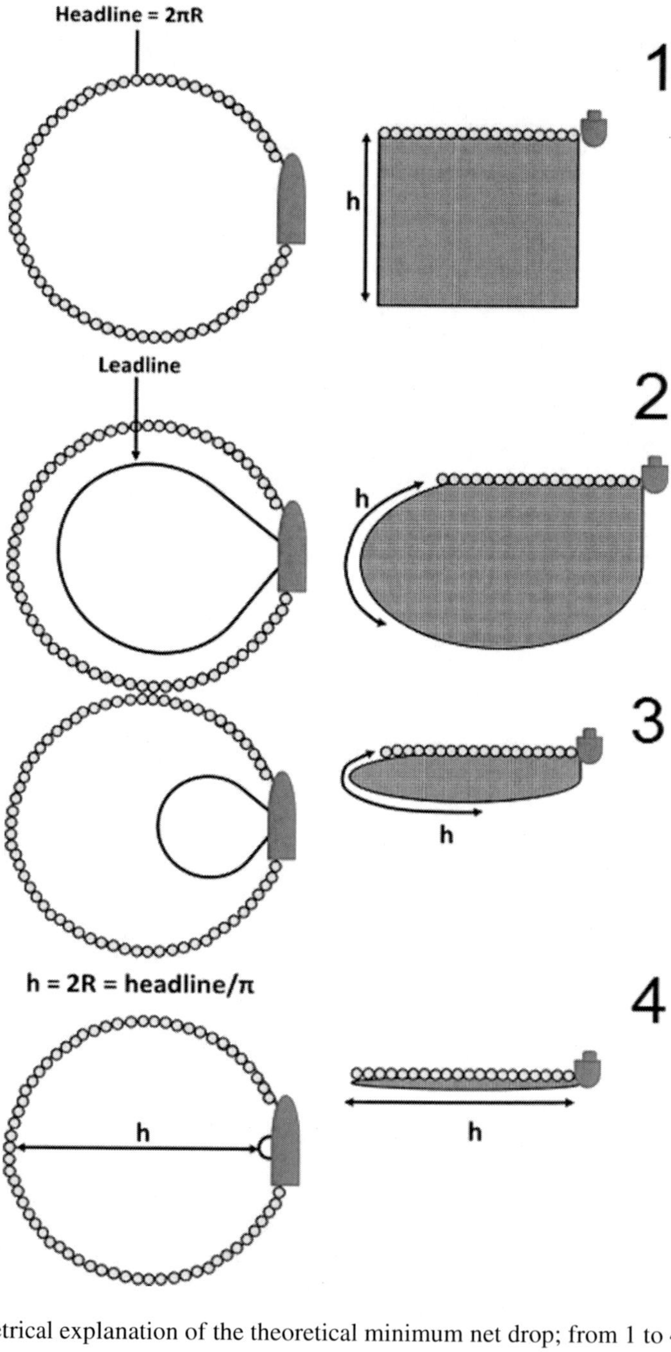

Figure 23. Geometrical explanation of the theoretical minimum net drop; from 1 to 4 the different hauling phases.

Surrounding nets can be used by a large range of vessel sizes, ranging from open boats up to large industrial vessels. Currently, surrounding nets are commonly used in the Mediterranean together with artificial "Fish Aggregating Devices" (FADs). FADs are permanent, semi-permanent or temporary structures or devices systematically used in some surface and mid-water fisheries to concentrate schools of fish. Buoys and floats close to the surface usually comprise the major part of the aggregating system. Some FADs are permanent structures (anchored) while others are moveable (free drifting). In the Mediterranean Sea FADs are mostly made of palm leaves used on the sea surface to aggregate pelagic species such as *Coriphaena hippurus, Naucrates doctor, Seriola dumerili.* However, there is a growing concern about the discard of small fish, sharks, rays, and some other species captured as bycatch when fishing near or under floating objects.

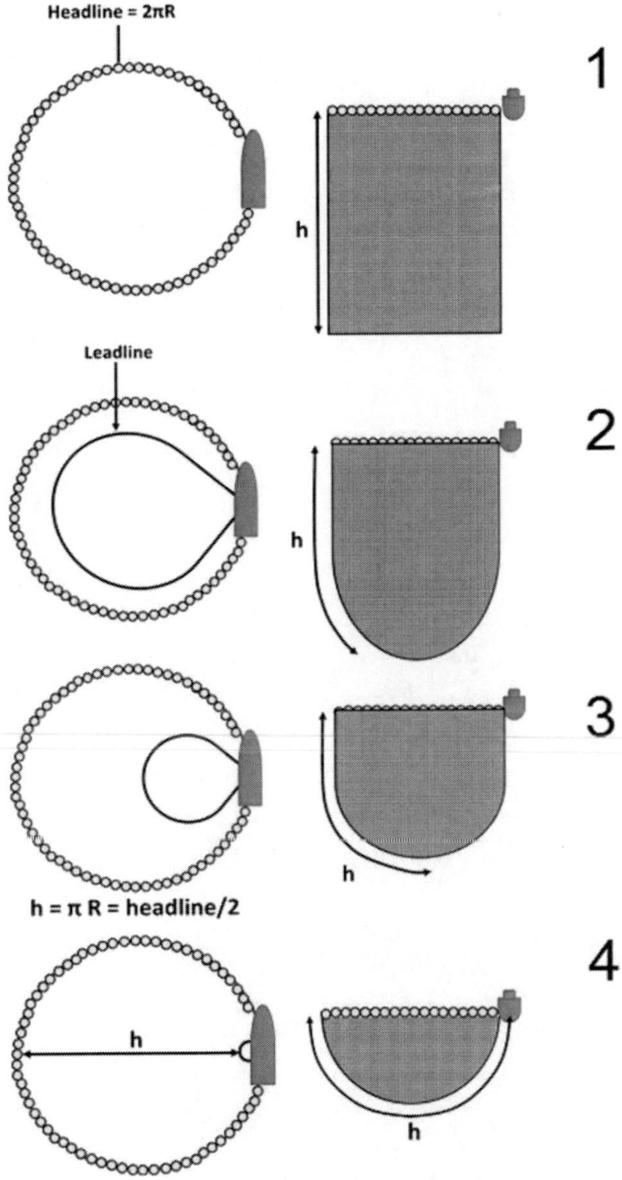

Figure 24. Geometrical explanation of the theoretical maximum net drop; from 1 to 4 the different hauling phases.

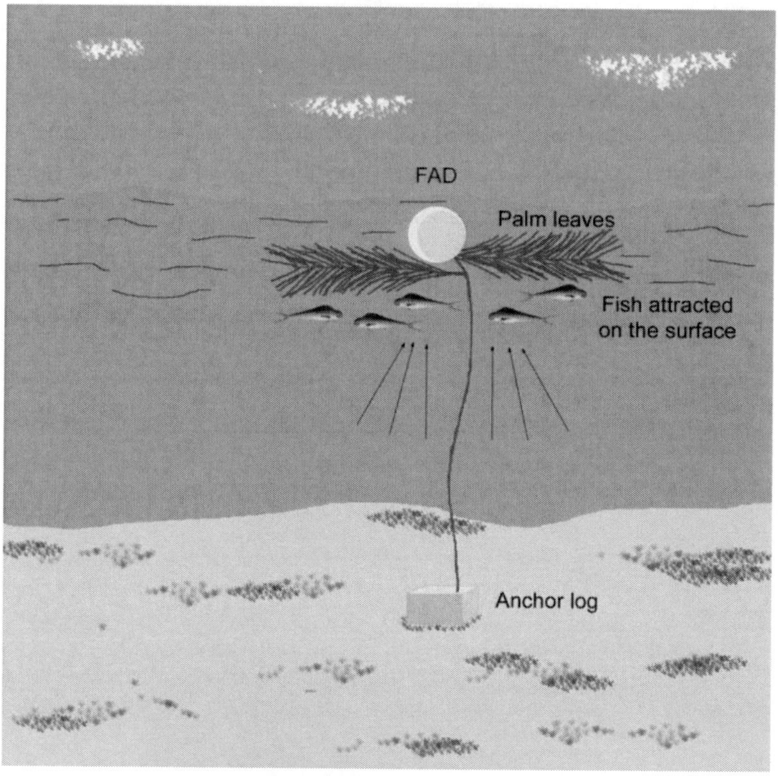

Figure 25. FAD scheme (top) and FAD commonly used in the Mediterranean Sea (down).

Surrounding nets without a purse line are used by artisanal fisheries to catch demersal and pelagic species. This type of net is hauled simply recovering firstly the ground rope and are quite similar to boat seines. The only limitation to the use of surrounding nets could be in too shallow waters where the water depth is less than the height of the surrounding net during the fishing operations making a risk of damage to the fishing gear.

2.4. BOAT AND BEACH SEINES

Seines are long nets usually operated by hand from the shore or by small boats. Thus, it is possible to identify two types of seines; the beach seines and the boat seines. The seine net design can be similar in some way to a bottom trawl. The net generally has a cone-shaped design, a groundrope with leads and a headrope with floats that defines the vertical opening of the net, two wings and a net "body" ending with a bag even though the two wings are quite long in comparison to other parts of the net. It is also possible to define seines with a bag (similar to a codend) and without a bag. In the seines without a bag, however, the central part of the net has the smallest meshes and has a slack since this is the netting where the fish are driven and retained.

The catching principle is to surround an area with the net. Two long ropes fixed to the net wings are used both for hauling the net and for herding the fish. Net hauling can be done either by hand or with a winch.

The beach seine can be set either by hand or by boat in shallow waters near the shore. The net is hauled from land by two or more fishers (Figure 26). The beach seine can be towed directly by the two wings.

In the Mediterranean, boat seines are operated from small boats without specific deck equipment (Figure 27). In some cases, hydraulic winches are used to recover the two ropes and rope drums can be used for storing them. The net is operated by dropping the rope into the sea on the external side of fish school with one end having a floater (marker buoy) and the other connected to the wing of the net. When the entire wing and the body of the net are shot into the water, the other wing is dropped followed by the second rope which circles the floater until reaching it. The two ropes are then retrieved on board and left so the net can drop until it almost brushes against the seafloor. The gear is then hauled with the vessel being held stationary by the anchor. The ropes are slowly pulled back on board, then the wings. Whereas the recovery of the ropes and wings can be mechanical, that of the body and bag is manual. The speed of the boat in circling and net dropping phase is always slow (1-2 knots) and completely null during the catch of the fish school. The conditions of the sea and weather strongly limit this fishing activity. In some areas, plastic bags or tatters are used at intervals along the two ropes to scare fish and drive them towards the net.

One of the main technical properties of seine nets is the prevalence of the weight of the groundrope on the buoyancy of the headrope. This means that nets touches the bottom during towing but the impact exerted by this type of net on the seabed is less destructive compared with a bottom trawl considering net dimensions and weight, and towing speed (Figure 28).

The seines are commonly used to catch demersal species having small size, such as fry of Sardine, transparent Gobiids, sandeel or to catch juveniles of commercial species that are used after that in aquaculture. Therefore, small meshes are commonly used in the codend even in violation of legislation.

Figure 26. Beach seines operated from the shore. This practice is often carried out illegally in the Mediterranean.

In the last 30 years, boat and beach seining have been commonly operated in the Mediterranean with nets having small meshes (3-5 mm mesh opening or even less), mainly around estuaries and within shallow waters where spawning, breeding and growing of most fish occur. The bycatch and the capture of juveniles is a reason of major concern for fishing practices carried out in the coastal areas with small meshes. Therefore, boat and beach seines have come under intense criticism in recent times by resource managers, policy makers and environmentalists due to its destructive effects on coastal and marine fisheries resources and to habitat degradation. Without a reasonable management limiting the gear, the areas and the seasonal activity fishing practice may reduce the recruitment potential of most fish species especially the small pelagic stocks shared by countries.

Figure 27. Boat seine operated by small vessel.

Figure 28. Bottom impact exerted by a boat seine.

In the last few years, there have been requests for restrictions of boat and beach seining and in some countries these gears have been banned. However, this action may be beneficial to improve the natural assets of the coastal environment and resources, but it would on the other hand increase the vulnerability of fishers' communities to poverty since beach seine fishery mainly provides employment and means of livelihood to several coastal rural folks.

2.5. Dredges

Dredges are active gears that are actively dragged on the bottom to catch bivalves (mussels, oysters, scallops, clams), gastropods, sponges or sea urchins. Dredges can be towed by the fishing vessel or completely controlled by hands (hand dredges). The dredges operated from a vessel can be either dragged by the main boat engine (boat dredge) or hauled by a motorized winch from an anchored vessel (mechanized dredge). Some dredges, the most important from a commercial point of view, are finally equipped with hydraulic gear (hydraulic dredges).

The dredges usually consist of a net bag or a metal basket mounted on a rigid frame or rod of variable size and shape whose lower part may carry a scraper blade or a toothed rake. The dredge may or may not be mounted on two lateral sledge runners to prevent it from digging into the substratum. The catching principle of all the dredges is that the gear (a frame or cage equipped with a rake or a chain) digs into the sea sediments to enable the escapement of organisms within it, and the catch is driven in a sort of bag or sieve which allows the sediment and water to run out. Average daily and annual production of the dredges varies greatly between zones and from one year to the next, since this depends on the condition of the resource, environmental conditions, number of fishing vessels and number of fishing days.

Dredges pulled by hand or by manual winch are normally used in shallow waters with or without a boat to catch bivalves, gastropods or sponges (Figure 29).

Figure 29. Hand dredge towed in shallow waters for the catch of clams: back and forth movement.

Figure 30. Boat dredges dragged on the bottom to catch mussels, sponges and sea urchins.

Boat and mechanized dredges have a quite simple design (Figure 30); the gear is usually a metal frame (rectangular or triangular shape) with a width no greater than 1.2-1.5 m. The net is no longer than 1.5 m and a total weight of the gear must not exceed 12-15 kg. The side of the frame touching the sea bottom can be equipped with a rake or a chain. Certain types of boat dredges consist of a box dredge of about 1.5 m wide weighing 30 kg, rigged with 5-7 cm teeth along the lower leading edge and a net bag to collect the catch. Two inclined wooden boards (150x30x2.4 cm) are fitted to the front of the metallic frame to act as a depressor, to keep the gear in contact with the seabed and, moreover, to press it on to the bottom to facilitate the penetration of the teeth into the soft sediment (Figure 30 c,d). The gear is equipped with 4 skids that enable the sliding of the gear on the seabed. The net, made of polyamide and polyethylene, consists of two panels, the upper and the lower panel 4 m long. This type of gear is commonly used to target bivalves such as the Pectinidae (variegated scallop, *Clamys varia*), the bearded mussel (*Modiolus barbatus*); these species are strongly firmed to the seabed so that the fishing practice implies more than one single passage along the same track to scrape the mussels from the bottom.

Figure 31. Boat dredge repeatedly towed along the same track to scrape mussels from the bottom.

Figure 32. (Continued).

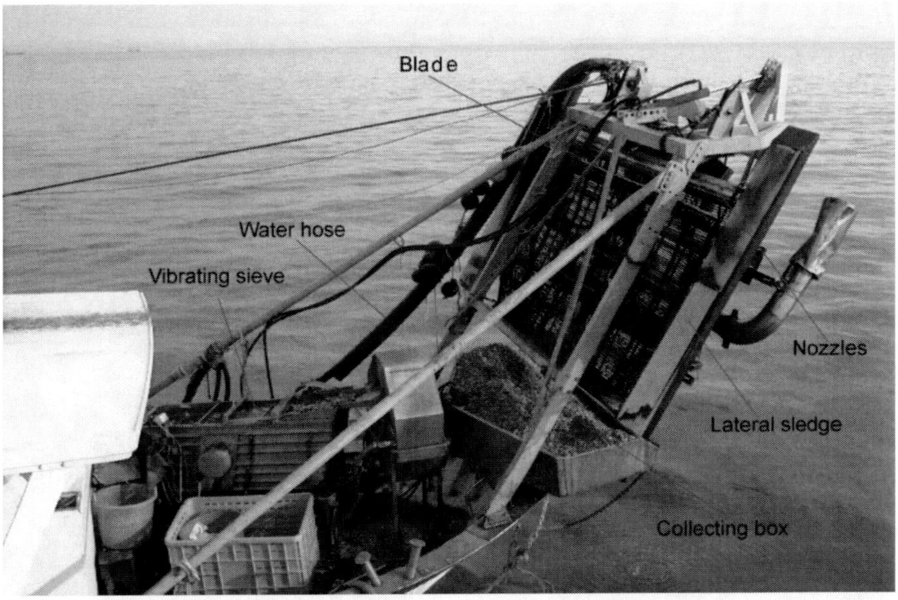

Figure 32. a) Detail of an hydraulic dredger used to catch mussels (Chamelea gallina, Ensis spp, Callista chione); detail of the dredge.

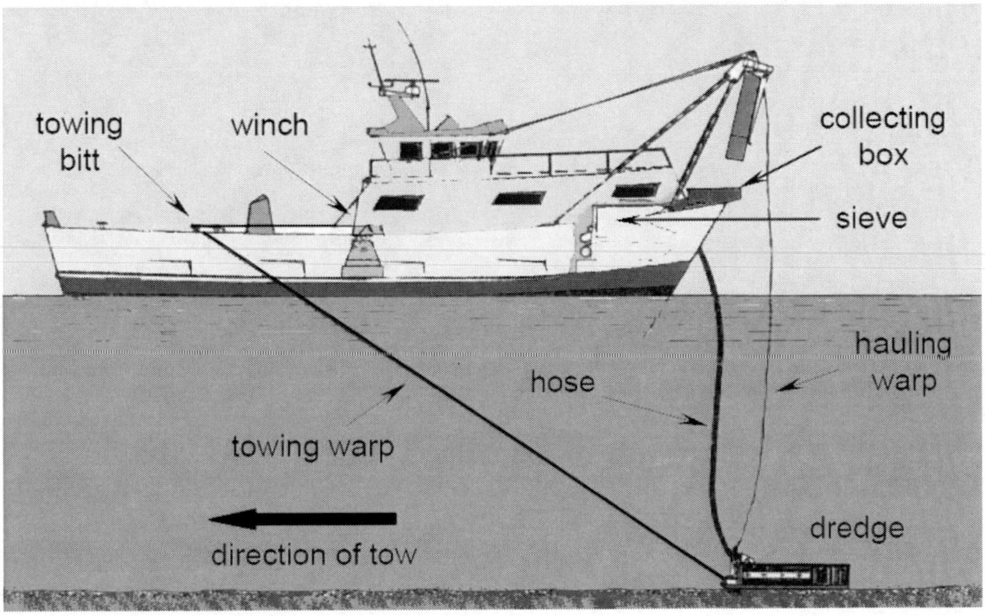

Figure 33. Scheme of the hydraulic dredge towing procedures.

Hydraulic dredges design and functioning is much more technologically advanced. In this type of dredge the front of the cage is connected by a hose to a centrifugal water pump that ejects water under pressure from the nozzles at the mouth of the dredge and inside the dredge cage (Figure 32). The water pressure is used to drive out the marine organisms living within the bottom sediments and to facilitate their catch. Once suitable fishing grounds are reached, the dredge is lowered from the bow of the vessel. Due to its weight the cage immediately sinks on the bottom and the vessel moves astern for a variable distance. At the end of the tow, the cage is hauled and its contents are spilled into a collecting box. The catch is finally sorted in different commercial classes by means of a mechanical vibrating sieve, sometimes composed of two or more sieving planes or grids (Figure 34). The unwanted material is directly thrown back to sea.

The catching efficiency of these gears strongly depends on the dimensions of the dredge and towing speed. While the selectivity is mainly affected by the characteristics of the dredge and the sieves, space between the rods, mesh side of the grid (if a grid is used as a sieve), diameter of the holes (if a metal sheet is used to sort the catch) and the mesh size of the bag etc.

Figure 34. Detail of the mechanical vibrating sieve (a) and sieving procedures to sort the catch in different commercial classes (b).

2.5.1. Dredges Impact

The dredges, mainly the hydraulic ones, are considered as highly impacting fishing gears; these gears are dragged on the bottom and deployed within coastal areas, in relation with the distribution of the target species. The most evident physical effects of trawling are the furrows produced by the dredge on the sea bed (Lucchetti and Sala, 2012). In area heavily impacted by dredging, side scan sonar studies showed that a relatively flat seabed can be transformed into a series of irregularly spaced deep, wide furrows produced by the combined actions of the deeply penetrating cutting blade and the water jets eroding the sediment. The density of the dredging marks can be really high with less than 2 m among the tracks (Figure 35). Studies show variability in the time required for dredge tracks to resolve but in most case trenches and tracks created by harvesting dissipate quickly. Time required for a trench to return to pre-harvesting conditions from dredging is usually very site specific and depends on natural sedimentary processes.

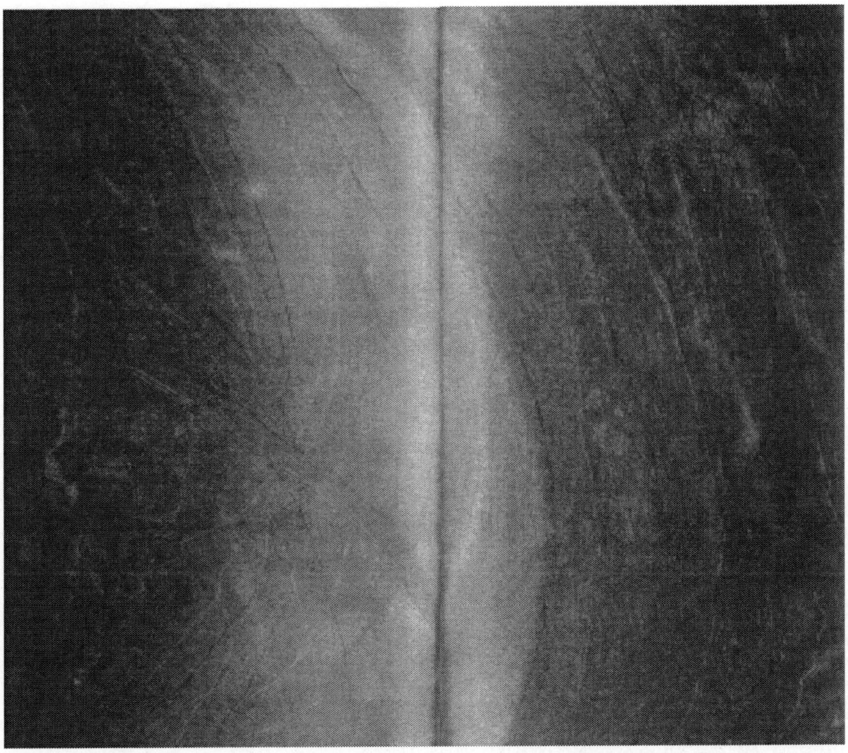

Figure 35. Side scan sonar sonograms: area heavily exploited by hydraulic dredge.

The use of dredges has been shown to destabilize and partial modify the sediment conditions, decreasing the habitat complexity and leading to fluctuations within the benthic communities. Intense and continued dredging trawling on the continental shelf has a noticeable effect on water turbidity, which must be considered in addition to natural processes (Palanques et al. 2001). This interaction may induce turnover of the sediment surface, oxygenation of deeper sediments, and release of organic matter and nutrients. These factors may locally have positive effects by stimulation production, however negative effects such as destruction of habitat and mortality in the benthos communities and the release of pollutants may be more important (Pilskan et al. 1999; Cloern, 2001; Duplisea et al. 2001). Scientific studies carried out in the Adriatic Sea showed that hydraulic dredging causes the destabilization and partial modification of sediment conditions, decreasing the habitat complexity and leading to fluctuations within benthic communities (Brambati and Fontolan 1990).

These factors can lead to an overall reduction in production. The turbidity can also negatively affect photosynthetic processes, mainly in coastal areas, by decreasing light intensity (Tudela, 2004). Under extreme conditions of turbidity, visual feeders and photosynthesizing plankton may be disturbed. Moreover sediment re-suspension can potentially reduce the survival of bivalves and fish from gill obstruction or by preventing their movements and/or burrowing activity.

Moreover when dredges move along the seafloor sediments, surface benthic organisms can be removed, crushed, buried, or exposed. Tubes and burrows may be scraped away and erect and sessile organisms removed from the substrate surface. Clam harvesting can result in direct mortality due to the impact with the dredge, and indirect mortality can occur when shellfish are caught in the dredge, washed out, or discarded

A radical reduction in the abundances of infaunal organisms in favour of few dominant opportunistic species has been widely demonstrated as a consequence of dredging activity. Moreover the impact on the eggs and larvae of other commercial species should be considered.

The impact of dredging can be summarized as follows.

- physical effects
- suspension of sediment, increased turbidity
- disruption of the sediment surface
- creation of dredge tracks (10-20 cm deep)
- mortality and damage in benthic organisms: shell damage to bivalves caused by dredging
- effects of suspended sediment on marine organisms
- reduction in benthic organisms
- Temporary Decline in Commercially-sized Shellfish
- Indirect Effects of Dredging Related to Habitat Disruption
- Effects of Dredging on Biodiversity
- Alterations of the Biochemistry and Physiology of Shellfish after Dredging
- Evidence of Reduced Shellfish Recruitment

Therefore, the management of dredging activities deserve a particular concern among fisheries managers.

The system adopted for the management of dredge fishing is highly structured. It features a combination of measures to control both output and input, and technical measures. A whole series of management measures are usually applied including:

- output control measures limiting the catches (daily catch quantities) of the fleets belonging to each harbour or region, by departure at sea and unit of time. One of the principal management measures traditionally implemented in some countries in the Mediterranean is the suspension of fishing activities for short or long periods to allow greater protection of the resources and the marine environment;
- closed areas;
- input control measures addressing production factors, through restrictions on the number of boats licensed to fish and the time (daily fishing hours) available for fishing;
- input control measures limiting the size and the engine power of the fishing vessels;
- technical measures regulating the relationship between input and output, governing the size of catches, the temporary or seasonal suspension of exploitation activity and the closure of fishing areas;

- technical measures on the characteristics and dimensions of the fishing gear and sieves.

Fishing with dredges over phanerogams beds (e.g., Posidonia oceanica) and hard substrates is generally forbidden. Technical characteristics prohibit dredge fishing on seabeds where there is a presence of marine phanerogams since the gear needs loose and not particularly compact sediment to move through the seabed, with the dredge acting as a sieve. This is impossible in the presence of marine phanerogams, whose long roots transform the sediment into an almost rigid carpet.

2.6. Passive Nets

Passive nets include the stationary or fixed nets (anchored to the bottom) and the driftnets (drifting with the currents).

Gillnets, trammel nets, combined nets, and small driftnets are currently among the most important passive nets largely used along the Mediterranean coasts for the catch of a great number of demersal, benthic and pelagic species.

Gillnet, trammel net and combined net are known as fixed nets because each end of the net is connected to an anchor, weight or stake to prevent the net moving with the current, and usually both ends are marked with a buoy. Gillnet consists of a single netting wall generally made of mono-filament nylon. Trammel net is made of three panels of nets with the inner panel being made either of twisted or mono-filament nylon, while the outer panels have a larger mesh size than the inner one and are generally made of twisted nylon filament. Combined net is a bottom-set gillnet combined with a trammel net that constitutes the lower part. The driftnet is made of a single netting panel. This kind of net is left free to drift with the current usually near the surface or not far below it.

All the above net types are kept vertical by a floatline (upper line/headrope) and a weighted groundrope (lower line/footrope). Small solid floats usually made of plastic and spherical, cylindrical or egg-shaped are attached to the headrope, while weights are evenly distributed along the footrope. A recent innovation implies the replacement of large number of floats by "swimming lines". Lead weights are evenly distributed along groundrope but the lower line can also be made of lead cored rope that negates the need for additional weight.

The nets can be fished alone but commonly in groups placed in a line (fleets of nets). The two ends of a net are fixed to the sea bottom by means of weights or anchors that maintain the gear in place. According to the target species, the gear can be designed to fish either close to the bottom or, more rarely, in the water column. Each weight or anchor is joined to a marker buoy or flag on the surface.

Fixed nets can be hauled by hand at least from shallow or moderate depth, but automatic net haulers are very common now.

Fish can be caught because gilled, entangled or enmeshed in the netting. Gilling and entangling are two different principles of catching:

- Gilling is where the mesh surrounds the fish just behind the gill cover;
- Wedging or enmeshing is where the mesh surrounds the body as far as the dorsal fin;

- Entangling is where the fish is held in the net by teeth, maxillaries, fins or other projections without necessarily penetrating the mesh; and
- Entrapping is where the fish is entrapped in bags or pockets of netting. This is the typical catching method of a trammel net.

The catching method strongly depends on the gear properties (Figure 36). Fixed nets can be designed to catch both demersal and pelagic species. They are, especially the gillnets, fishing gears with a high degree of selectivity that's regulated by the mesh size.

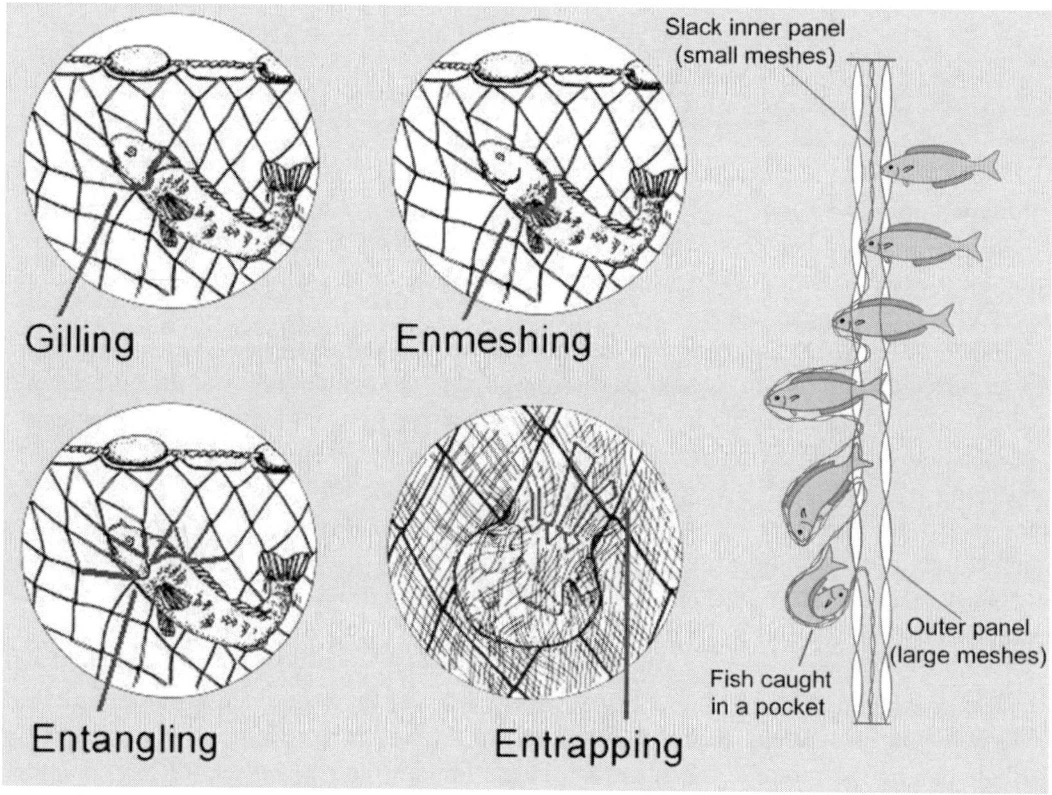

Figure 36. Catching processes in passive nets. On the right a detail to show how fish are entrapped in a trammel net (Adapted from Clark, 1930).

There are three main types of fixed nets: trammel net, gillnet and combined net:

Gillnets (FAO Code: GNS) are strings of single netting walls held vertically in the water by floats on the headrope and weights on the footrope (Figure 37). Fish can be caught gilled, entangled (by the spines, teeth etc.) and enmeshed.

Gillnets are highly selective gears and the mesh size is usually decided on the basis of the target species. Therefore, it is possible to say that gillnet are species-specific gear.

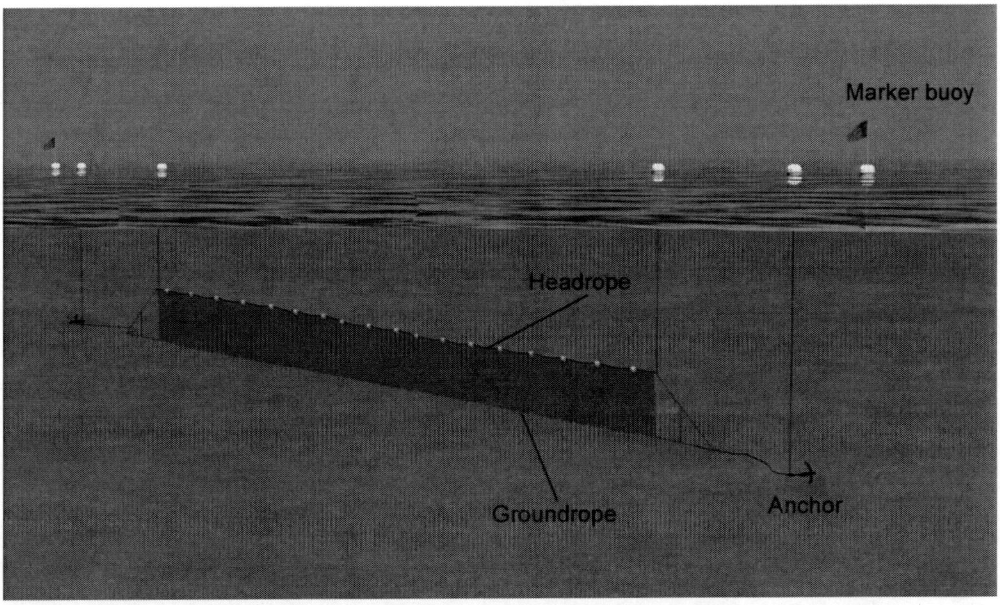

Figure 37. Main features of a gillnet (Adapted from FAO, 2011).

Trammel nets (FAO Code: GTR) are fixed nets consisting of three layers of nets; a slack, small mesh, inner panel of netting and two outer layers of netting that are tight and have a larger mesh size (Figure 38). The three panels are hung jointly in parallel on a single headline and leadline. The catching process implies that the fish entangle themselves in a pocket of small mesh webbing between the two layers and large meshed walls (Figure 36). This type of capture is exclusive of the trammel net. Fish can be caught also because gilled, entangled or enmeshed in the netting.

Therefore, the trammel net is considered to be less selective than the gill net.

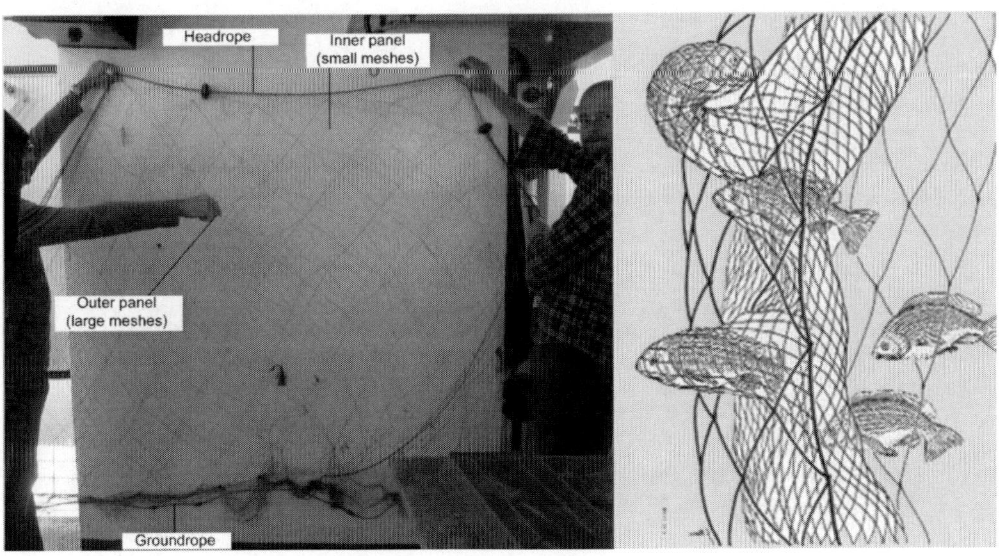

Figure 38. Main features of trammel net and catching properties.

Combined nets (FAO Code: GTN) are nets fixed on the bottom composed of two parts; a trammel net in the lower part, and gillnet in the upper part (Figure 39). This type of nets combines the catching methods of trammel net with those of gillnet. In addition to these main fixed net types, some other ones are locally used.

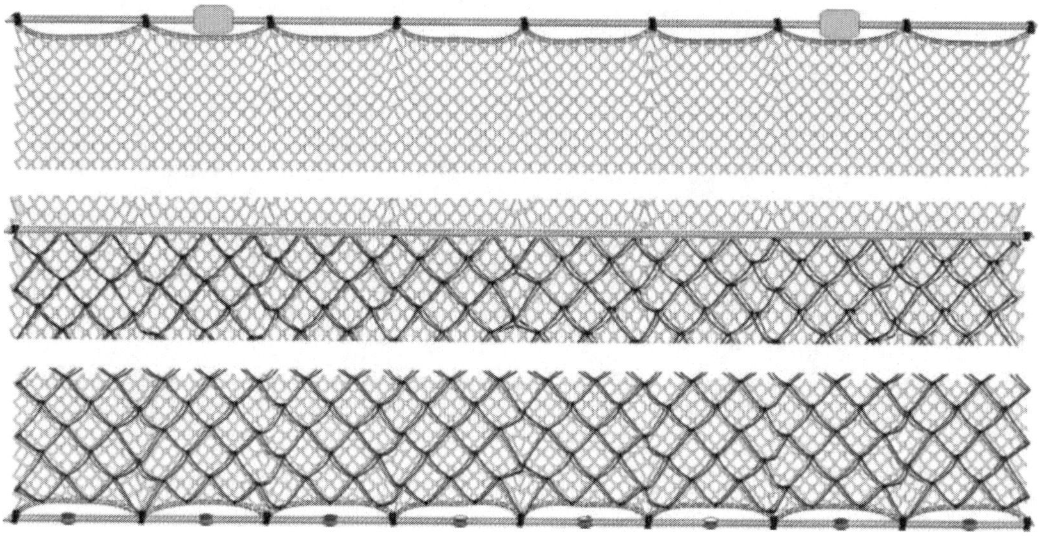

Figure 39. Main features of a combined net.

Encircling gillnets (FAO Code: GNC) are gillnets set vertically in shallow waters or in lagoons encircling school of fish (Figure 40). After the fish has been surrounded by the net, noise or other means (frequently, by fishers moving themselves in the water, or by beating the sea surface with an oar) are used to scare the fish and force them to gill or entangle themselves in the netting. It is a technology commonly used by groups of small-scale fishermen.

Figure 40. Encircling gillnets used in shallow waters (by FAO, 1980).

Gillnet on stakes (FAO Code: GNF) is a fixed gillnet stretched between a series of stakes inserted into the bottom in inter-tidal area along the cost, estuaries, lagoons and shallow waters (Figure 41). These nets are usually hauled by hand. The catching process is analogous to a simple gillnet.

Figure 41. Gillnet on stakes commonly used in shallow waters and in the lagoons.

The driftnet (FAO Code: GND) is a form of gillnet (one single netting panel) that's held on sea surface or at a certain distance below it by floats to drift with the current either independently or attached to a boat (Figure 42). Driftnets are mainly used in the Mediterranean to catch both small and large pelagic species. However, the use of driftnets for the catch of large pelagic species such as Swordfish, Bluefin tuna and like tuna species is forbidden in several Mediterranean countries. The driftnets targeting large pelagic species can extend up to 50 km in length, and 20-30 m in drop.

This gear has proven to be quite unselective, catching a great number of non-target species (bycatch) such as sharks, rays, dolphins, whales, turtles and sea birds.

In 2005 the GFCM (General Fisheries Commission for the Mediterranean) endorsed an ICCAT (International Commission for the Conservation of Atlantic Tunas) decision that rendered a total ban on driftnet fishing, irrespective of net size, on large pelagic species (including tuna, sharks, and Swordfish) applicable to all Mediterranean States. Small driftnets are used by artisanal fishers to catch dense schools of fish, such as mackerel, Anchovy, and Sardine.

The GFCM recommended that "*no vessel flying the flag of a Contracting Party of GFCM may keep on board, or use for fishing, one or more driftnets whose individual or total length is more than 2.5 kilometers*" (REC.CM-GFCM/22/1997/1 Limitation of the use of driftnets in the Mediterranean).

Figure 42. Main features of a driftnet.

2.6.1. Passive Net Impact

The catching process in passive nets depends on intercepting fish as they move, therefore these gears are considered as more selective than active gears such as a bottom trawl. Size-selectivity is generally good with gillnets, but the trammel nets are generally less selective since both small and large specimens can be caught. In general, the species-selectivity is often poor in passive gears since most of the species can be caught. Moreover, it should be considered that the direct mortality is really high since most fish die during the capture so there is little potential benefit from releasing bycatch.

Recently, the entanglement of endangered and sensible species such as sea birds, turtles and marine mammals has aroused concerns. In these cases the survival rate is also dramatically low. In general, these species get entangled in the tentative of feed from the prey previously caught in the net. In the case of driftnets marine mammals and turtle can be caught simply because they are unable to see the net. Driftnets and passive nets used in deep-water s are responsible for considerable mortality in pelagic and demersal elasmobranch species which frequently get entangled in them.

Important commercial catches of *Alophias vulpinus* and blue shark, as well as minor discards of *Dasiatys violacea*, have also been reported for the driftnet fishery operating in the Ligurian Sea (Di Natale and Notarbartolo di Sciara, 1994). Little attention is paid nowadays to the impact of Mediterranean fisheries on seabird populations even if this type of bycatch should be considered with more attention.

2.7. Pots and Traps

Traps and pots are gears in which marine organisms enter voluntarily either for sheltering or attracted by a bait but are hampered and retained from coming out. They are designed in a way that the entrance/s itself becomes a non-return device allowing the fish to enter the trap but making it impossible to leave the catching chamber. Pieces of fish or squid are often used as bait but artificial baits are also in use.

A pot is designed in the form of small or large cages or baskets with dimensions ranging from around half a meter to two meters. They consist of a rigid or semi-rigid frame made of various materials (wood, wicker, metal rods, wire netting, plastics) that may or may not be covered with netting (Figure 43). Pots shape, design and type of bait are developed on the basis of tradition and target species ethology: cuttlefish are attracted by pots for shelter during reproduction season, fish and crustaceans are attracted by bait etc.

The size of pots may vary from small octopus pots (Figure 44) to large basket pots such as the tubular fyke nets for eels and cuttlefish. Traps are usually set in rows where 20-30 or even more pots are normally joint to a main line (Figure 45). Fishermen visit traps regularly for collecting catch and replacing bait if used leaving the gears set in the same place for several days. Like pots, traps can more easily be moved from one fishing location to another. Pots are hauled either by hand if the depth is not too large and there are only a few pots to be retrieved, or with a hauler (for deep water fishing or hauling a series of pots).

Pots are normally used for catching a wide range of species including fish, lobster, crabs, shrimps, octopus, and cuttlefish. Regarding the target species, pots are generally operated in a very wide range of depths either in inland, estuarine, or in marine waters and they are usually set on the bottom but few models targeting fish and prawns are designed to be set in mid-water.

Certain types of traps are provided with large guiding panels made from netting to lead the fish into the catching chamber (Figure 46). These traps usually consist of cylindrical or cone-shaped netting bags mounted on rings or other rigid structures (fyke net). The fyke nets are fixed on the bottom by anchors, ballast or stakes. The fyke nets may be used either separately or in groups. Fyke nets are generally operated in coastal zones and shallow waters either in inland, estuarine, or marine waters. They are usually set in contact with the bottom in areas with strong currents.

Figure 43. Different types of pots used in the Mediterranean.

Figure 44. Pots used in the Mediterranean to catch Octopus.

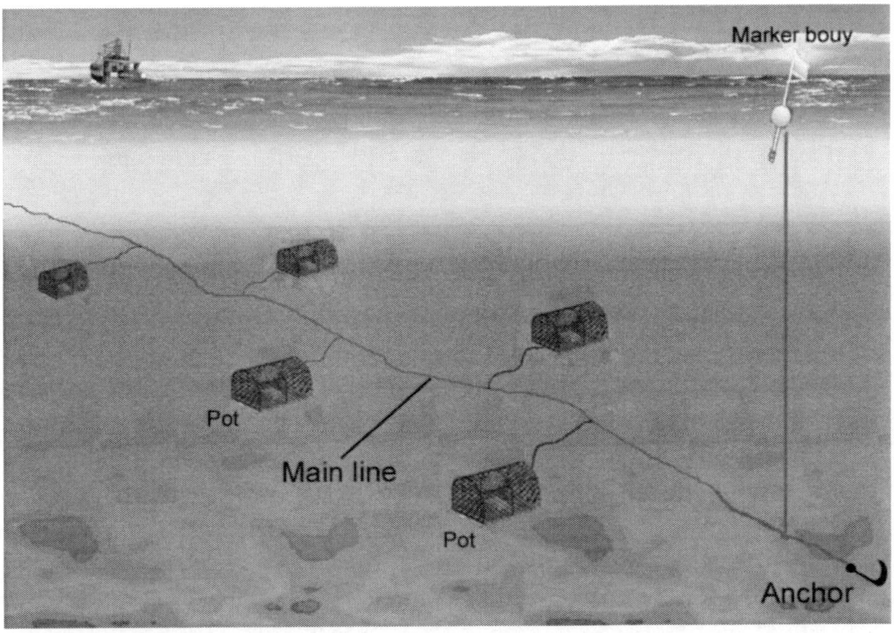

Figure 45. Schematic of pots set in row.

Pots and traps are in general considered as highly size and species selective gears.

Marine species that are caught with traps are normally still alive and uninjured when the gear is hauled on board, so that generally unwanted bycatch organisms can be released with a good chance of survival although factors such as on-deck injury, and air exposure, and thermal shock may influence the survival of released organisms. Thus, traps offer the potential for low bycatch mortality in comparison with many other fishing methods.

Figure 46. (Continued).

Figure 46. a) Fike net used in lagunes; b) Fike net setting c) Fike net hauling.

Large stationary nets or barrages are finally used in certain areas to catch migrating fish. In the past large "tonnara" traps were widely used to catch Bluefin tuna during the reproduction season between May and June (Figure 47), but presently these large and selective traps are worldwide replaced by purse seines. The tonnara trap comprises a long netting panel (leader) used to drive the migrating tunas towards the catching system composed of 3-5 chambers until the last chamber (death room). Similar traps ("tonnarelle") were also used to catch migrating pelagic species such as *Seriola dumerili*. Really large traps made by a complex system of barrages are also used in the channels between the lagoons and the sea to catch demersal species such as *Dicentrarchus labrax*, *Sparus auratus*, Mugilids, *Anguilla Anguilla* etc.

Figure 47. Large "tonnara" used to catch Bluefin tuna.

Figure 48. Stationary traps used in the channels between the lagoons and the sea to catch demersal species.

3.8. Longline

The longline is a fishing gear that comprises a main line carrying numerous hooks on branch lines (snoods) of variable lengths and spacing depending on the target species. It may be set either at or near the bottom (bottom-set longline; Figure 49), drifting in mid-water, or near the surface (surface longline; Figure 50).

Longlines are used in many areas of the world to catch a variety of species. Bottom longlines are commonly used to catch demersal species like hakes, groupers, sea basses, sparids, flatfish etc. Surface longlines comprise a main line many kilometers long (up to 60-80 km) and they are mainly deployed to target large pelagic species such as Bluefin tuna (*Thunnus thynnus*), Swordfish (*Xiphias gladius*), and Albacore (*Thunnus alalunga*). Sometimes, sharks are considered as target species. The success of this type of fishing activity is to set the longline for optimized overlap between the distribution of hook depths and the supposed preferential vertical habitat of the target species.

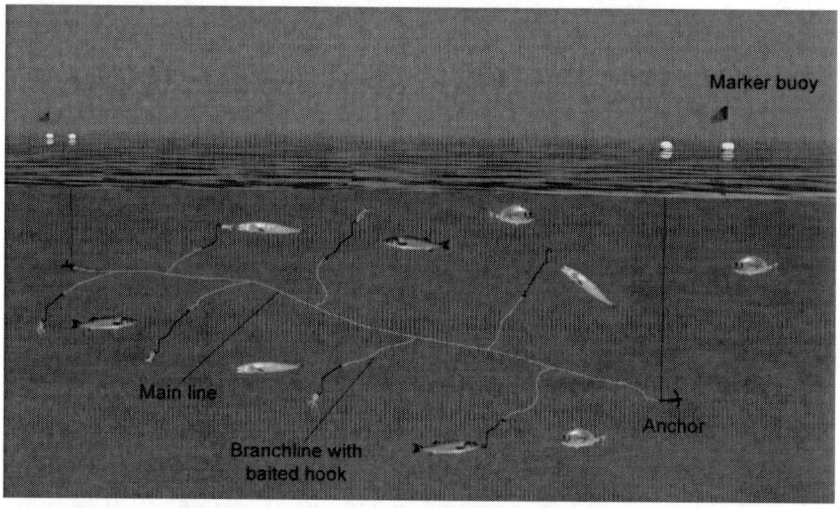

Figure 49. Scheme of a bottom longline.

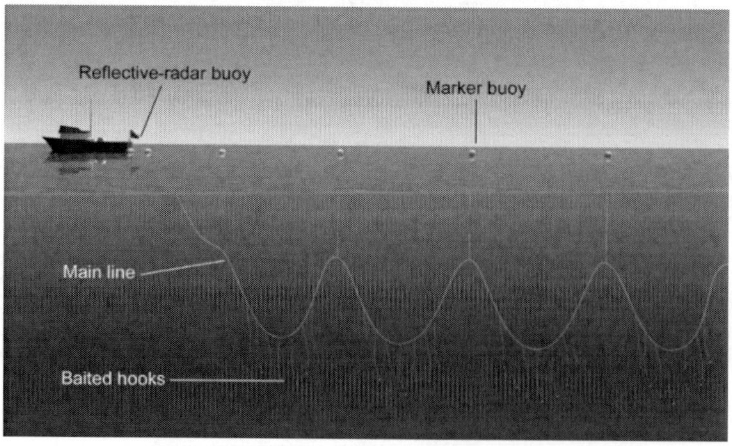

Figure 50. Scheme of a drifting longline.

The hook is the core of the longline system. The hook mainly consists of an eye, shank, length, width, point, gap and throat (Figure 51).

The main factors affecting the catch ability of a longline are:

- The number of hooks;
- The hook shape and material;
- The bait type (squid baits, fish baits, artificial bait, lightsticks etc.);
- The longline rigging: main line diameter, branch lines length, space between snoods etc.;
- The bait colour;
- The fishing depth: depth at which the branch line is positioned;
- The location of fishing grounds in relation to the topographic and oceanographic features, sea temperature etc.;
- The location of fishing grounds in relation to the target species migration routes; and
- Total catch: higher catches make the longline more visible.

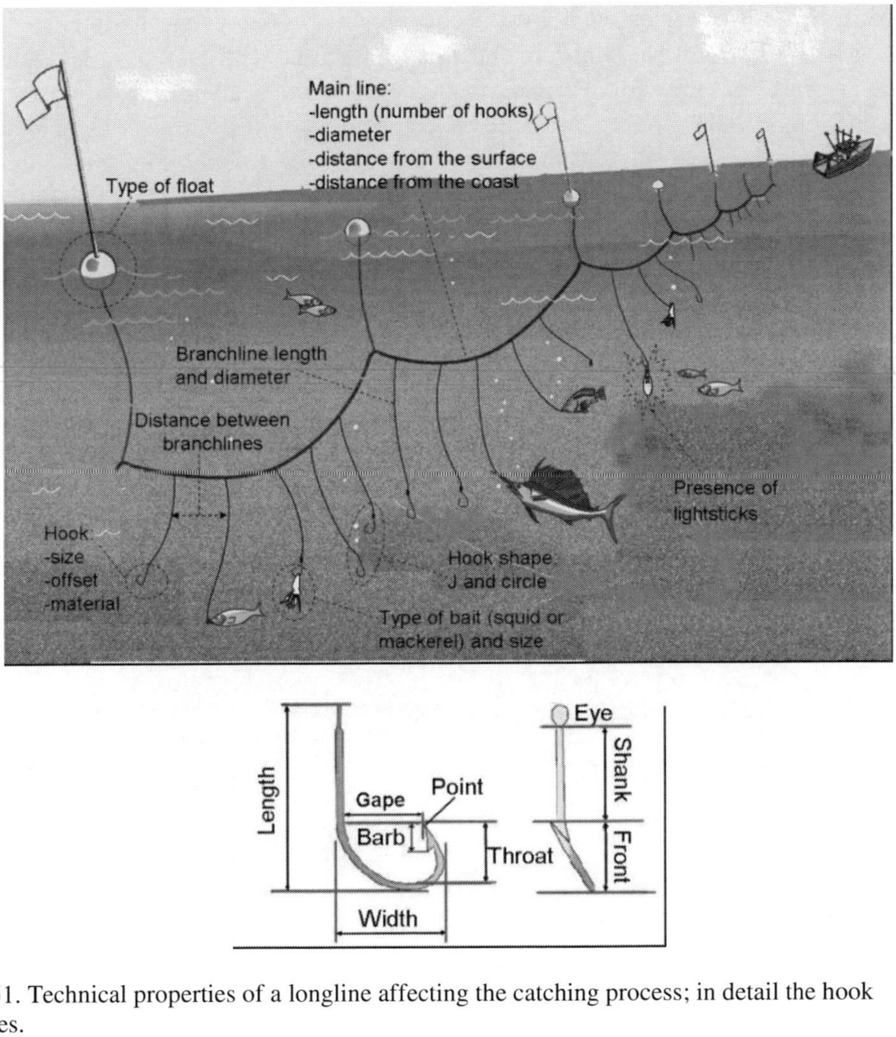

Figure 51. Technical properties of a longline affecting the catching process; in detail the hook properties.

Species- and size-selectivity of a longline can be modified by bait and hook. Hook size strongly influence the size that can be captured (Alòs et al., 2008). Only large fish can bite a big hook and that's way one of possible management measures in the longline; to set a minimum hook size. Hook shape (J or circle) influences the location of hook in the prey; circle hook seems to be more easily pierced in mouth of the prey, while J hook is more easily swallowed leading to high direct mortality rate.

2.9. Longline Impact

In addition to the target species, the longlines are responsible for the incidental catch of non-target species mainly turtles, elasmobranches and even seabirds and in many cases these are simply thrown dead or dying back into the sea. In some longline fisheries, fishes are usually alive when hauled on board and many may survive if released carefully. However, shock due to change in temperature and pressure may strongly influence the survival of released fish.

The drifting longlines deployed over the continental shelf (for tuna-like species) or offshore waters (for Swordfish, Albacore and Bluefin tuna) are considered as the main threat to marine turtles in the Mediterranean (Lucchetti and Sala, 2010). The GFCM Recommendation 35/2011/4 on the incidental bycatch of sea turtles in fisheries in the GFCM Competence Area stressed that "*Contracting Parties and Cooperating non-contracting Parties of GFCM (CPCs) should ensure the implementation of fisheries management measures that strongly mitigate or eliminate the risk of incidental taking of sea turtles in fishing operations and/or the mortality associated with those incidental takings. Specimens of sea turtles accidentally taken in fishing gears shall be safely handled and released unharmed and alive to the extent possible.* [...] *Any event of incidental taking as well as releasing or discarding shall be recorded by the ship-owner/master in the logbook*".

Figure 52. Small Elasmobranches caught in a bottom longline.

Moreover, the catch of drifting longlines often includes undersized or juvenile fish of targeted species. This is mainly because the juveniles of Bluefin tuna can bite and swallow the small hooks used to catch for example the Albacore (*Thunnus alalunga*).

Large and small pelagic elasmobranches are regularly caught in the surface and bottom longlines (Figure 52). Some of these species are landed and marketed. The blue shark (*Prionace glauca*) is perhaps the most impacted species. Pelagic or benthopelagic stingrays (*Dasyatis spp.*) and marine turtles are also important victims of certain pelagic Mediterranean fisheries.

Chapter 3

TRAWL NET DESIGN

Several different net designs are used in the Mediterranean relevant to sea bottom characteristics, traditions, target species etc. Two main types of net can be distinguished:

- The two panels/faces trawls which are normally fit for the capture of benthic animals living close to the bottom like flatfish, red mullet, shrimp in particular but also demersal fishes like hake. Most of the traditional Mediterranean trawls belong to this type including the Italian "*volantina*" or the Spanish "*huelvano*" with a vertical opening that does not exceed 2 meters in general; and
- The four panels/faces trawls which normally have higher vertical opening reaching up to 5 meters. Such trawls are sometimes known as "*semi-pelagic*" trawl or "*rete francese*" "*GOV: Grand Ouverture Verticale*".

Based on Eigaard et al. (2011) and personal communications with Mediterranean net makers as well as knowledge of how trawl geometry, ground gear and mesh sizes are best matched to the behaviour and size of target species, five conceptual trawl typologies can be defined in Mediterranean. Each inventory observation will be assigned to one of the five trawl typologies according to the nominated target species:

1. *Demersal/bottom 2-panel trawls (OTB2).* They tend to have long sweeps and bridles with large wingend spread and low vertical opening (1-2 m). Typical Italian OTB2-trawls are the "Tartana" and "Volantina" or the Spanish Cadenero, Huelvano, Minifalda, etc. They are usually used to target mixed demersal species, such us European hake (Merluccius merluccius), red mullet (Mullus barbatus), whiting (Merlangius merlangus), poor cod (Trisopterus minutus capelanus), monkfish (Lophius spp.) and Nephrops (Nephrops norvegicus). They are commonly entirely made up of knotless PA-netting (Figure 53).
2. *Demersal/bottom 4-panel trawls (OTB4).* OTB4 trawls are generally used to target crustaceans such as deep-water rose shrimp (*Parapenaeus longirostris*), deep-water red shrimp (*Aristaeomorpha foliacea*), and Nephrops (*Nephrops norvegicus*). All shrimp species are predominantly captured by a process of filtration. These trawls are characterised by having two bridles of 10-15 m to increase the vertical opening, which can be of around 2-4 m. Typical OTB4 are the Italian *Americana* trawl; the

Spanish *Cuadrado, Dos Bocas, Espada, Tangonero, Semitangonero*; and the French *Jumeaux, Filet a chains*. Manufactured mainly with knotted polyethylene netting, can have sometimes knotless polyamide netting in the lower panel (Figure 53).

3. *Pelagic 4-panel trawls (PTM4).* They are designed to catch shoaling pelagic fish that are off the bottom, such as anchovies (*Engraulis encrasicolus*), sardine (*Sardina* pilchardus), mackerel (*Scomber scombrus*), and horse mackerel (*Trachurus trachurus*). PTM-trawls are generally large trawls (e.g., *"Volante"* in Italian), with a high vertical opening, constructed with very big meshes (600-3200 mm) or ropes in the forward part of the trawl that herd fish towards the centre of the body of the trawl constructed in much smaller mesh size (Figure 53).
4. *Semi-pelagic 2-panel trawls (OTM2).* They are designed for the Tyrrhenian fisheries to catch demersal fish such as gilthead sea bream (*Sparus aurata*), sea bass (*Dicentrarchus labrax*), and silver scabbardfish (*Lepidopus caudatus*) which are generally close to the seabed but exhibit an upward migration. The volume swept by the relatively large meshes (120-1600 mm) is what determines the capture efficiency and OTM2-trawls are typically constructed for having a mean vertical opening around 3-4 m. This trawl typology has always a 4-cable rigging (Figure 53).
5. *Semi-pelagic 4-faces Trawls (OTM4).* As for the OTM2, these are designed for the Tyrrhenian fisheries to catch demersal fish. Compared to the OTM2 it has a higher vertical net opening and is more efficient for catching pelagic species. OTM4-trawls are designed to have a mean vertical opening around 4-10 m. This trawl typology has always a 4-cable rigging (Figure 53).

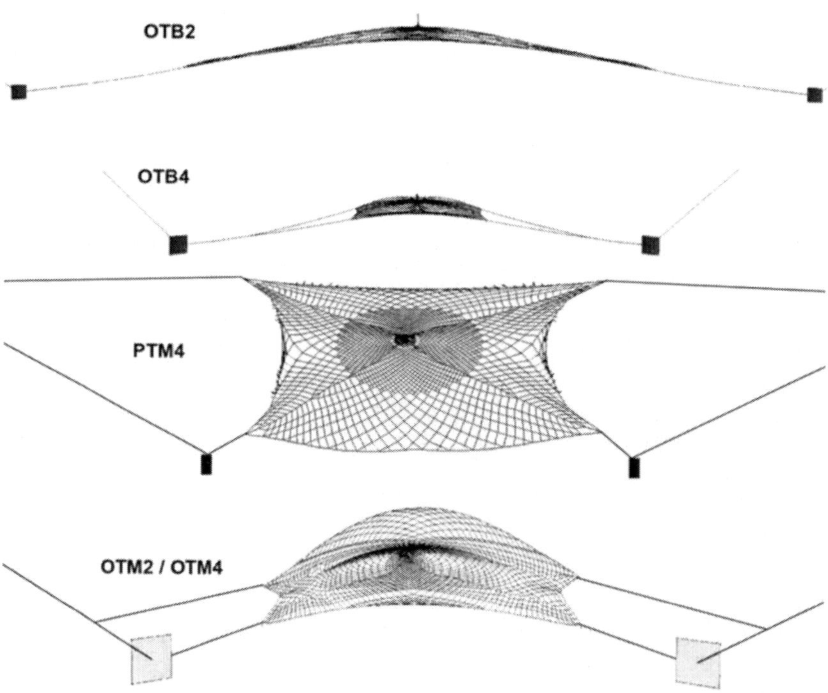

Figure 53. The five trawl types defined: Demersal/bottom 2-panel trawls (OTB2), Demersal/bottom 4-panel trawls (OTB4), Pelagic 4-panel trawls (PTM4), Semi-pelagic 2-panel trawls (OTM2), Semi-pelagic 4-panel trawls (OTM4).

3.1. Demersal/Bottom 2-Faces Trawls (OTB2)

The left side of the drawing below represents the Upper Panel (UP), while the right part represents the Lower Panel (LP). From the arrow down, UP and LP are equivalent and, therefore, only one panel is showed.

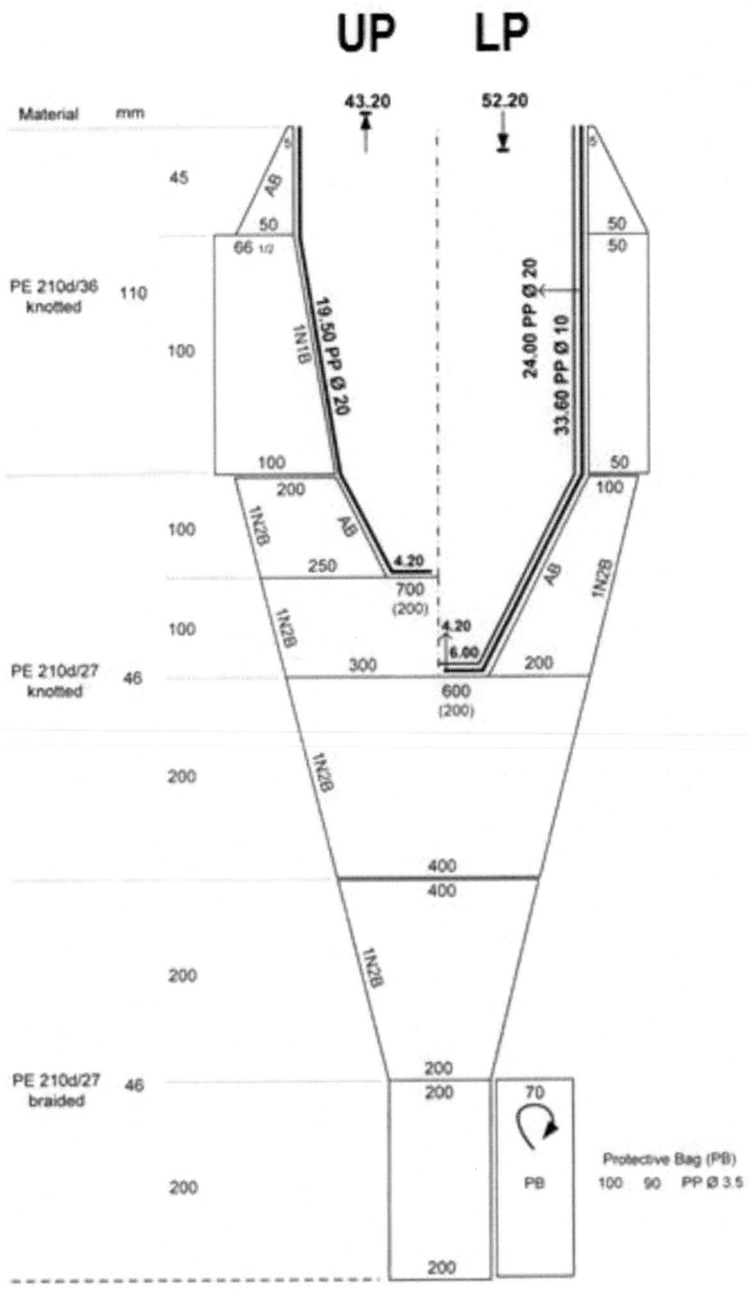

Figure 54. Scheme of a Demersal/bottom 2-faces Trawls (OTB2).

3.2. Demersal/Bottom 4-Faces Trawls (OTB4)

The drawing below represents a general 4-panels trawl, left panel is the Upper Panel (UP), right panel is the Lower Panel (LP) and the panel in the middle represents the two Side Panels (SP).

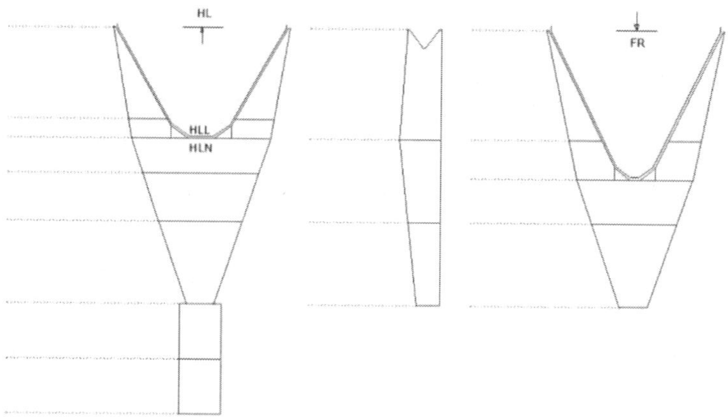

Figure 55. Scheme of a Demersal/bottom 4-faces Trawls (OTB4).

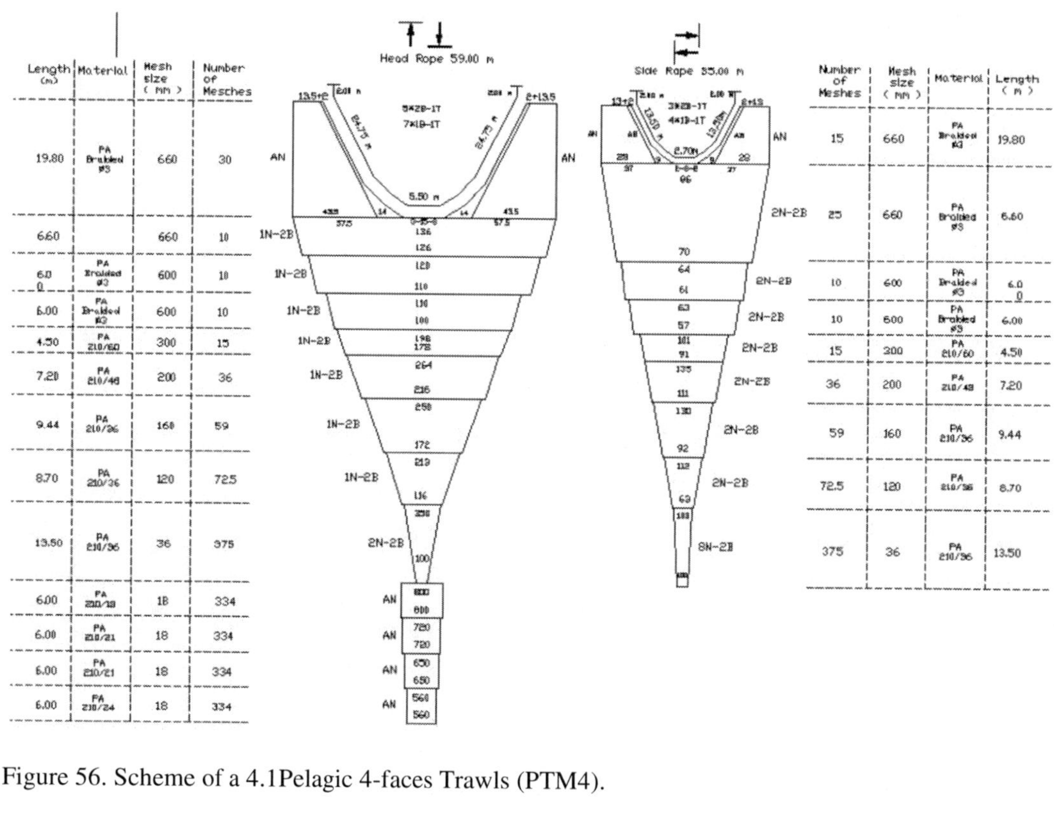

Figure 56. Scheme of a 4.1Pelagic 4-faces Trawls (PTM4).

3.3. Pelagic 4-Faces Trawls (PTM4)

The drawing below represents a general 2-panel trawl: left panel is the Upper Panel (UP) and the right panel is the Lower Panel (LP).

3.4. Semi-Pelagic 2-Faces Trawls (OTM2)

The left side of the drawing below represents the Upper Panel (UP) while the right part represents the Lower Panel (LP). Typical OTM2 trawls are the French High Opening (GOV) trawls.

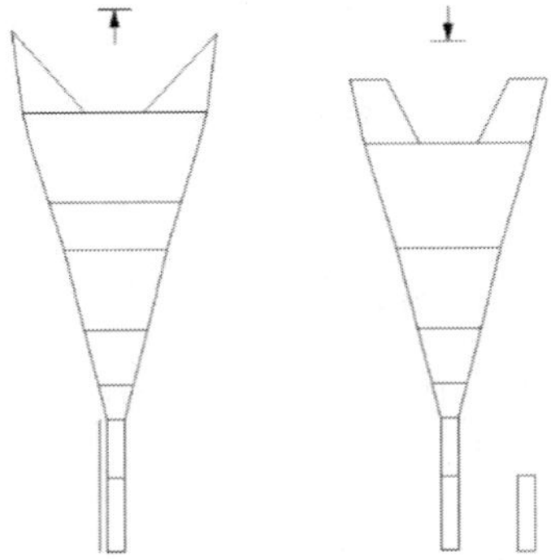

Figure 57. Scheme of a Semi-pelagic 2-faces Trawls (OTM2).

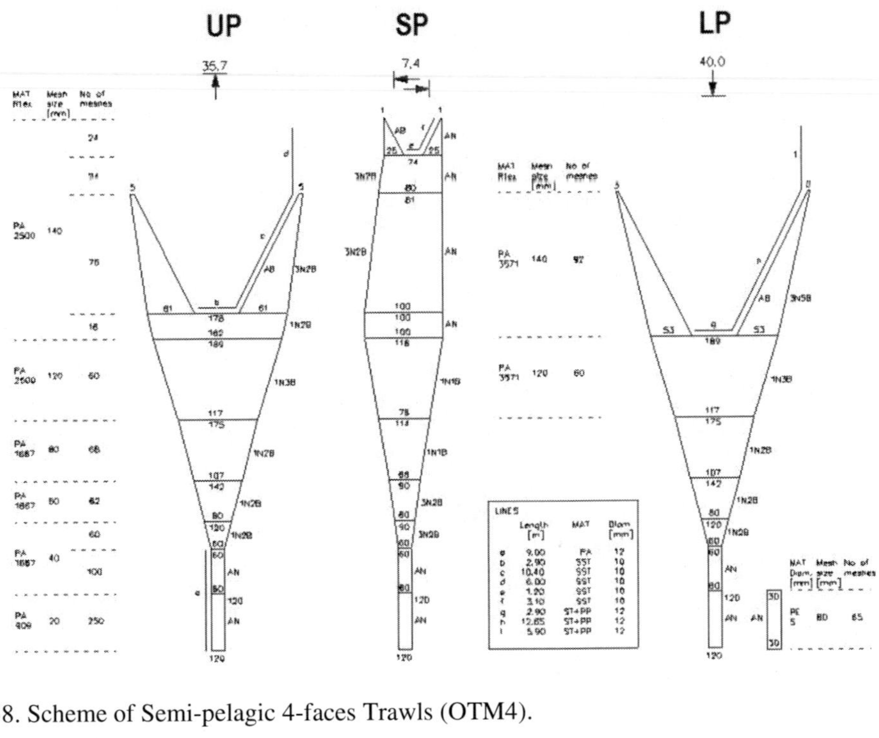

Figure 58. Scheme of Semi-pelagic 4-faces Trawls (OTM4).

3.5. Semi-Pelagic 4-Faces Trawls (OTM4)

The left side of the drawing below represents the Upper Panel (UP), the panel in the middle represents the two Side Panels (SP), while the right part represents the Lower Panel (LP). A typical OTM4 is the GOC73 trawl used in the International Mediterranean Stock Assessment programme (MEDITS).

1 - How to recognize main parts of the net: Top Wing, Lower Wing, Top Belly, Lower Belly, Top Codend, Lower Codend.

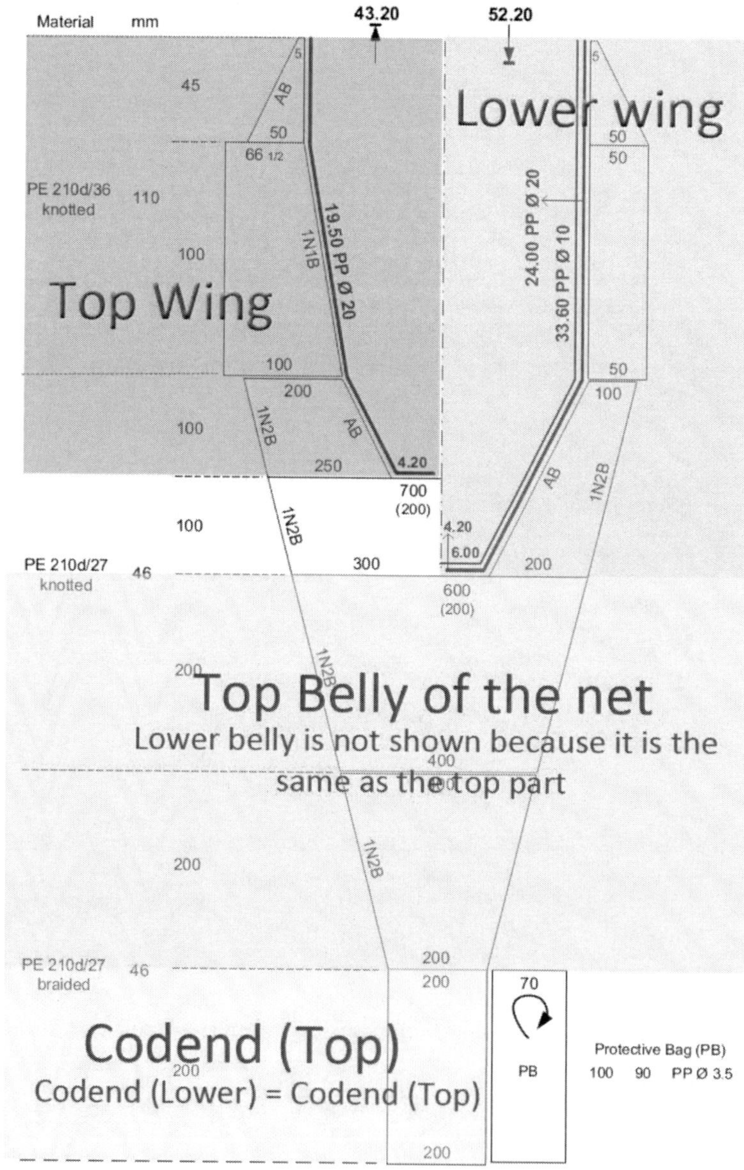

Figure 59. Main sections of a trawl net.

2 - How to identify mesh size of every panel MS [mm]

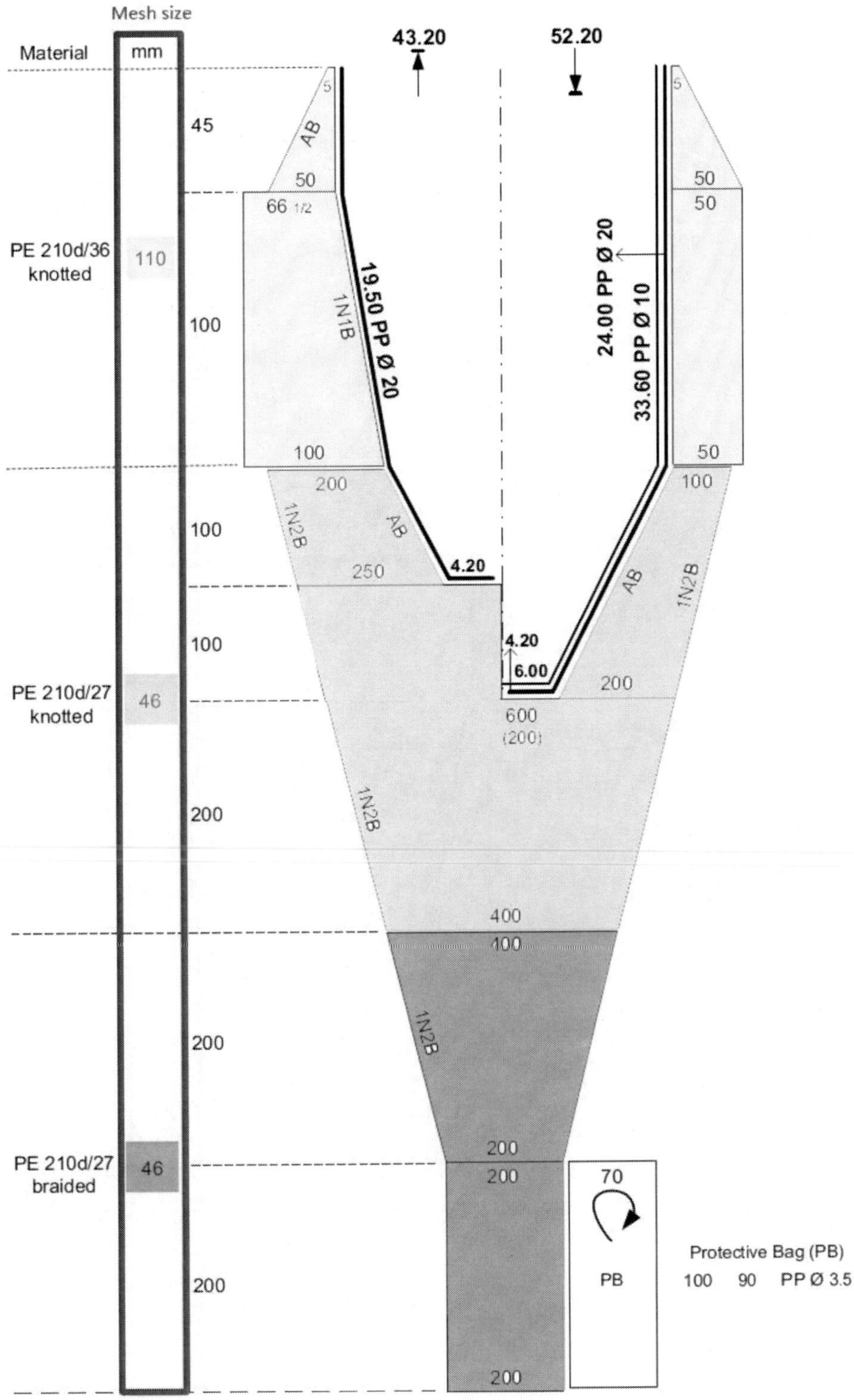

Figure 60. Details of the mesh size of different net sections.

3 - How to read and/or calculate headrope and groundrope length HL and FR

From the picture:
Total length of the headrope = *43.2* m

If total length is not shown on the plan of the net, then it can be calculated as following:
19.5 x 2 + 4.2 = 43.2 m

Total length of the groundrope = *52.2* m

If total length is not shown on the plan of the net, then it can be calculated as following:
24 x 2 + 4.2 = 52.5 m

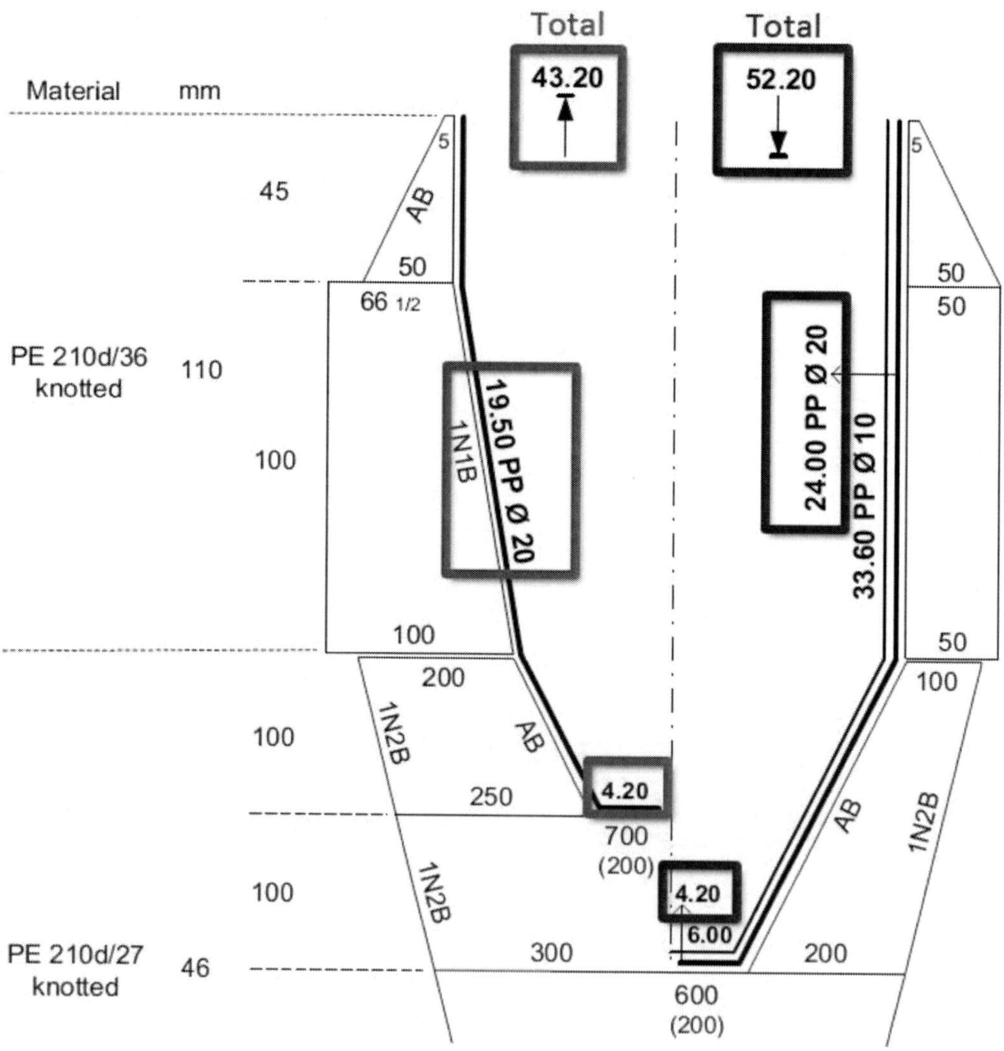

Figure 61. Headrope and groundrope calculation and reading.

NOTE: *19.50 PP ⌀ 20* means: length of the part of the headrope is *19.5*m, it is made of *PP* (Polypropylene) with a diameter of (⌀) *20*mm.

4 - How to determine the number of meshes in each panel (length of the panel) NMH

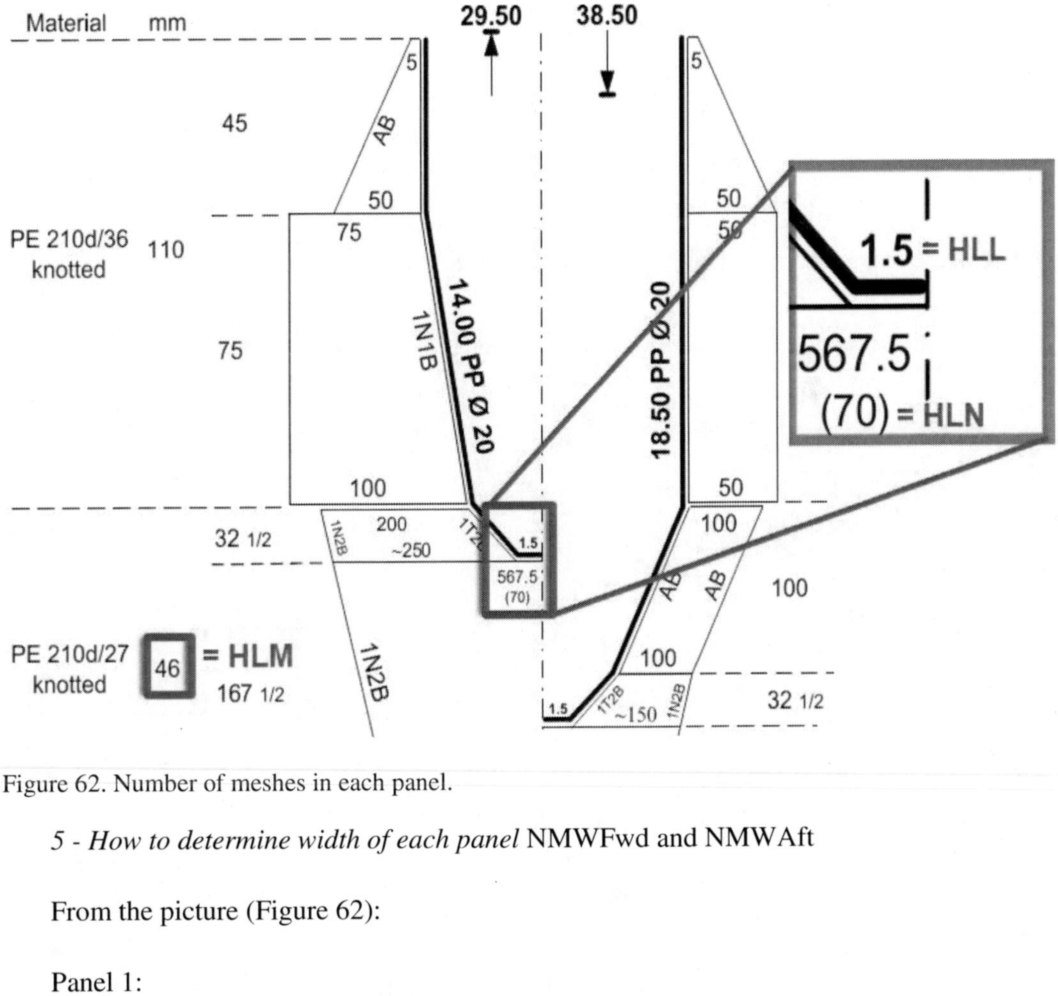

Figure 62. Number of meshes in each panel.

5 - How to determine width of each panel NMWFwd and NMWAft

From the picture (Figure 62):

Panel 1:
Number of meshes on the top of the panel (**NMWFwd**) = 5
Number of meshes on the bottom of the panel (**NMWAft**) = 50

Panel 2:
Number of meshes on the top of the panel (**NMWFwd**) = 200
Number of meshes on the bottom of the panel (**NMWAft**) = 250

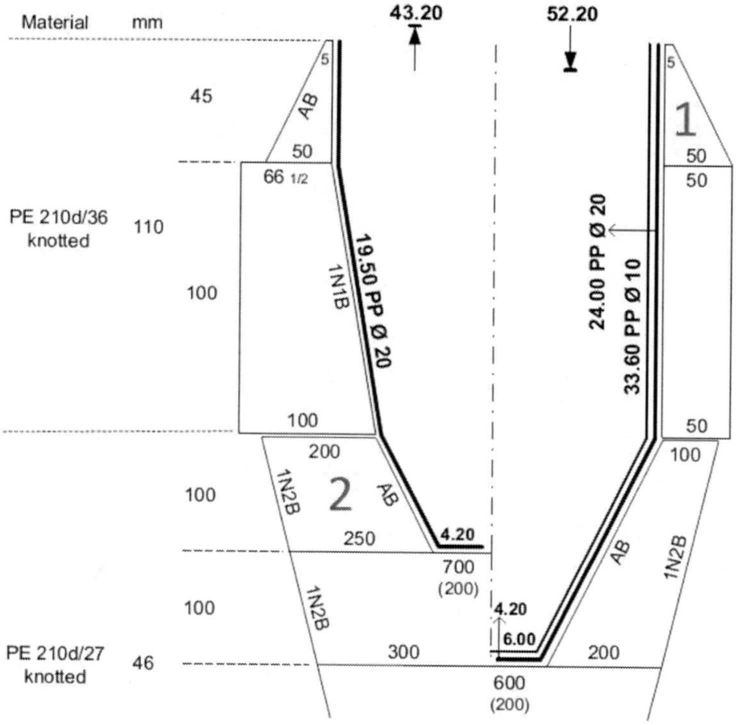

Figure 63. Width of each panel.

6 - Cuts used for shaping the netting T, N, B, AB.

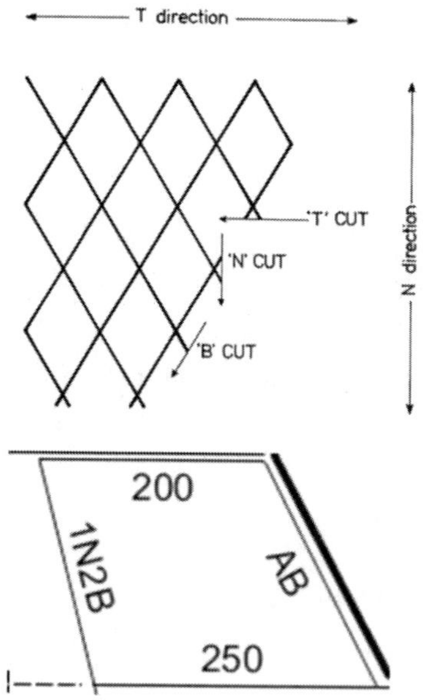

1N2B means: 1 cut in N direction and 2 cuts in B direction.
AB means: All cuts are in B direction.

Figure 64. Cuts used for shaping the netting.

7 - Difference between diamond mesh (DM) and square mesh (SM) netting. DM, SM.

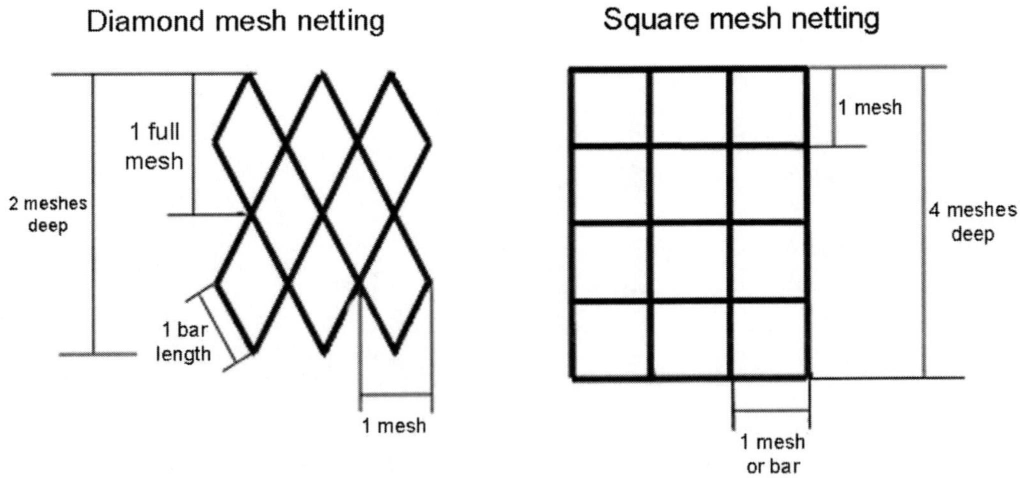

Figure 65. Diamond and square mesh panels.

IMPORTANT: some countries still uses mesh bar when defining trawl mesh, so it is important to emphasize which measurement are you inserting.

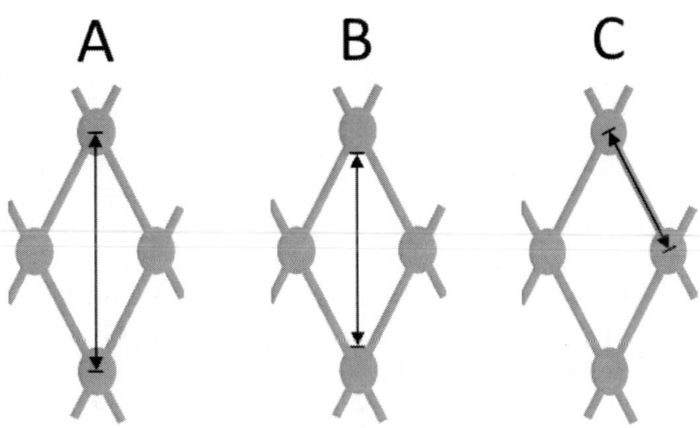

Figure 66. Mesh size parameters: A - Mesh Length; B - Mesh Opening; C - Mesh Bar or Mesh Side.

8 - Hanging ratio. Database reference E.

Hanging ratio (E) is commonly defined as:

Length of rope on which a net panel is mounted *(L) / Length* of stretched netting hung on the rope *(Lo)*

$$E = \frac{L}{L_0}$$

Example: 200 meshes of 50 mm stretched hung on a rope of 8 m.

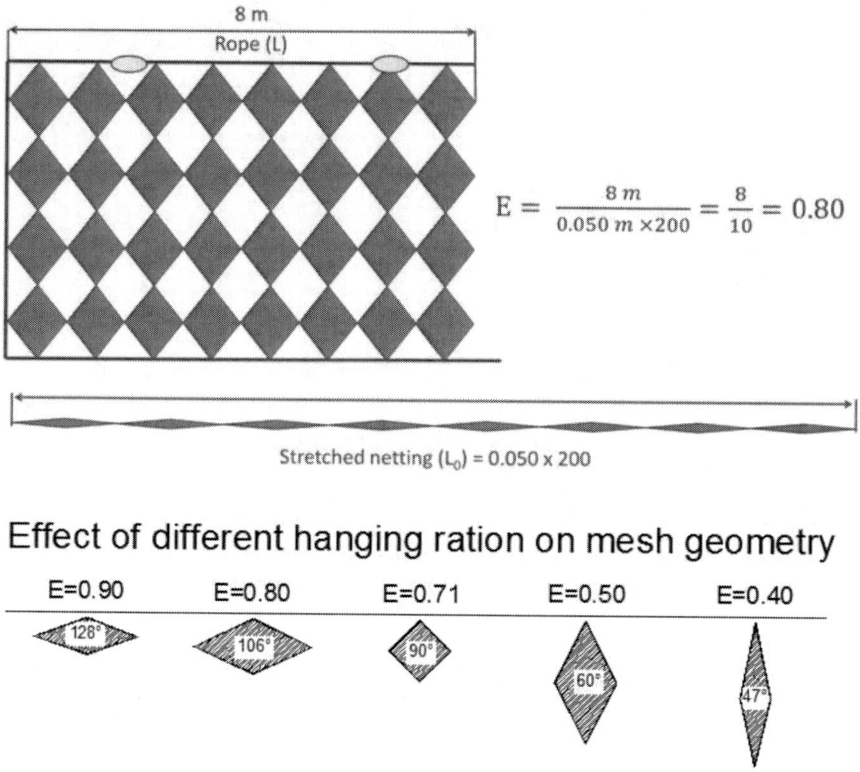

Figure 67. Hanging ratio explanation.

9 - Netting material. den, tex, dtex, Rtex, PA, PES, PE, PP, PVA.

Units of textile measurement (**den** and **tex**):

Denier (den) – is a unit of measure for the linear mass density of fibers. It is defined as the mass in grams per 9,000 meters.

Tex (tex) - is a unit of measure for the linear mass density of fibers and is defined as the mass in grams per 1000 meters. The most commonly used unit is actually the decitex, abbreviated *dtex*, which is the mass in grams per 10,000 meters. Sometimes on the net drawings Resultant linear mass density of the rope is given (*Rtex*). Most commonly used synthetic fibers are:

Nylon, polyamide (**PA**)
Polyester (**PES**)
Polyethylene (**PE**)
Polypropylene (**PP**)
Polyvinyl alcohol (**PVA**)

10 - CoefHLL, CoefHLN, CoefHLM.

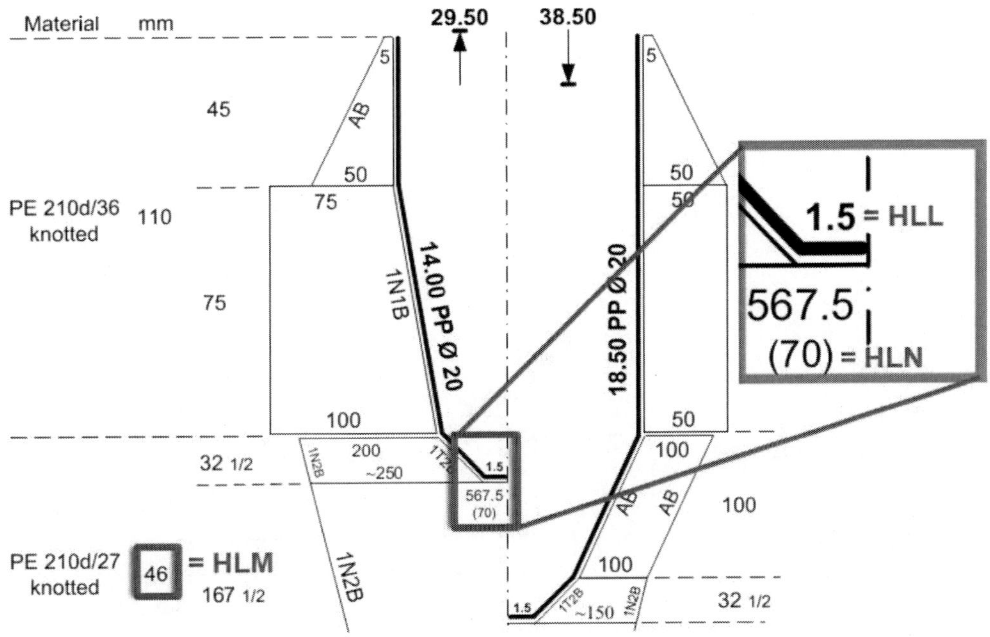

Figure 68. Exaplanation of CoefHLL, CoefHLN, CoefHLM.

11 – Rigging.

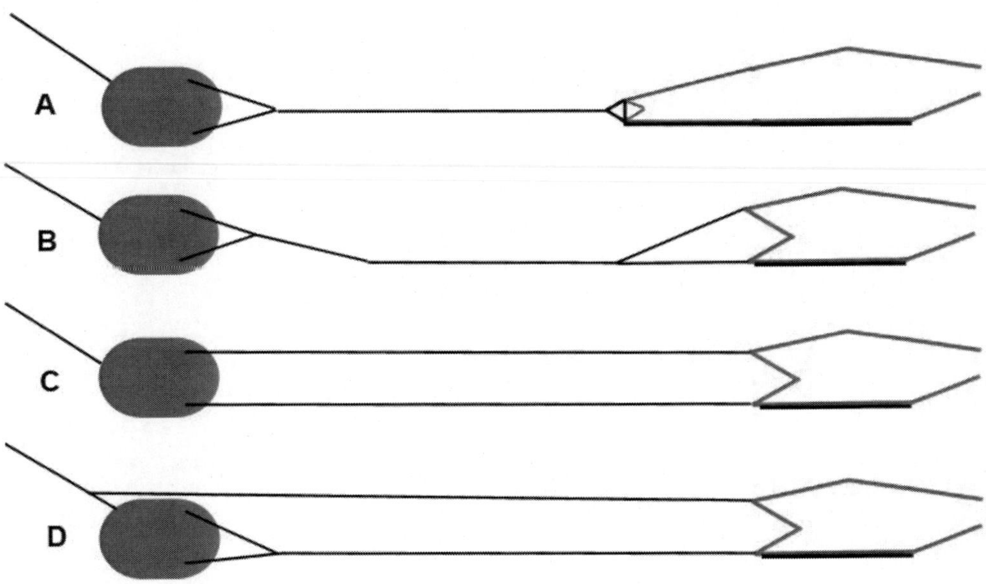

Figure 69. Different rigging between net wing and door.

Chapter 4

FISHING GEAR PERFORMANCE AND MONITORING

The technology for the control of trawl performance consists of acoustic sensors that measure the physical distance between trawl doors, the horizontal net opening (defined as upper net wing end spread) and vertical net opening (defined as the height of the headline center above the seabed), water temperature, catch, symmetry, door pitch angle, roll angle of the doors as well as depth of the trawl door etc. (Figure 70, Figure 71 and Figure 72). The sensors receive the different underwater parameters and transmit the information back to the vessel using wireless acoustic telemetry. The data is finally received by a hull mounted hydrophone. The acoustic signals are decoded and converted into digital data that is transmitted to a bridge computer for graphical display. Each sensor can be equipped with one or more transducers. Sensors together with the receiver (hydrophone) are the most important elements in trawl control system.

The control system of the distance between doors or between the two wings of the net consists of a master sensor which communicates with slave sensors using a transversal wireless acoustic frequency (generally between 100 kHz or 144 kHz). The master sensor sends an acoustic signal to the slave sensor which receives the signal and re-sends it to the master sensor. The time-lag between sending and receiving the signal from and to the master sensor is proportional to the distance between master and slave sensors. Distance information as well as optional pitch, roll, temperature and depth etc. measurements are wirelessly transmitted to the vessel using broadband transducers.

With the use of acoustic sensors placed on the trawl gear, fishermen and fishing technologists can have a much better understanding of what is happening underwater. Due to current resource problems facing the fishing industry, fishermen want to maximize their productivity and increase overall efficiency, while scientists want to be able to standardize their data. In fact and for scientific purpose, the acoustic sensors can be used to standardize methodologies, the performance of the sampling gears, the explored areas, and finally the catch data (Catch per Unit Effort). The sensors are very useful in the shooting phase because one can discover problems that arise early.

In the towing phase, the sensors give a real picture of whether the doors have the desired distance or have changed due to changes in the towing speed as a result of underwater currents or changed force. Changed warp lengths can also affect the doors distance.

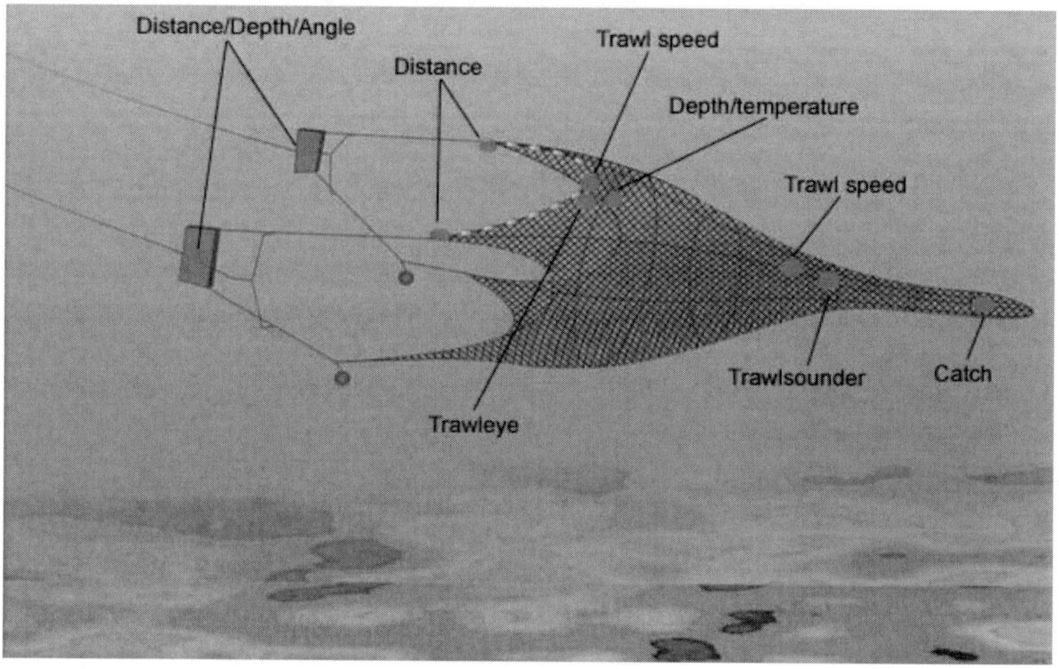

Figure 70. Scheme of different sensors that can be mounted on a trawl net for the monitoring of different parameters.

Figure 71. Sensors mounted on a pelagic trawl that are essential in the echosurvey.

When hauling in difficult conditions, information about doors distance can be important in avoiding problems. The Doors Distance sensor is perhaps the most important for the majority of fishermen.

Depth sensors are necessary in a pelagic trawl to ensure that the trawl is at the exact depth it should be at. This is very important during the echo-survey in order to monitor the desired layer.

The trawl sounder gives the distance from the headline to the bottom, the distance from ground rope to the bottom (clearance) and the vertical net opening. With this sensor it is possible to know the height of the trawl, if the bottom gear is touching the seabed, and the distance to the seabed (Figure 71).

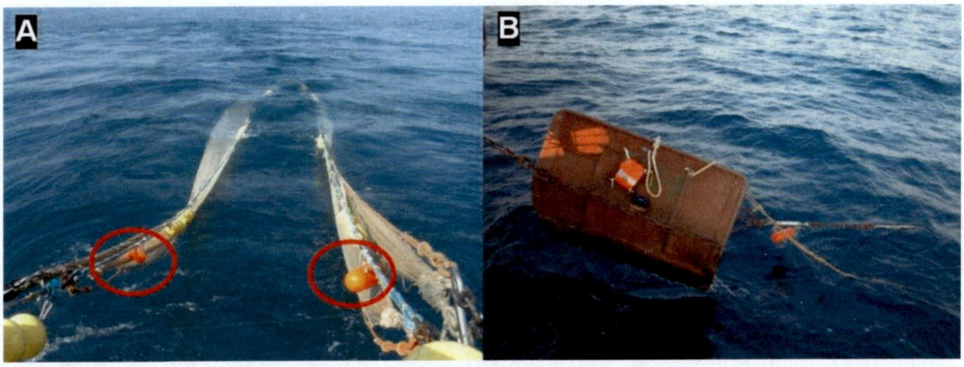

Figure 72. Gear sensors mounted on net wings (A) and door (B).

One of the most common inquiries from fishermen to fishing technologists is whether they are spreading their gear correctly. Without trawl sensors empirical formulae can be applied in order to estimate the door spread and the distance between wings, even if this estimate should be considered with care.

5.2. Use of Gear Control System for Trawl- and Echo- Surveys

Marine fisheries scientists and management agencies routinely conduct long-term demersal trawl surveys and pelagic echosurveys to estimate the condition of exploitable fishery resources, generate recruitment indices for exploitable fish stocks, or track changes in the fish community over time. Fish abundance indices obtained from annual research trawl- and echo-surveys are integral parts of fisheries stock assessment and management in the Mediterranean.

In early national stock assessment experiments, the standardization of catch data obtained from different national trawl and echo-surveys made with different gears/boat size, was only possible through the comparison of dimension of some compartments of the gears used. In such a context, the need of standardized biological information was considered of fundamental importance by the scientific community in order to achieve a proper assessment of fisheries resources and their management. A key aspect of standardization in trawl and echo-surveys is the monitoring of the net deployment in the field. Even if all institutes were able to deploy identical nets, it would still be possible to introduce considerable variation due to different net performance. As already explained net performance is now routinely monitored using a variety of instrumentation; e.g., for headline height or wing spread.

The development of underwater electronic instrumentation enabled to carry out the real time measurement of the main parameters that affect the functioning and the performance of trawled gears (Fiorentini et al., 1999).

The acquisition of such data is essential to calculate the filtered water volume and/or the swept area, and inter-calibrate the catch data obtained from different scientific units involved in the stock assessment program. For this purpose, some authors developed a correction coefficient to apply to different catch yields obtained from different surveys with different vessels. The knowledge of gear performance is essential to standardize the catch data in terms of number of individuals/m^2 or kg/m^2 in the case of bottom trawl, or Nr./m^3 and kg/m^3 in the case of pelagic trawl. For example in the case of echosurvey, fishing samplings are conducted along the survey routes for biomass allocation into species and to know mean lengths and weights of the pelagic fish. This is not the right document to explain in detail the echosurvey. In brief, however, the echosurvey or acoustic survey is essential for the evaluation of highly variable and widely distributed pelagic stocks that create challenges for fishing industry and management. Currently, multi-frequency echosounders are used to record acoustic backscatter data beneath the survey vessel and along prefixed transects spanning the pelagic species habitat. Pelagic trawl samplings are randomly conducted to collect catch information and to assign these data to the different sound scatters recorded in the area. The total backscatter from each species is divided by the backscatter representative of an average individual of that species to estimate and map their respective biomass densities. The total biomass of each species is estimated by multiplying their mean biomass density by the survey area. To collect biological data on the target species, it is therefore essential to tow the net at a depth that the scientist wants to investigate in relation with the information on the fish school obtained from the echosounder (Figure 73).

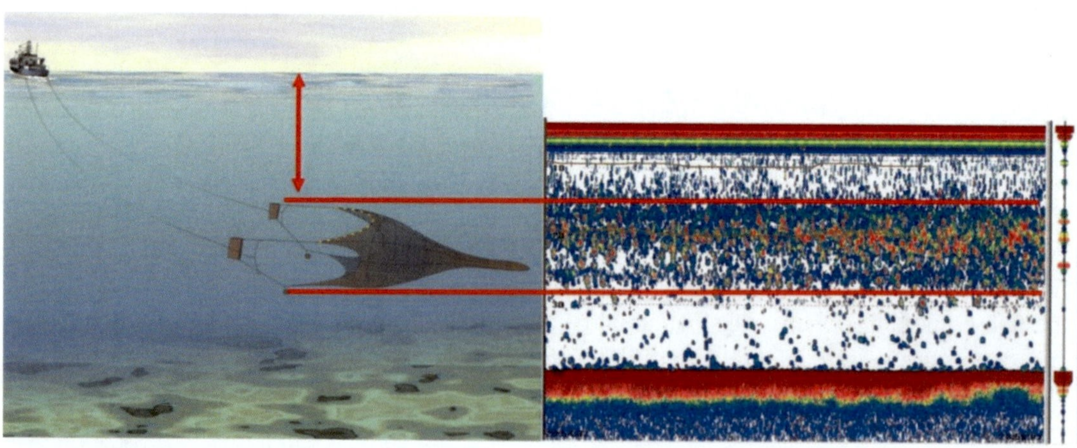

Figure 73. The net should be towed at a depth at which the echosounder shows the school of fish.

When trawl surveys are carried out by multiple vessel–gear combinations, systematic error may be introduced in combining catch-per-unit-effort (CPUE) data across vessels. Maintaining comparability of survey methods over time and across geographic regions poses particular problems for multi-vessel surveys. Differences in vessels, gear, or gear configuration can result in differences in catchability which, in turn, can confound interpretations of fish abundance, recruitment indices, and community change when time series data are combined (Beverton and Holt 1957; 1996). This bias is mainly associated with relative differences in catchability among vessel–gear combinations. The potential for differences in catchability is due to differences in survey vessels and captains that cannot be standardized even if the gear and procedures can be standardized (Gulland 1956; Byrne and Fogarty 1985). In addition, most long-term surveys face the potential of vessel replacement, which could introduce bias into the time series (Pelletier 1998).

Therefore, it is essential to inter-calibrate the different combinations in order to eliminate this systematic bias. Intercalibration trawl survey is essential to develop correction factors for CPUE data associated with different vessel–gear combinations. Failing to account for these differences can result in biased mean catch-per-unit-effort (CPUE) estimates and inaccurate time series for abundance when data are combined.

4.1.1. Sampling Gear

Trawl and echosurveys results are strongly affected by the differences in catchability of the sample vessels used. Differences in catchability among different sampling vessels can be minimized by standardizing the fishing equipment (trawls net design and dimensions), sampling design and procedures; i.e., tow duration and trawl speed, and rigging; i.e., warp length).

Therefore, trawl surveys and echosurveys should be theoretically conducted using a standard gear in all the different areas investigated. The different research units should use sampling gears having the same design and rigging.

The following characteristics should be considered for the ideal standard survey gear design:

Basic Design: An uncomplicated gear design is essential to enable ease of handling, deployment and repair on different vessels. The trawl design must be compatible with the required towing (ground) speed and the actual speed through the water to maintain the geometry, stability and ground gear contact.

Rigging: Rigging adjustment should also be as simple and steady as possible to avoid differing adjustments leading to differences in trawl performance.

Ground gear contact: A good contact of the groundrope with the ground is essential for most of the species considered, particularly for shrimps, Norway lobsters, anglers and flatfish. Nevertheless, the ground gear must also be adaptable to different seabed conditions. In any case, the same amount of weights should be used in all sampling trawls to fish at the same conditions.

Vertical opening: It is essential for some target species that the vertical opening is high enough to collect a representative sample. To do this, the same amount of floats should be used in all trawls. Generally, the vertical opening should be around 2-2.5 m for bottom trawl.

Horizontal opening: It must be adequate to collect sufficient samples, and compatible with the vertical opening for the stability of the net.

Mesh size: Theoretically, an unselective gear should be used in order to collect the entire population in the swept area. Therefore, mesh size must be small enough to catch shrimp,

Norway lobster, flatfish as well as the juveniles. To maintain geometry and efficiency of the trawl, it is recommended to use larger meshes in the upper wings. However, to maintain good water flow in the body of the trawl, the meshes in the top panels must diminishes gradually to equal the meshes in the lower panel before the extension piece. The net should have minimal mesh selection and also ground gear selection.

Robustness and durability: The material used in construction of the trawl must be chosen to ensure the strength and minimize the damage to the trawl. The same material should be used in the different trawls because the twine thickness and material can strongly affect the gear performance. The best solution is that the same net maker should produce all the nets, in order to reduce the bias caused by different constructions.

Towing speed: The towing speed must be previously decided on the basis of the behavior of the different target species. Towing speed should remain constant for the duration of the survey tow. The herding effect of rigging must remain constant at all times. The sweep angle and length must be chosen with reference to the behavioral characteristics of the target species.

Stability: Geometry of the trawl gear must be maintained for different water depths, water flow on the trawl, sea state and seabed conditions to ensure a stable catchability of the sampling trawl.

Extra trawl net: The auxiliary net should have the same characteristics of the sampling net: design, material, groundrope properties (weight/meter, chain etc.), headrope (number and type of floats etc.), rigging (bridle diameter, bridle length etc.).

Chapter 5

MONITORING OF BOTTOM TRAWL IMPACT BY SIDE-SCAN SONAR TECHNOLOGY

The increased sensitivity of the International scientific community toward the exploitation of fishery resources promoted the development of new technologies to study the physical impacts of fishing gears on seabed. Several techniques are available for the rapid assessment of trawling impacts on seabed. At the moment, the “acoustic imaging technology” mainly through side-scan sonar is often used for the evaluation of trawling physical effects (Franceschini et al., 2002; Humborstad et al., 2004; Smith et al., 2007; Lucchetti and Sala, 2012). Furthermore, Friedlander et al. (1999) showed that side-scan sonar can be used to assess trawling impacts over a wide area.

In side-scan sonar images, physical changes caused by trawling are visible as tracks across the seabed (Schwinghamer et al., 1998; Friedlander et al., 1999; Humborstad et al., 2004; Malik and Mayer, 2007; Lucchetti and Sala, 2012).

Therefore, side-scan sonar technology can be used for:

- Fishing effort assessment by detection and quantification of the furrows traced by bottom trawls on a geo-referenced area of the seabed;
- Evaluation of trawl impact;
- Real time analysis of the behaviour, geometry and performance of different fishing gears in addition for qualitative and quantitative evaluation of their impacts on seafloor.

Side-scan sonar is a kind of sonar device that transmits conical or fan-shaped pulses down toward the seafloor and analyzes the returned signal (echo) since seabed and other objects reflect some of the sound energy back in the direction of the sonar (known as backscatter). The acoustic fan sweeps the seafloor directly from under the towfish to either side, typically to a distance of 100 meters. The strength of the return echo is continuously recorded, creating a "picture" of the ocean bottom.

Side-scan sonar is used to efficiently create an image of large areas of the sea bottom. It usually consists of a "towfish" towed behind the research vessel at slow speed and connected by a cable to the portable PC on board (processing unit).

The sound frequencies used in side-scan sonar usually range from 100 to 500 kHz; higher frequencies yield better resolution but less range. For a deeper explanation of the side-scan sonar functioning and image interpretation (see Figure 74).

Side-scan sonar is used extensively for many commercial, military and leisure applications. It is commonly used to detect stones, debris, and other obstructions on the seafloor that might be hazardous to shipping or to seafloor installations by oil and gas industry.

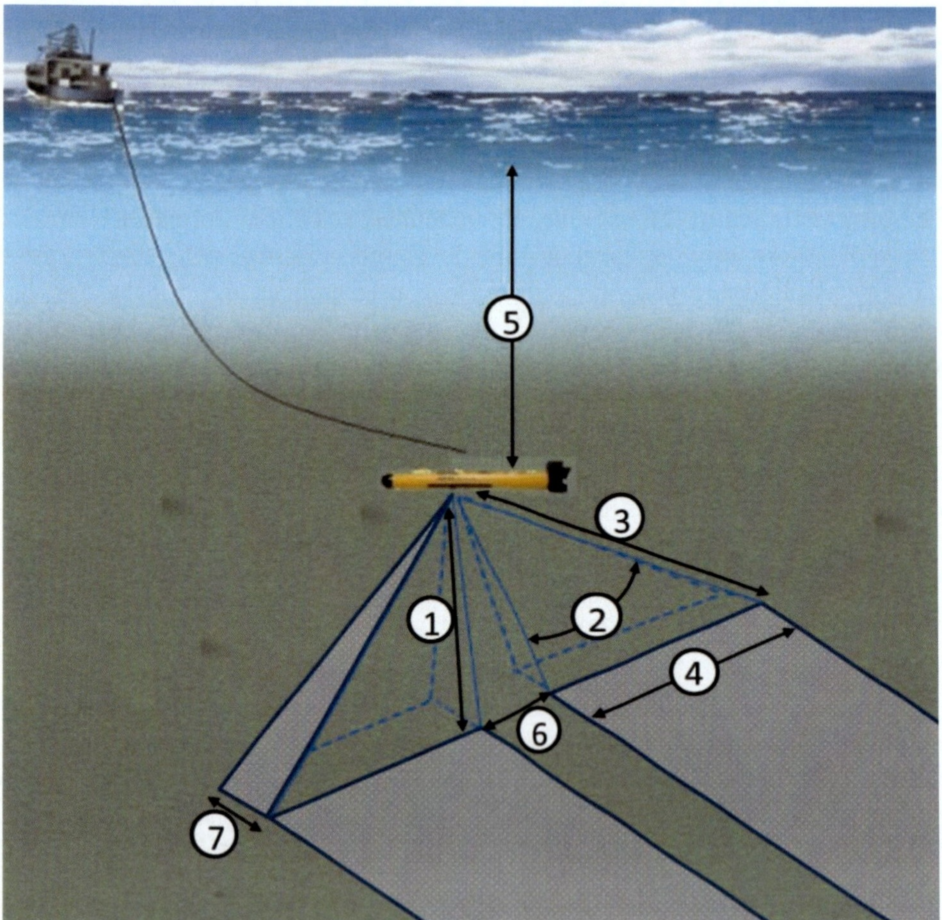

1. Depth to inside of acoustic path.
2. Vertical beam angle.
3. Range setting in software (maximum acoustic range).
4. Swath width accross seafloor.
5. Tow depth of side scan sonar.
6. Port and starboard channel separation.
7. Horizontal beam width.

Figure 74. (Continued).

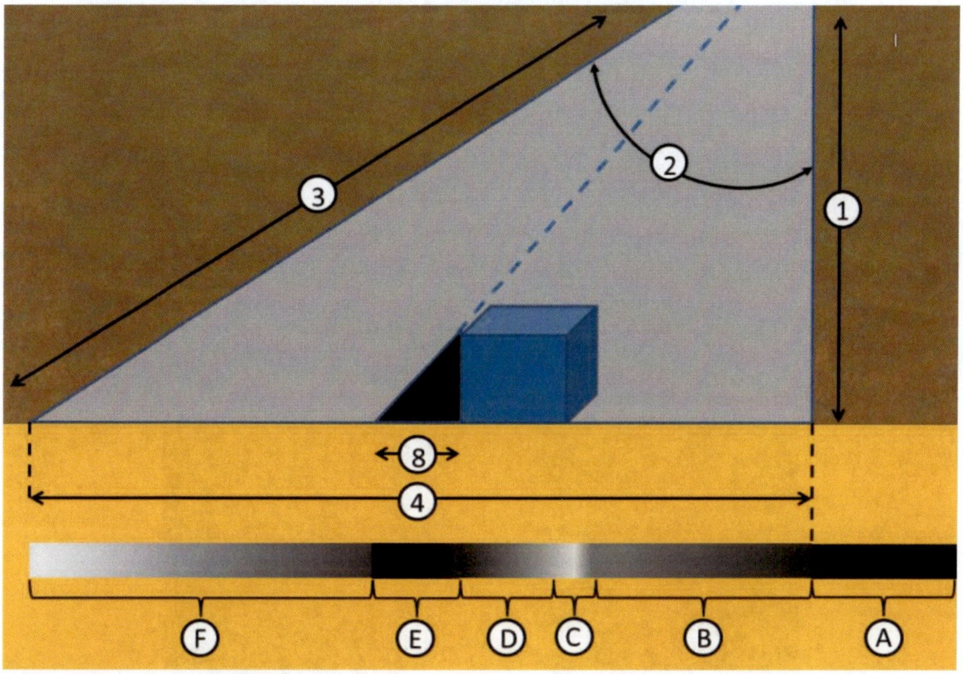

1. Depth to inside of acoustic path.
2. Vertical beam angle.
3. Range setting in software (maximum acoustic range).
4. Swath width accross seafloor.
8. Acoustic shadow length, corresponding to height of target.
A. Area before first "bottom" return (no sound = black).
B. Seabed texture.
C. Very reflective corner of object (brightest intensity).
D. Reflective object (target).
E. Acoustic shadow of target (no returns in here).
F. Seabed texture.

Figure 74. Side scan sonar functioning and interpretation of images.

Side-scan sonar sonograms allow to easily recognize the passage of different towed gears in a fishing ground making it possible to classify the fishing areas on the basis of the type of furrows generated by different gears; something that is almost impossible with a towed video because of its narrow field of view and its limited discerning of different gear impacts.

Sidescan sonar can also be used to monitor in real time the performance and impact of towed gears. Normally the towfish is towed at 2-2.5 knots at an operating height of around 10 m above the seabed. By towing the towfish in parallel with a bottom trawler at this speed the research vessel is easily overtaken along the axis of towing the towfish by the fishing vessel, which usually operated at speed of about 3-3.5 knots (Figure 76). Such a parallel pass survey design allows the fishing gear to be detected directly during fishing operations (Lucchetti and Sala, 2010). Such a technique has made it possible to link each recorded furrow on the seabed to a specific towed gear and to classify each fishing ground according to the gear with the greatest impact. This approach also provides additional information on the net 84 rigging: it is possible to have semi-quantitative information about the headrope, the floats, the tickler chain, and the groundrope etc (Figure 77).

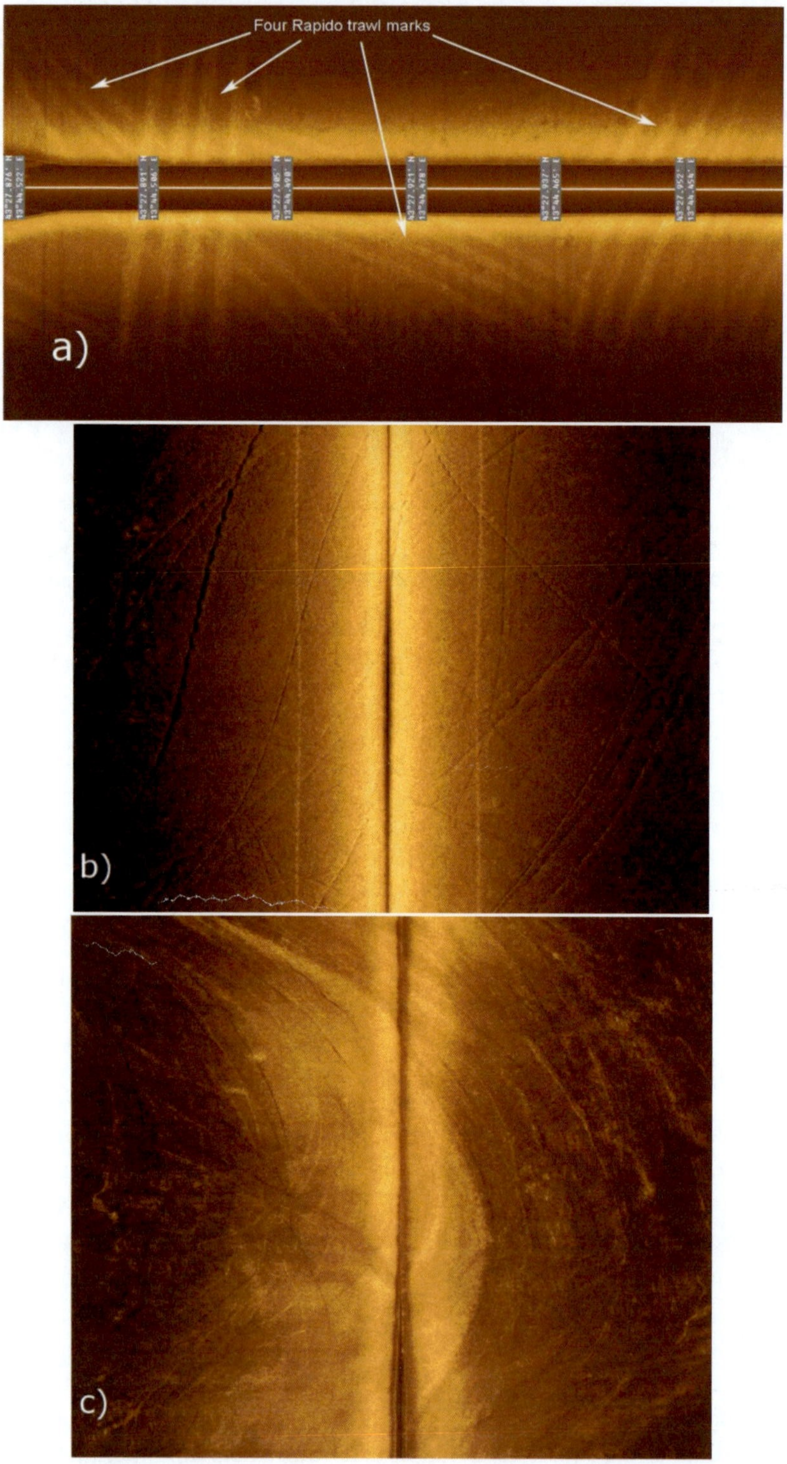

Figure 75. Side-scan sonagrams obtained in different fishing grounds: A) fishing grounds impacted by beam trawling (4 gears are towed by the same vessel); B) fishing grounds impacted by otter trawling (the marks left by the otterboards are clearly showed); C) fishing grounds impacted by dredges (from Lucchetti and Sala, 2012).

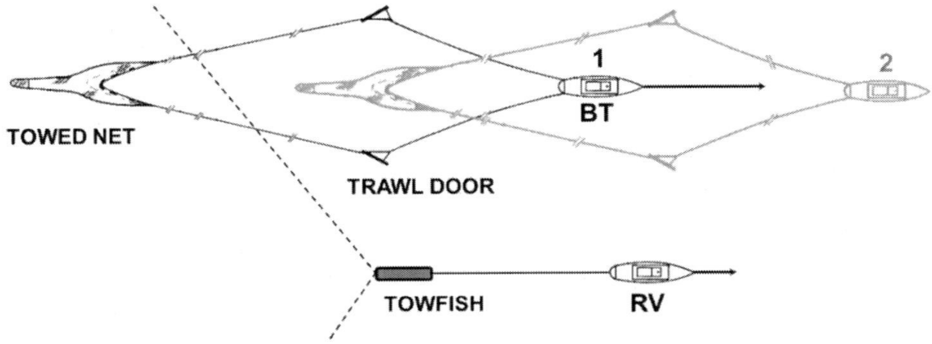

Figure 76. Scheme of the experiment method used for detecting, quantifying and mapping the fishing gear impact and performance by side scan sonar. BT: commercial bottom trawler (higher towing speed); RV: Research Vessel (lower towing speed; from Lucchetti and Sala, 2012).

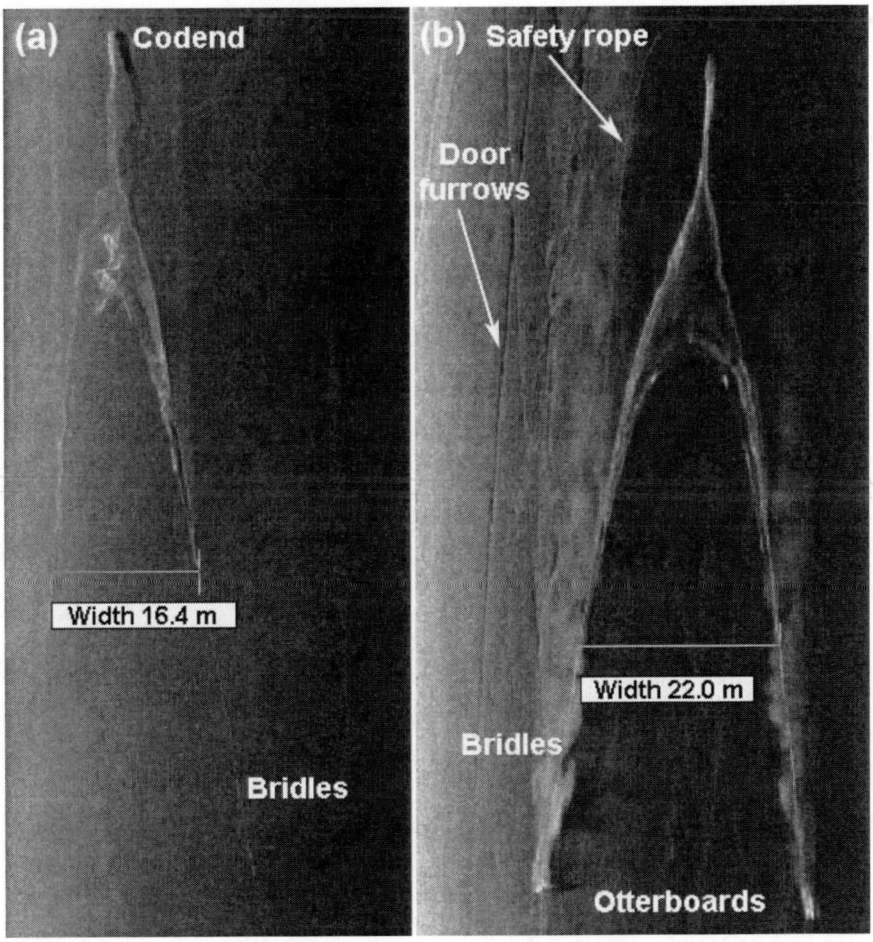

Figure 77. Details of the net rigging of a typical Italian "Volantina" trawl (angel) and of an "Americana" trawl (beer). The horizontal openings (e.g., distance between the wings) are also reported. Bridle: the rope between otterboard and net wings; width: distance between the two wings of a net; safety rope: the rope used for hauling the codend in case of net breaking.

Chapter 6

GHOST FISHING

Ghost gears are fishing gears that have been abandoned or accidentally lost in the sea by fishermen. Ghost fishing normally occurs with passive fishing gear such as longlines, gillnets, entangling nets, trammel nets, traps and pots. These gears, although outside of the control of fishermen can continue entangling fish, marine mammals, sea turtles, sharks, crabs and other creatures (Figure 78). The use of synthetic materials for the netting of different gears strongly raised the problem of ghost fishing. The use of natural fibres for like cotton in the past made the lost gears inoffensive in a short period of time.

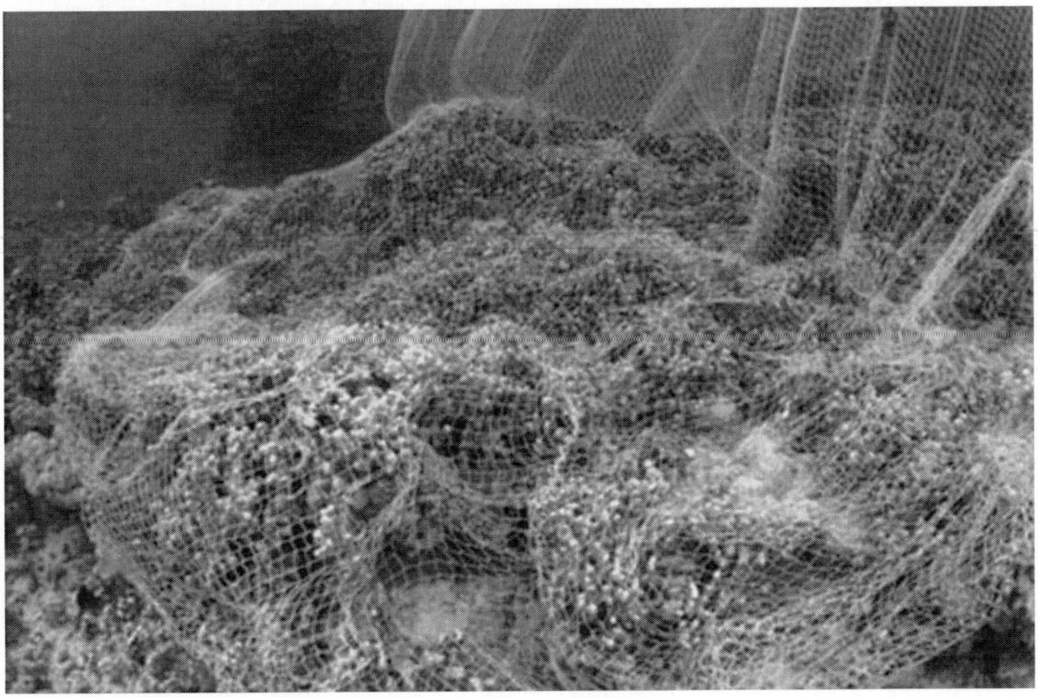

Figure 78. Ghost net lost on the seabed.

Acting as designed, the nets especially restrict movement causing starvation, laceration and infection, and suffocation. Ghost gears can be also dangerous to passing vessels when the gear is lost on the surface because the net may entangle in the propeller disabling the vessel.

The ability of ghost gear to catch fish and crustaceans usually declines over time. This decline depends mainly on environmental conditions. Moreover, marine fouling on different gear's components also increases the visibility of nets, thus, reducing their catching efficiency over time.

The reasons for ghost fishing can be briefly summarized as follows:

- Conflict with other fishing sectors. Passive nets, pots and longlines can be carried far from the area where they have been set by towed gears such as dredges and trawls;
- Working in deep-water ;
- Working over hard bottoms;
- Working in bad sea-weather conditions;
- Using very long passive nets or longlines;
- Bad marking system on the surface for passive gears; and
- Working more gear than can be hauled regularly.

A good advice for fisheries management is that fishing vessels should be equipped with appropriate devices to retrieve lost gear, and the master of a fishing vessel that has lost a gear or part of it should attempt to retrieve it as soon as possible.

A major problem with ghost gears can also be related to the fact that animals like marine mammals can incidentally ingest huge quantities of lost nets. This is responsible for a high rate of mortality.

Chapter 7

SELECTIVITY

7.1. DEFINITION

A basic definition of selective fishing refers to a fishing practice's ability to target and capture organisms by size and species during the fishing operation allowing non-targets to be released unharmed. This means that all fishing gears are only able to catch a certain portion of the total fish community present. In a fishing area, a range of species and sizes normally exist together. When encountered by a fishing gear, they will be captured at different rates depending on gear design and its mode of operation, on fish body characteristics, on swimming behavior etc. The selectivity of a fishing gear is a measurement of the selection process. It describes the relative likelihoods that different sizes and species of fish would have been caught by the gear if there are equal numbers of each in the population (Wileman et al., 1996). Recently the Regulation EU 1380/2013 has defined a "selective fishing" as *fishing with fishing methods or fishing gears that target and capture organisms by size or species during the fishing operation, allowing non-target specimens to be avoided or released unharmed.*

Important contributions in the last two decades made a better understanding of the principles of selection by the different types of fishing gears. Various gears are already known to be more selective than others. Gillnet is considered as a quite selective gear since the selection process mainly depends on the mesh size. However, no gear is known to be one hundred percent selective for a given species or size range of individuals. Most gears and methods fall into a group that has some selective performance, and their ability to select targets can be modified through changes of design and operation methods.

In the Mediterranean, discarding and bycatches occur because most of fishing gears and fishing practices are not selective enough for the species and sizes being targeted and because target species also inhabit areas that are occupied by a wide range of other species (multispecies).

Fisheries resources in the Mediterranean are mainly managed by defining technical measures such as minimum landing sizes (MLS) and minimum mesh sizes. The protection of juveniles through the definition of MLS has been widely recognized as an important measure for regulating those fisheries in which the amount of fishing is not otherwise controlled. The most practical approach to preserve fish above the MLS is wordwide recognized as the setting of an appropriate minimum mesh size.

The main goal of setting a legal mesh size (LMS) is to protect juveniles until they are large enough to spawn and, in this way, the size of the spawning stock can be protected. From an economical point of view, this means to protect juveniles so as to give them time to grow to an economically useful size before they are harvested. LMS also aims at reducing bycatch and minimizing gear impacts on benthic communities. In spite of its basic theoretical simplicity, however, the determination of a legal mesh size can be associated with a number of practical difficulties and management problems (FAO, 1984). An objection to an increase in the mesh size is usually that there is an immediate loss and that it could need several years before long-term benefits are likely to be obtained.

Moreover, determining the best mesh size for the Mediterranean fisheries, in which more than one species is caught by the same gear at the same time, is complicated because the mesh size has to be a compromise between the best mesh sizes for the individual species. For example, there is rarely a single minimum codend mesh size for towed nets that is appropriate for all the species caught in an area in the Mediterranean, due to differences in body shape and size at maturity. A mesh size appropriate for one species will be unsuitable for many others. Some species are long and thin, whereas others are short and fat and others may be small-bodied when fully grown. Thus, it is impossible to define an ideal mesh size and the best solution is to fix the best compromise-mesh size for many of the species individually. A certain mesh tends to catch only fish of a fixed size or of a limited size range like in the case of gillnet for example, or over a certain size as in the case of bottom trawl. Therefore, the concept of "mesh size" is strongly correlated with the "selectivity" of a fishing gear. Size selectivity can be defined as the probability of fish being retained in a fishing gear as a function of the length of the fish.

In general, all fishing gears have peculiar species and size selection properties. This implies that the species and size compositions of an observed catch composition differ from that of the population fished. Thus, selectivity can be affected by several parameters related to fishing gears and operations. However, mesh size and configuration (diamond-, square- and T90-meshes) are some of the main factors affecting the selectivity of fishing net.

There is a considerable interest and activity now throughout the world to improve the selectivity of fishing gears to reduce the capture and discarding of unwanted sizes and species. The Code of Conduct for Responsible Fisheries (FAO, 1995) stated that "*selective and environmentally safe fishing gear and practices should be further developed and applied, to the extent practicable, in order to maintain biodiversity and to conserve the population structure and aquatic ecosystems and protect fish quality*".

7.2. Trawl Selectivity

Bottom trawl is considered as the gear having the lowest selectivity performance. A trawl can be thought as a filtering device where all organisms entering the gear are sorted by the meshes and some, in reason of their size or shape, are retained in the net.

Once a fish enters the mouth of the trawl and although there is a large mesh size in the wings and a substantial portion of the front part of the trawl net, the sharp angle of attack of meshes during towing makes escapements from this section unfeasible. Therefore, fish entering the net are directly driven towards the codend and in traditionally designed trawls, most of the selectivity occurs in the codend.

The majority of undersized fish that are unable to escape and collected in the codend are typically dead when hauled on deck at the end of the tow and they are sorted out and discarded at sea. The poor size selectivity of the trawl results in the discarded bycatch of undersized species.

Historically, research into trawl selectivity around the world has focused mainly on the codend as this is where the majority of fish escapes take place. The codend is the final region of the net where the catch of exhausted fish builds up. Nevertheless, it is recognized that there is rarely a single minimum mesh size for the codend of towed nets that is appropriate for all the species caught in an area of multispecies fisheries due to differences in body shape and size at first maturity. A mesh size appropriate for one species could be unsuitable for many others (Stewart, 2001).

Traditionally, a diamond mesh codend with a mesh opening of 40 mm is used in the Mediterranean though the results of selectivity studies there demonstrated that this codend (40D) is not selective enough for protecting the juveniles of most of species. As catch volume increases, these codends adopt a bulbous shape and tension in the meshes is increased. Then, meshes in the front part of the codend become almost totally closed with only a narrow band of meshes immediately ahead of the catch open sufficiently for small fish to escape. If catch rates are high, fish pressed up against the netting may quickly block these meshes. In some areas, the high market request for fry fish strongly incentivizes the use of small meshes at the codend.

In general, the larger the codend mesh size the greater the number of small fish that can escape. However, the ability of fish to escape from the codend is also affected by catch rate and volume, mesh orientation (diamond, square or T90), and codend design.

7.2.1. Methods for the Study of Selectivity

Experiments for the study of selectivity performance with towed nets can be conducted by different methods. The most commonly used are the covered codend, and paired gear techniques with the latter involving four different approaches (Figure 79). These methods are characterized by the collection of fish in two compartments one of which is assumed virtually non-selective. The purpose of the non-selective compartment is to collect information about the local abundance of fish that entered the combined gear.

Covered-codend Method

The "covered codend" method is the most widely used technique to estimate the codend selectivity curves of towed fishing gears (Beverton and Holt, 1957; Pope et al., 1975).

The cover is a netting bag that completely surrounds the codend. The role of the cover is to collect all the organisms escaped from the codend. In fact, the selectivity study is based on the knowledge of the proportion of fish retained in the codend which also implies the knowledge of the population investigated. To fulfil that, the mesh opening of a cover must be considerably less than that of the codend to retain small fish and is typically 10-20 mm (stretched) in the Mediterranean. Moreover, the dimensions of the cover should be at least 1.5 times those of the codend.

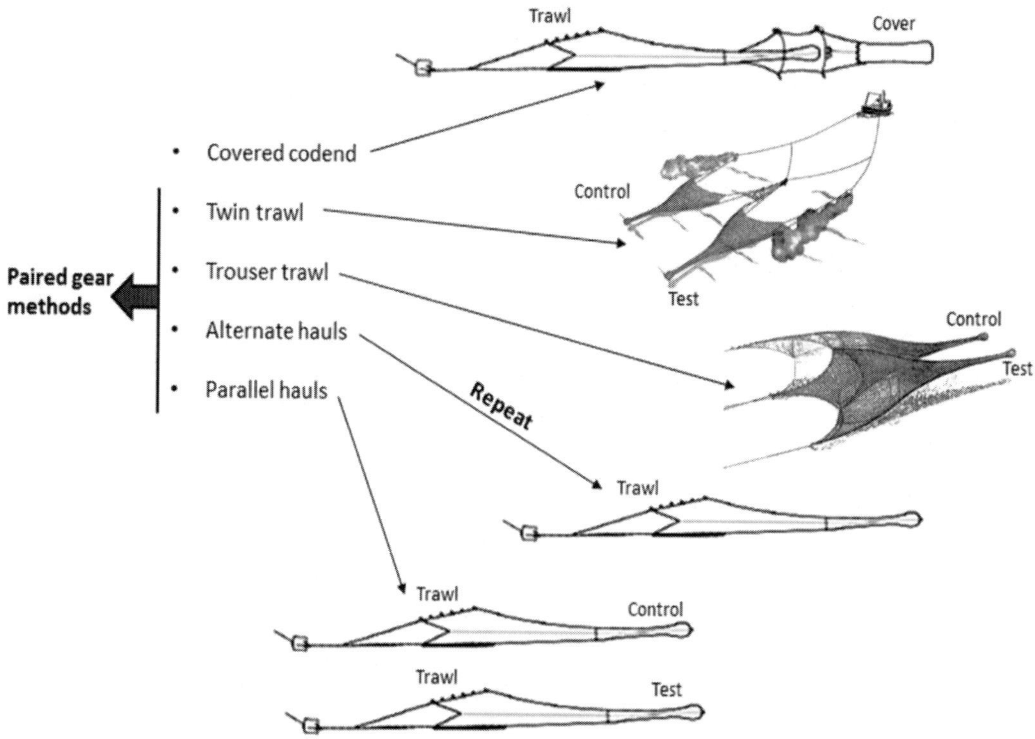

Figure 79. Scheme of the different methods for the study of trawl selectivity.

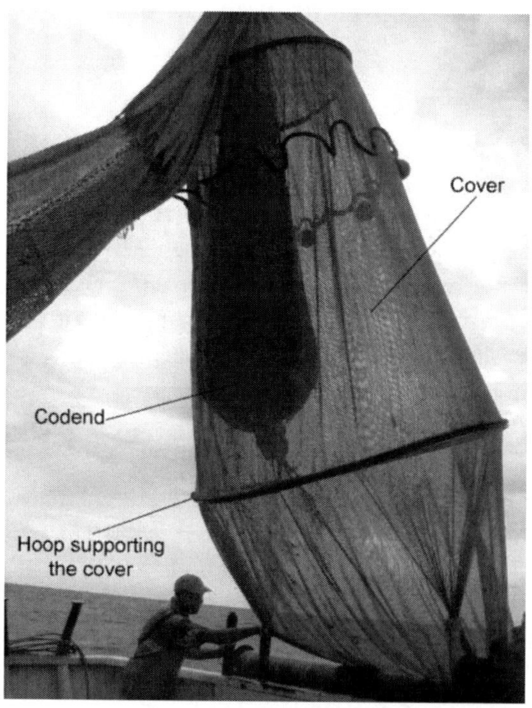

Figure 80. (Continued).

Figure 80. Covered-codend experiment carried out in the Mediterranean sea.

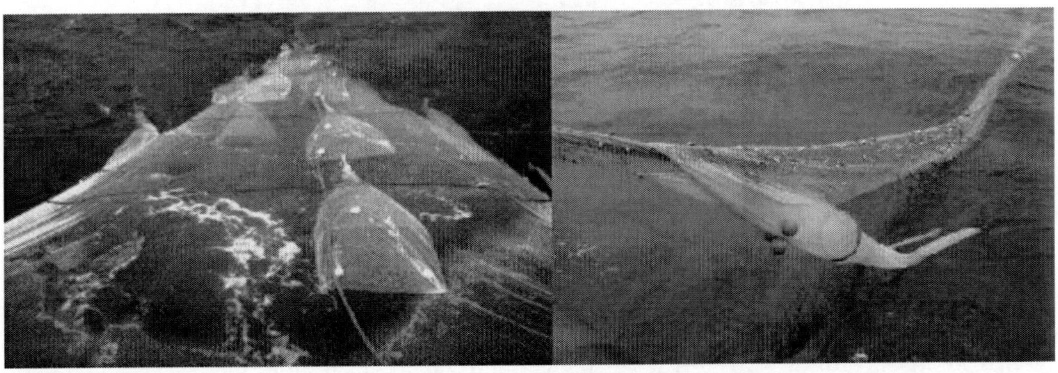

Figure 81. Particulars of the net bags mounted on the different parts of the trawl body in the Mediterranean sea research carried out to evaluate current gear selectivity of a pelagic trawl.

In case of pelagic nets, the dimensions of the net make it unfeasible to rig a cover around the codend. In these cases, the escapement through the body or through the codend can be assessed by mounting small net bags on different parts of the two nets (Figure 81). Similar bags can be mounted to assess the escapement from a selective window (Figure 82).

Small mesh codend cover enables scientists to obtain immediately the fish population in the area swept by the trawl. The advantage of the method is that the codend selectivity can be estimated directly because the fish escaping the codend will be collected in the cover.

The use of such covers has been criticized because the cover might affect fish escape process and, hence, bias results. The most serious potential problem is that the small mesh netting may obstruct the codend meshes; i.e., masking effect in which the cover comes in physical contact with the codend. Another problem is that the cover might restrict water flow through the codend, thus influencing the swimming performance of fish and limiting their opportunities to escape (Davis, 1934; Pope et al., 1975). A cover effect might also occur if the cover has a visual effect that deters or stimulates the escape of fish.

To limit these problems, it is usually recommended to use covers supported by hoops (Figure 80) to keep cover netting clear of the codend (Wileman et al., 1996).

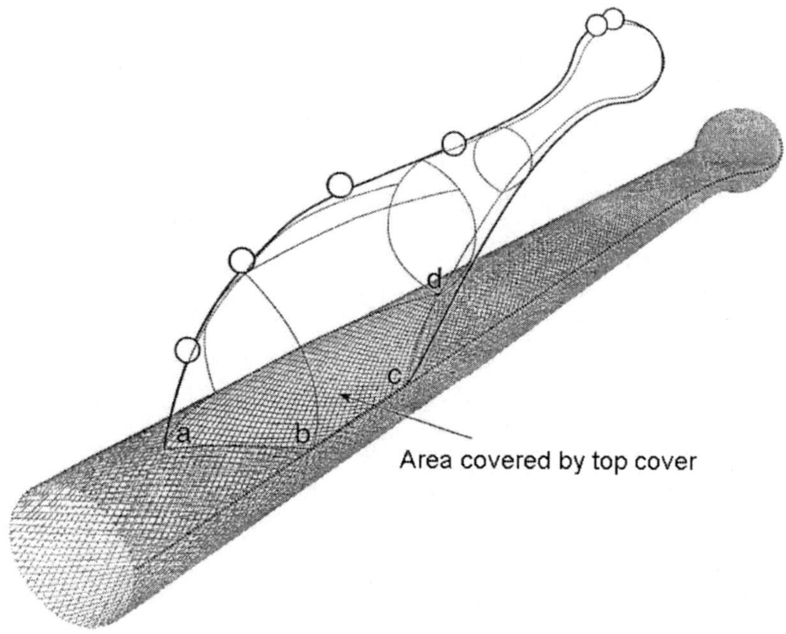

Figure 82. Scheme of a cover bag mounted over a selective window.

More recently, a kite cover concept has been developed and successfully used during sea trials (Madsen et al., 1999; Tschernij and Holst, 1999). The use of kites attached to the top and sides of the cover and chain weight attached to the bottom of the cover generates spreading forces that keep the cover well clear of the codend (Figure 83). The cover should be attached to the selvedges well in front of the codend which ensure that the drag of the cover does not influence the codend meshes. A mixed configuration between hoop-cover and kite-cover has been also tested with promising results (Figure 84).

The weak point of the covered-codend method is that this approach allows the estimate of selectivity of the codend and not the selectivity of the whole net. However, it also well known that the selection process takes place mainly in the codend.

Normally, it is recommended that the cover should be 1.5 times the fully extended length and width of the codend (Stewart and Robertson, 1985). However, this may be too small to collect huge catches. Therefore, it is now recommended that the cover should be two or more times the fully extended length of the codend. The recommended fully extended width is at least 1.5 times the codend width.

Escaped fish rarely approach the forward end of the cover during the tow but fall back into the area behind the codend catch where the flow is reduced.

The cover behaviour can be inspected using an underwater camera (Figure 85). There is also the possibility that the meshes in the extension just ahead of the codend may be opened wider than normal due to the presence of the cover. If this happens and an estimate of the true mesh opening in the extension can be made, then a constrictor rope might be used. As already mentioned, supporting hoops are recommended to keep the cover netting clear of the codend, and it is usually recommended to reduce the first hoop diameter or position it further aft of net.

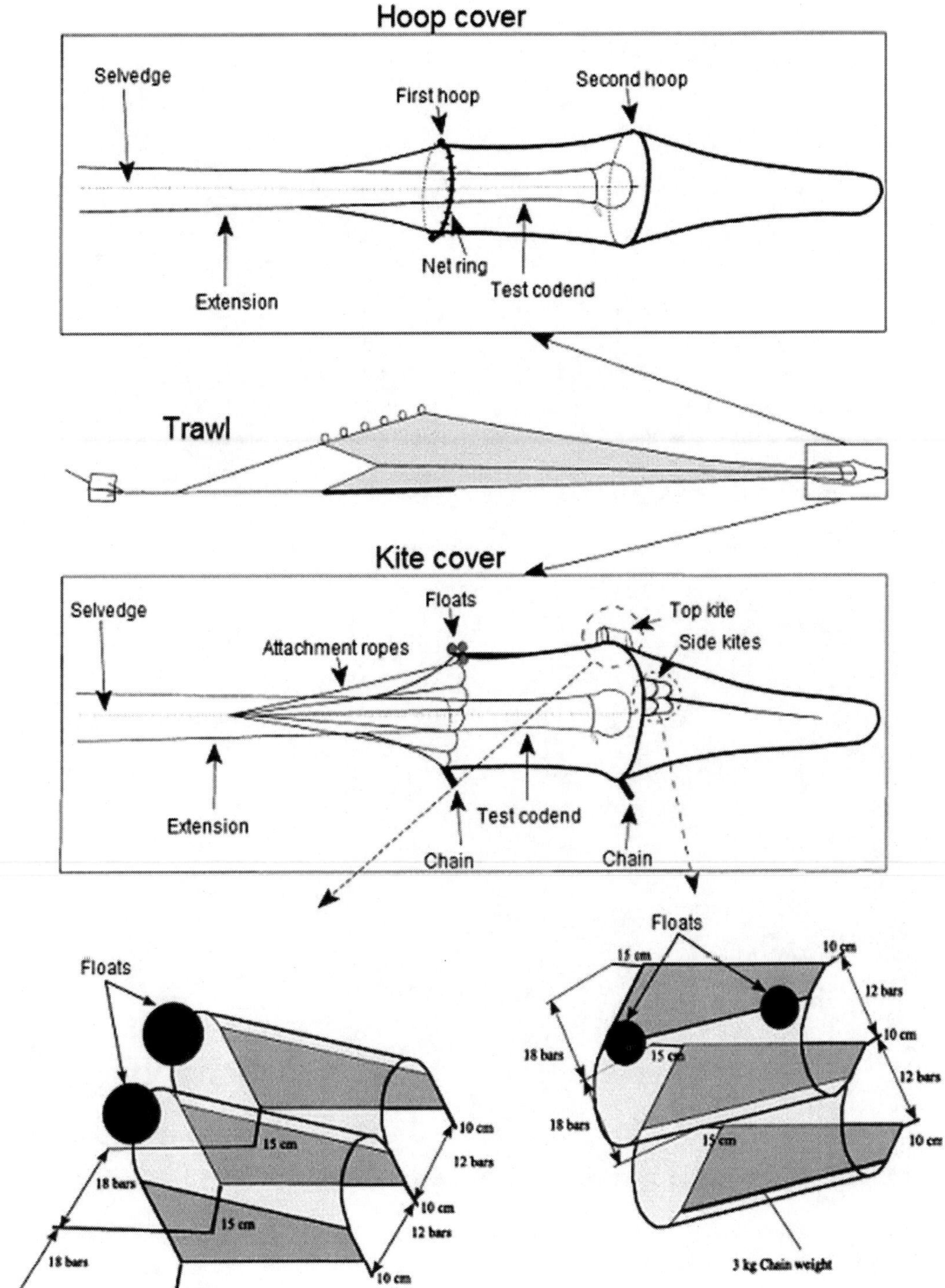

Figure 83. Hoop and Kite-cover rigging: kites are attached to the top and sides of the cover and chain weight attached to the bottom of the cover (Madsen et al., 2001).

Figure 84. Mixed configuration: kite cover supported by hoops.

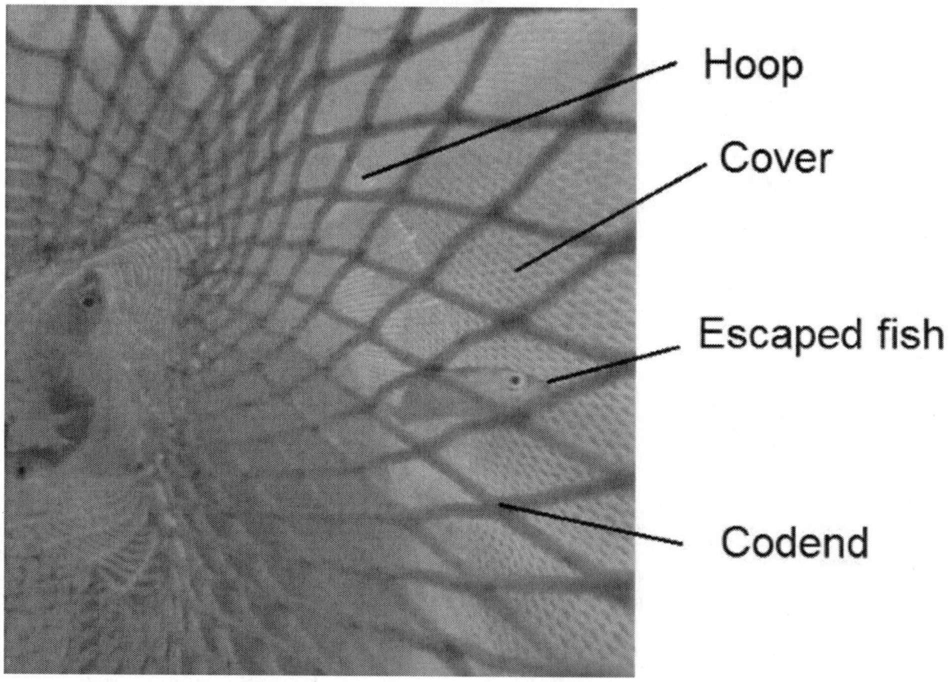

Figure 85. Underwater camera used to monitor the right setting of the cover; it is possible to observe that cover is well detached from the codend during towing.

The colour of cover netting could affect the escaping process. Therefore, this color should be chosen to reduce the contrast of netting to background and make the cover as invisible as possible for fish in the codend. This is mostly important for the upper panel as the majority of escapes are made through it (Glass et al., 1995).

On a vessel where the codend and cover will be hauled up in the air for unloading, zips can be used to simplify emptying the codend catch.

Paired Gear Methods

Paired gears methods include the alternate haul; parallel haul; twin trawl; and trouser trawl methods. In each of these methods, two gears (nets) of equal overall dimensions are towed alternately or alongside each other. In one gear, the rear sections of netting whose selectivity is to be determined are made in small mesh so that all fish are retained. These methods are, therefore, suitable for determining both codend selectivity and whole gear selectivity.

Alternate Haul Method

This method involves alternating one haul with the gear the selectivity of which is to be determined with a haul of the same gear with a small mesh codend. This second haul is carried out in order to obtain an estimate of the fish population entering the test net (Figure 86). This means that two hauls with the same gear have to be done in order to calculate one single selection curve.

This method enables obtaining the selectivity of the whole net or part of it, usually the codend selectivity. If this is the target, the second set of gear should be made in small mesh throughout. It is essential that the pairs of hauls should be similar in every respect except for the mesh size in the part of the gear whose selectivity is being measured.

In fact, the test codend/net is fished as in normal commercial fishing, The main advantage of this method is that it allows avoiding any bias caused by cover., while its main disadvantage is that larger numbers of hauls are needed to obtain good results (two hauls are needed to construct a single selection curve for one codend) which, in turn, increases the costs of the experiment. On the contrary, a selection curve can be constructed in each haul by applying the covered-codend method. Another disadvantage is that the two nets are fished at different times, which means that the population encountered by the two nets might not be the same and irremediably compromises the experiment.

Parallel Haul Method

The principle is similar to that applied in alternate hauls method. However, here two vessels fish in the same ground and at the same time and speed as well with one of them towing the test net and the other towing the "virtually unselective gear" (Figure 87). The only difference between the two gears is the part of the net whose effect on selectivity is to be determined. If the objective is to determine codend selectivity for example, the experimental gear whose selectivity is be determined is towed by one vessel and a gear of identical design but with a smaller mesh codend is towed by the other vessel to obtain an estimate of the population of the target species entering the test codend. Fishing practice; i.e., fishing area, time, and speed, should be as similar as possible. Like in alternate haul method, the test codend/net is fished as in normal commercial fishing method.

Figure 86. Alternate haul method: the same vessel carries out one haul with the gear the selectivity of which is to be determined (test gear, usually test codend) and another haul with the same gear with small meshes (control).

The positive sides of this method are that it allows for the study of selectivity of the whole gear and negates the bias caused by using a cover.

Disadvantages of the method include the doubled cost of the experiment due to the use of two vessels. It is also possible that the two nets will not catch the same population even though the boats tow the nets at a close distance. Finally, bias can arise from different properties; e.g., engine powers, which can affect the towing performance of the two vessels.

Twin Trawls Method

In this method, the trawler tows two identical trawls ("twin") simultaneously side by side using special rigging. The horizontal opening is maintained by a single pair of otterboards that are attached to the trawl close to the wings.

Usually, the objective of this study is the evaluation of selectivity of the codend of the test net while the other net has the same characteristics but small meshes in the codend. Small meshes are used to collect the total fish population entering the test codend (Figure 88).

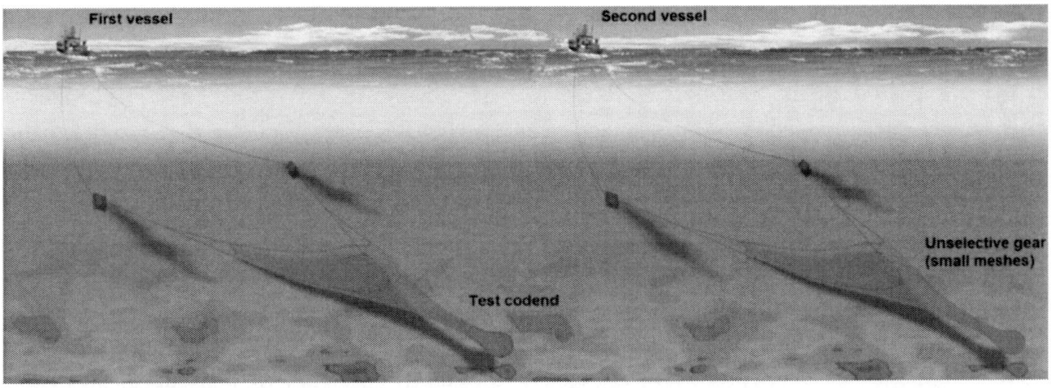

Figure 87. Parallel haul method: one vessel tows the test gear (test codend, generally) while the other vessel (towing in parallel with the test vessel) tows a non-selective gear with small meshes (control).

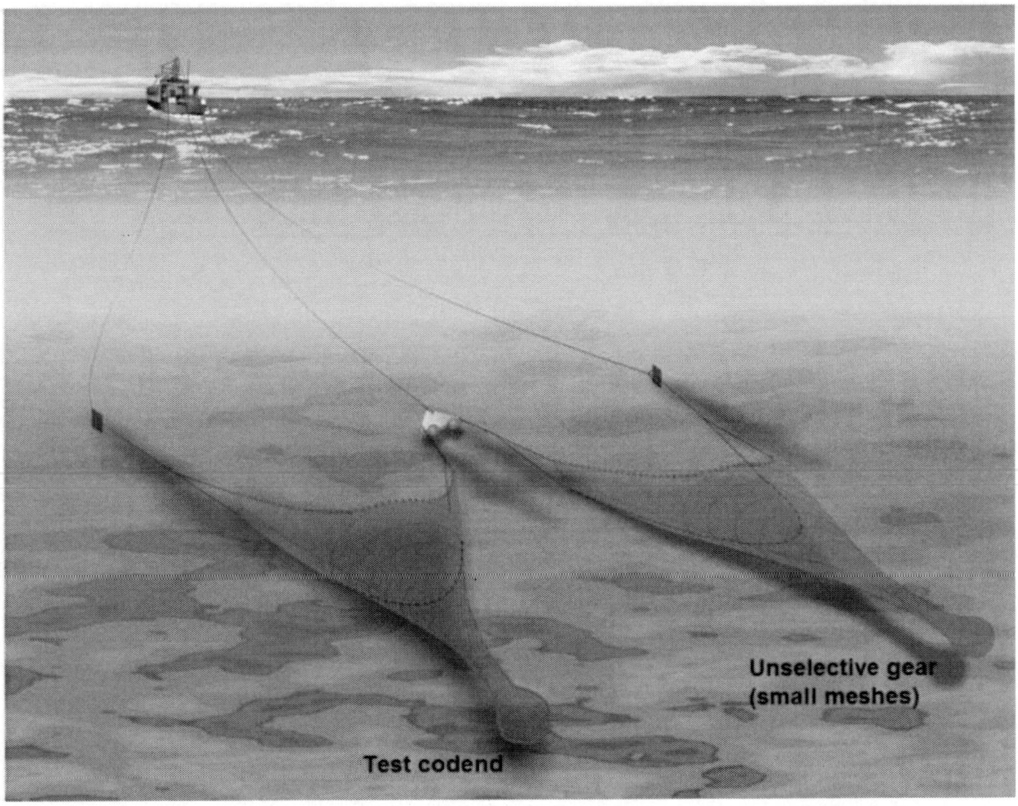

Figure 88. Twin trawl method. Two identical nets are towed by the same vessel: one net is the test gear (test codend, generally) the other net is the non-selective gear having small meshes (control).

Taking into account the difficulties in using this gear, this method is only recommended for fisheries in which twin trawls or beam trawls are commonly used.

This method enables the evaluation of selectivity of the whole trawl net and to conduct catch comparison trials. In this case, the net used to collect population in the swept area should be entirely of small meshes.

The twin trawl method has several advantages. Firstly, it is free from any bias caused by the use of a cover. Secondly, it makes possible the estimation of selectivity of the whole net and the population in the two nets is likely to be the same. Finally, one haul is needed to construct one single selectivity curve. However, some researchers believe that although the two trawls are working close to each other, there is no certainty that the same population of fish will enter each trawl.

On the other hand, twin trawl rigging is different from traditional demersal trawl and the catching process as well as the behaviour of the fish ahead of the trawl can be affected by the change in rigging. Moreover, the performances of the two nets could not be exactly the same.

Generally, variations in haul-to-haul catches can be observed and a somewhat larger number of hauls are usually needed to achieve the same precision of estimation as in the covered codend.

If the objective of the study is to study the selectivity of two codends, the effects of haul-to-haul variability are likely to be reduced if covers are used over both codends.

Trouser Trawls Method

In trouser trawl method, one vessel tows a standard trawl divided down the middle by a vertical panel with two codends attached to the aft end, one on each side of the panel. One codend is the test the other one is the control codend made of small meshes.

The assumption of this method is that an equal number of fish will enter each side of the trawl.

The advantage of the method is that the trawl can be managed in a similar manner to a standard trawl. Moreover, there is no need for using covers that can affect the selective process. The method also implies that one haul is needed to calculate a single selection curve. It also does not have any special rigging that might affect the behaviour of fish in front of the trawl though the rigging of the two codends is quite different from the commercial situation; a fact that can affect the behaviour of fish inside the net. Strong currents can also affect the performance of the two sides.

In order to reduce the potential of unequal catches in the two codends and, hence, a greater variance in calculated selectivity parameters, it is recommended that the panel should extend the full length of the trawl from the footrope to the codend openings. However, it might be difficult to design it so that it maintains a vertical position without distorting the upper or lower sheet. Thus, a shortened panel may be the best solution in this case.

7.2.2. Trawl Selection Curve

In trawl selectivity studies, the comparison between the size frequency distribution of specimens caught in the codend and those living in the area investigated; i.e., specimens in the codend plus cover, enables the construction of selectivity curve. The easiest mathematical model used to describe the selection curve of a trawl net is the "*logistic curve*" (Figure 89):

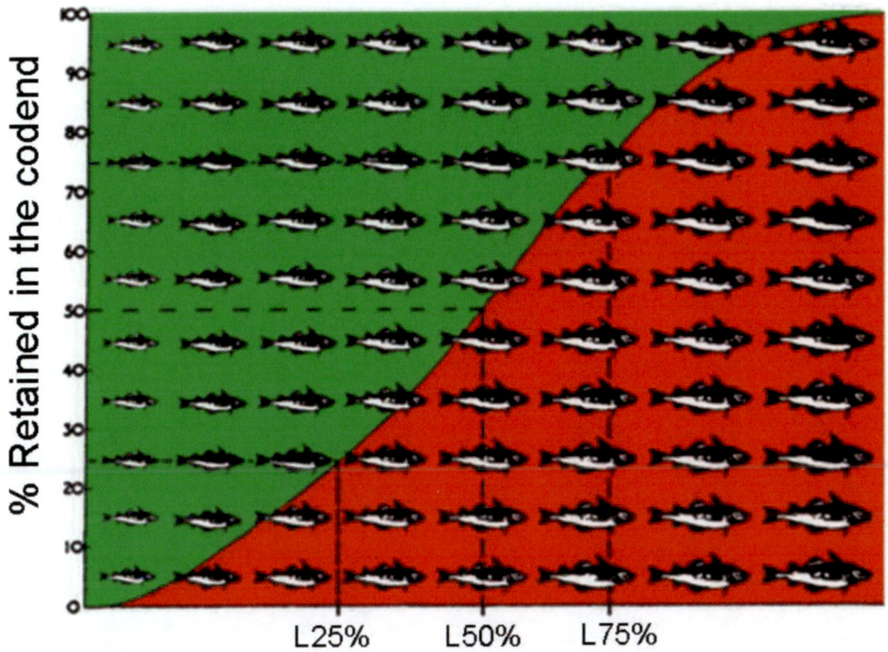

Figure 89. In green: the specimens escaped; in red: the specimens caught.

$$S(L) = \frac{1}{1+\exp(S_1 - S_2 L)}$$

Where $S(L) = \dfrac{\text{Number of fish having length } L \text{ in the codend}}{\text{Number of fish having length L in the codend and in the cover}}$

L is length mean interval point,
S_1 and S_2 are constants.

In Figure 89, the specimens and lengths escaped from the codend are in the green zone while those caught are in the red one. The smallest specimens are able to escape easily from the meshes (retention probability very low) because they are able to pass through the meshes. The retention percentage increases when fish size increases, and the percentage of fish escaping from meshes decreases. It is very difficult for the biggest to pass through the meshes and almost 100% of them are retained into the codend (asymptotic part of the curve).

The flex point of the curve is the 50% retention length (*L50*) that is the length at which 50% of the fish are retained in the codend.

The *L25%* and *L75%* are the lengths at which respectively the 25% and 75% of the fish are caught in the codend.

The length range between the *L25* and *L75* is the so called "*selection range*" (*SR*). This length range is symmetrical around the L50. *SR* is the factor affecting the slope and shape of the curve and it expresses the efficiency of the selection process.

By applying few algebraic manipulations to the *logistic curve*:

$$\frac{1}{S(L)} - 1 = \exp(S_1 - S_2 L).$$

The logistic curve can be transformed as: $\ln(\frac{1}{S_L} - 1) = S_1 - S_2 \cdot L$

This is a straight line where $S_1 = a$, and $S_2 = b$.

If we consider the 50% retention length (*L50*), *S(L)* will assume the value = 0.5 (that is the 50% retention). The equation can be rewritten as:

$$\ln(\frac{1}{0.5} - 1) = S_1 - S_2 \cdot L_{50}$$

$$\ln(1) = S_1 - S_2 \cdot L_{50}$$

$$0 = S_1 - S_2 \cdot L_{50}$$

And finally: $L_{50} = S_1 / S_2$.

In the same way, it is possible to obtain the following equation for *L75* and *L25*:

$$\ln(\frac{1}{0.75} - 1) = S_1 - S_2 \cdot L_{75}$$

$$\ln(\frac{1}{3}) = S_1 - S_2 \cdot L_{75}$$

$$L_{75} = \frac{-\ln(\frac{1}{3}) + S_1}{S_2}$$

$$L_{75} = (S_1 + \ln 3)/S_2$$

The same approach can be applied to the L_{25}.

$$\ln(\frac{1}{0.25} - 1) = S_1 - S_2 \cdot L_{25}$$

$$\ln(3) = S_1 - S_2 \cdot L_{25}$$

$$L_{25} = (S_1 - \ln 3)/S_2$$

Therefore, the selection range can also be expressed as:

$$SR = L_{75} - L_{25} = 2 \cdot \ln 3/S_2$$

Resuming:

L_{50}, 50% retention length	$L_{50} = S_1 / S_2$
L_{75}, 75% retention length	$L_{75} = (S_1 + \ln 3)/S_2$
L_{25}, 25% retention length	$L_{25} = (S_1 - \ln 3)/S_2$
Selection range: $SR = L_{75}\text{-}L_{25}$	$SR = 2 \cdot \ln 3/S_2$

If the L50 fits with the Minimum Landing Size (MLS) the selectivity curve help us to understand the trawl catching process: only in hypothetical situation (knife edge curve) is it possible to clearly cuts between undersized and legal specimens (marketable fish landed). In a real situation a portion of the catch can be composed by undersized specimens and discarded (discard), while a portion of the legal catch can be lost due to selectivity (marketable fish lost; Figure 90).

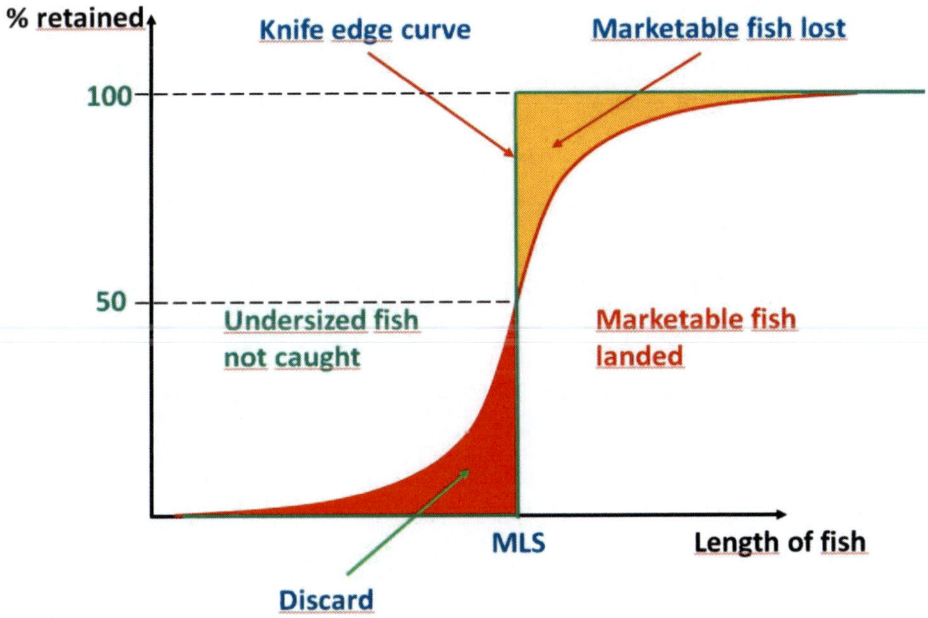

Figure 90. Scheme of the trawl catching process; in green the knife edge curve.

The *L50* can also be used to calculate the *Selection Factor* as:

$$SF = \frac{L50\ [mm]}{MO\ [mm]}$$

where *MO* is the mesh opening.

The *SF* is a parameter that enables comparing the selectivity results obtained for a certain species. Theoretically, using the same codend characteristics (mesh size, codend circumference, netting twine) would almost result in the same *SF*. Thus, *SF* can be considered as a species-specific parameter, because the *L50* for a certain species will increase upon increasing the mesh size.

This factor should be taken carefully into account when management measures have to be decided.

For example, the European hake (*Merluccius merluccius*) has *SF* around 3 with a diamond mesh codend. The Minimum Landing size for this species is fixed at 20 cm. Applying the *Selection Factor* equation:

$$SF = \frac{L50\ [mm]}{MO\ [mm]} = 3 = \frac{200\ [mm]}{MO\ [mm]}$$

Thus, the mesh opening that would protect the hake specimens below the minimum landing size should be:

$$MO\ [mm] = \frac{L50\ [mm]}{SF} = \frac{200\ [mm]}{3} = 66.7[mm]$$

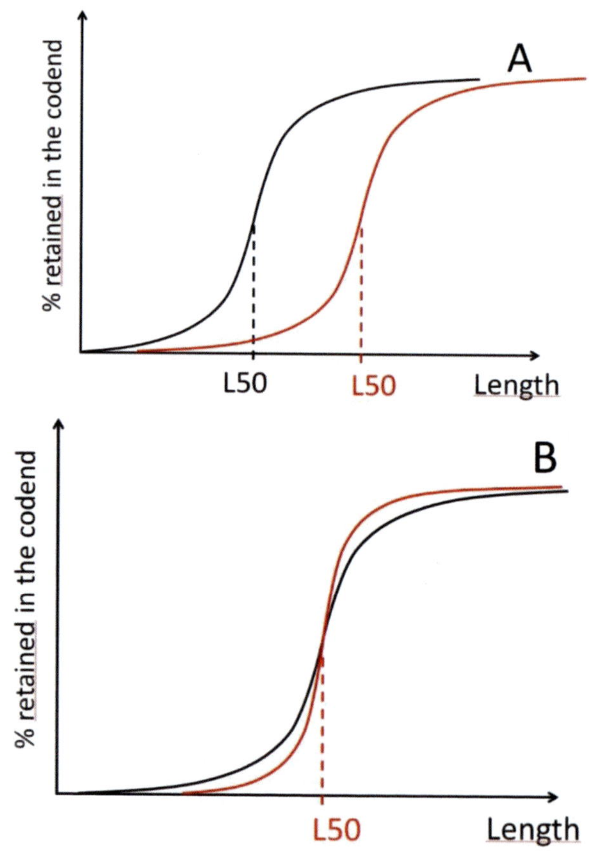

Figure 91. Trawl codend selectivity curves; A) same shape but different L50's; B) same L50 but different shapes.

So, a mesh opening of 66.7 mm should applied to the diamond mesh codend in order to protect the hake specimens below the MLS.

Figure 91 shows two different scenarios. In Figure 91A, selectivity curves show the same shape (same *SR*) but different *L50's*. In Figure 91B, selectivity curves show the same *L50* but different shapes (different *SR*). In both cases, the best selection performance is obtained with red curve. The red curve in case A shows higher *L50*, while the red curve in case B has a lower *SR* compared to the black curve.

In the following section, a selectivity study carried out for *Mullus barbatus* species is described (Table 2 and Figure 92).

$$SF = \frac{L_{50}[mm]}{AM[mm]} = \frac{88.4}{40} = 2.21$$

Actually several softwares are available for the study of trawl selectivity.

Table 2. Selectivity parameters obtained in a covered-codend experiment in the Mediterranean. Data is for Mullus barbatus. A 40 mm diamond mesh codend was used

Length Class [cm]	Number in the codend	Number in the cover	Total number	Observed Retained fraction $S_L$obs	ln(1/S_L-1) (y)	Mean Lenght Class (L_1+L_2)/2 (x)	Estimated Retained fraction $S_L$est
4-5	0	2	2	0.00	-	4.5	-
5-6	1	5	6	0.17	1.61	5.5	0.15
6-7	2	9	11	0.18	1.50	6.5	0.23
7-8	5	15	20	0.25	1.10	7.5	0.33
8-9	9	10	19	0.47	0.11	8.5	0.46
9-10	17	8	25	0.68	-0.75	9.5	0.58
10-11	22	8	30	0.73	-1.01	10.5	0.70
11-12	24	5	29	0.83	-1.57	11.5	0.80
12-13	30	5	35	0.86	-1.79	12.5	0.87
13-14	35	3	38	0.92	-2.46	13.5	0.92
14-15	25	1	26	0.96	-3.22	14.5	0.95
15-16	18	1	19	0.95	-2.89	15.5	0.97
16-17	14	0	14	1.00	-	16.5	-
17-18	7	0	7	1.00	-	17.5	-

$L_{25} =$ 6.7130762

$L_{50} =$ 8.845702

$L_{75} =$ 10.978328

Intercept = a = S_1 = 4.556822331

\- Slope = -b = S_2 = 0.515145359

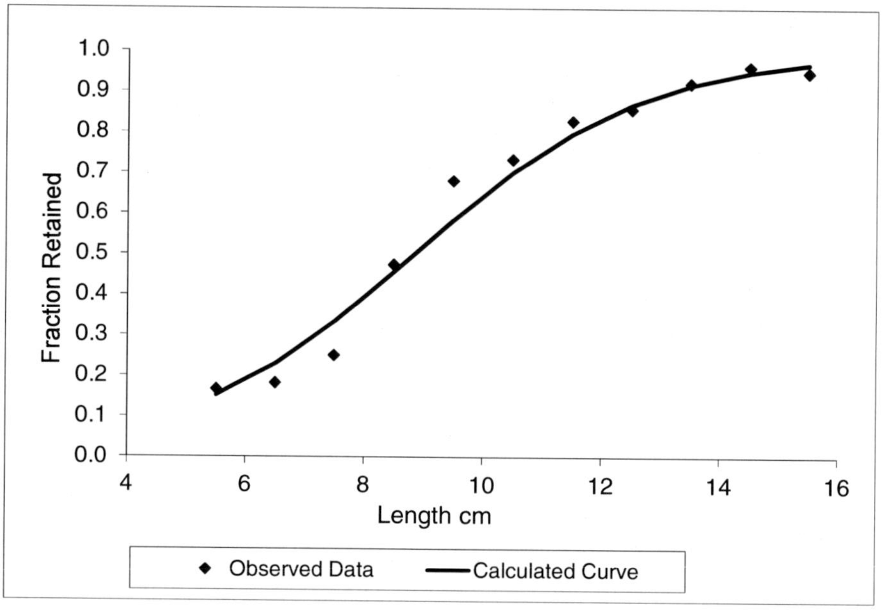

Figure 92. Selection curve for Mullus barbatus on the basis of data in Table 2.

7.2.3. Factors Affecting Trawl Selectivity

In general fish are able to avoid the catch because:

- They swim far from the bottom so that they can avoid the trawl mouth;
- They are able to swim out of the trawl mouth;
- They are able to escape under the ground rope; or
- They can pass through the meshes of the trawl body; they can pass through the meshes of the codend.

Therefore, the selectivity performance of a net is primarily affected by parameters related to fish, gear, environment, and different variables during tests.

Fish Parameters

- Fish size;
- Fish shape (girth at different body points);
- Presence of spines, teeth etc;
- Fish abundance;
- Fish availability to the net; and
- Fish behavior in the net.

Gear parameters

- Codend mesh opening;
- Codend mesh rigging (usually diamond, square or T90);

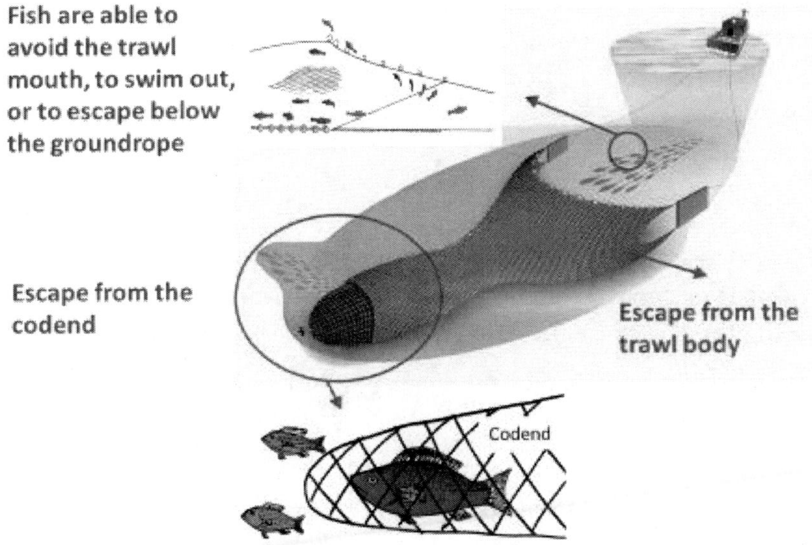

Figure 93. Escape of fish from the trawl net.

- Codend circumference: Increasing codend circumference relative to extension piece circumference decreases selectivity; and
- Additional devices like strengthening bag, bottom-side chafer can mask and reduce the escape possibilities.

Environmental parameters

- Temperature can affect the swimming performances of fish and, hence, their susceptibility to be captured;
- Sea conditions can affect gear performance. Induced pulsing movement of the codend at the end of tow when the net is pulled has been shown to affect fish escapement ahead of the catch. When the net is hauled by the winch, it moves from the bottom to the boat and in this path fish can have good possibility of escape;
- Type of bottom can affect both gear performance and turbidity;
- Water transparency can affect the ability of fish to see the meshes; and
- Depth since the higher the depth, the longer the cable used to tow the net and this can affect gear performance.

Haul variables

- Haul duration since the increase of catch inside the codend usually reduces selectivity. This is mainly due to not only the fact that fish already caught masks codend meshes but also the weight of the catch makes the codend meshes stretched with catch increase; and
- Vessel speed since increase in vessel speed usually stretches the codend meshes.

However, parameters are not equally important and the main technical factors affecting codend selectivity are:

- Mesh size;
- Mesh configuration (diamond, square, T90);
- Twine diameter;
- Codend circumference; and
- Strengthening bag.

7.2.3.1. Mesh Size

Mesh size is considered the main factor affecting codend selectivity and this is why all the Mediterranean legislations have set a fixed minimum mesh size for the codend.

Mesh size of the net can be defined by three different technical parameters:

- Bar length which is the longest distance between two opposite knots or joints in the same mesh when fully extended (it is equal to the sum of two bars; Figure 94A);
- Mesh opening which is the distance between the centers of two opposite knots or joints in the same mesh when fully extended in the N direction (Figure 94B); and
- Mesh side or bar length which is the distance between two sequential knots measured from center to center when the yarn between those points is fully extended (Figure 94C). The bar length measure is often used by commercial fishermen and by net manufacturers.

Mesh sizes must be carefully examined during net construction as the actual mesh often differs from nominal values given by the net maker.

Therefore, one of the three parameters is usually monitored in different legislations and fisheries inspectors should pay attention to which dimension has to be controlled. In any case, it is recommended to avoid the use of the generic term "mesh size" in legislation. The term "mesh opening" is the most commonly used parameter for the definition of mesh size, mainly because it is possible to define standard procedures for its measurement. However, "mesh bar" is also used in some countries.

Basically, the increase of mesh size usually means an increase in the 50% retention length (Figure 95). The use of greater mesh opening implies an increase in the mean size of fish caught.

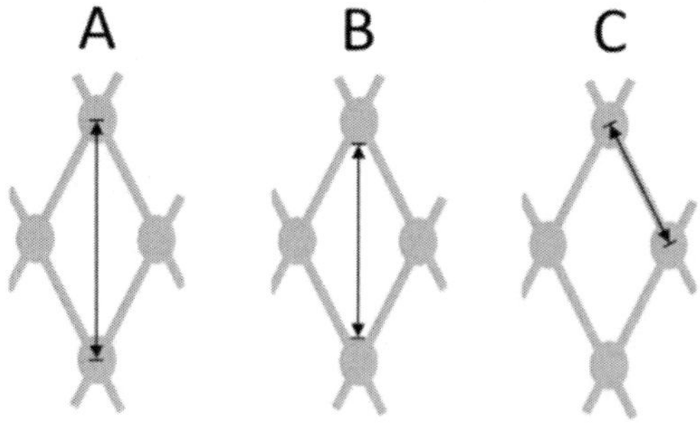

Figure 94. Technical parameters defining mesh size of net: A) Mesh Length; B) Mesh Opening; C) Mesh Bar or Mesh Side.

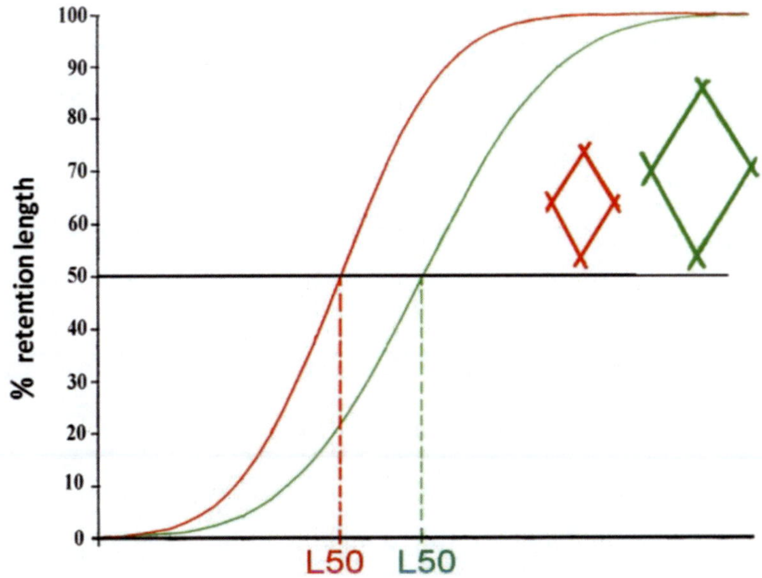

Figure 95. The effect of an increase in mesh size on L50 is well documented in the review of selectivity studies carried out by Sala et al., (2008, 2014; unpublished data), Figure 67 and Table 3).

Table 3. Review of selectivity studies in multi-species Mediterranean demersal trawl fisheries

SPECIES	MLS [cm]	Reference	Area	Cover	MI	MS [mm]	CMC	L50 [cm]	SR [cm]	SF
AL	-	Current paper	AS	HCC20	FM	39	DM	**8.30**	**1.18**	**2.15**
						39	SM	**7.61**	**0.77**	**1.97**
IC	-	Current paper	AS	HCC20	FM	39	DM	**4.90**	**0.97**	**1.27**
						39	SM	**8.38**	**1.90**	**2.17**
MM	20	Petrakis and Stergiou, 1997	WAS	CC20	PD	40	DM	**13.79**	**7.06**	**3.45**
						40	SM	**15.10**	**5.68**	**3.78**
		Ordines et al., 2006	WM	CC20	PD	40	DM	**10.60**	**3.30**	**2.65**
						40	SM	**15.20**	**3.30**	**3.80**
		Sardà et al., 2006	NWM	AH40	PD	36	SM	**18.47**	**5.07**	**5.13**
		Bahamon et al., 2006	NWM	HCC15	FM	40	DM	**10.10**	**3.10**	**2.51**
						40	SM	**16.00**	**3.20**	**3.97**
		Guijarro and Massuttí, 2006	WM	CC20	PD	40	DM	**11.60**	**0.80**	**2.90**
						40	SM	**15.30**	**2.20**	**3.83**
		Current paper	AS	HCC20	FM	39	DM	**8.26**	**1.74**	**2.13**
						39	SM	**14.17**	**3.64**	**3.67**
MB	11	Tokaç et al., 1998	AS	HCC20	PD	36	DM	**11.02**	**1.76**	**3.06**
						36	SM	**11.82**	**1.58**	**3.28**
						40	DM	**12.19**	**2.15**	**3.05**
						40	SM	**13.20**	**1.85**	**3.30**
						44	DM	**13.50**	**2.65**	**3.07**
						44	SM	**14.67**	**2.89**	**3.33**
		Sardà et al., 2006	NWM	AH40	PD	36	SM	**10.40**	**3.90**	**2.89**
		Current paper	AS	HCC20	FM	39	DM	**7.76**	**1.86**	**2.00**
						39	SM	**10.91**	**1.43**	**2.83**

Table 3. (Continued)

NN	2	Bahamon et al., 2006	NWM	HCC15	FM	40	DM	-	-	-
						40	SM	**2.20**	**0.65**	**0.55**
		Guijarro and Massuttí, 2006	WM	CC20	FM	40	DM	-	-	-
						40	SM	**2.66**	**0.34**	**0.67**
		Current paper	AS	HCC20	FM	39	DM	**1.46**	**0.28**	**0.38**
						39	SM	**1.91**	**0.37**	**0.49**
PE	12	Ordines et al., 2006	WM	CC20	PD	40	DM	-	-	-
						40	SM	**10.40**	**2.00**	**2.60**
		Current paper	AS	HCC20	FM	39	DM	**7.56**	**2.43**	**1.95**
						39	SM	**9.67**	**1.36**	**2.50**
PL	2	Guijarro and Massuttí, 2006	WM	CC20	FM	40	DM	**1.66**	**0.38**	**0.42**
						40	SM	**2.02**	**0.23**	**0.51**
		Current paper	AS	HCC20	FM	39	DM	**1.20**	**0.24**	**0.31**
						39	SM	**1.49**	**0.26**	**0.39**
TM	15	Ordines et al., 2006	WM	CC20	FM	40	DM	**13.70**	**2.10**	**3.43**
						40	SM	**15.20**	**3.00**	**3.80**
		Current paper	AS	HCC20	FM	39	DM	**9.71**	**2.75**	**2.51**
						39	SM	**13.12**	**2.43**	**3.40**
TMC	-	Petrakis and Stergiou, 1997	WAS	CC20	PD	40	DM	**13.73**	**5.50**	**3.43**
						40	SM	**11.85**	**5.95**	**2.96**
		Bahamon et al., 2006	NWM	HCC15	FM	40	DM	**9.20**	**3.00**	**2.28**
						40	SM	**13.00**	**3.00**	**3.23**
		Current paper	AS	HCC20	FM	39	DM	**8.11**	**2.07**	**2.10**
						39	SM	**11.26**	**1.65**	**2.92**

Definitions of selection parameters: The 50% retention length (L50) of a codend for a particular species is the fish length at which 50% of the species entering the codend is retained.

a) The selection range (SR) is the length difference between the 75% retention length and 25% retention length.

b) The selection factor (SF) of a codend for a particular species is the ratio of the 50% retention length and the mean measured mesh opening.

Selection parameter estimates: mean values of the retention length at 50% (L50), Selection Range (SR) and Selection factor (SF) are reported for diamond- (DM) and square-mesh (SM) codends. AL: Arnoglossus laterna (scaldfish); IC: Illex coindettii (broad-tail shortfin squid); MM: Merluccius merluccius (European hake); MB: Mullus barbatus (red mullet); Nephrops norvegicus (NN, Norway lobster); PE: Pagellus erythrinus (common pandora); PL: Parapenaeus longirostris (deep-water rose shrimp); TM: Trachurus mediterraneus (Mediterranean horse mackerel); TMC: Trisopterus minutus capelanus (poor-cod). MLS: Minimum Landing Size. MI: method for the estimation of the selectivity (FM=Fryer's method; PD=pooled data). MS: Codend mesh size. CMC: Codend Mesh Configuration (DM=Diamond-Mesh; SM=Square-Mesh).

7.2.3.2. Mesh Configuration

Traditionally diamond mesh netting panels are used for the construction of codend in the Mediterranean trawl. In recent years, many efforts have been made to study the selectivity of fishing gear used in the capture of commercial species as well as developing new gears with improved selective properties.

Several studies have demonstrated that a different configuration (square or T90) can greatly improve the selective performance of the trawl (Ordines et al., 2006; Lucchetti, 2008; Sala et al., 2008). In this regard, the GFCM Recommendation 33/2009/2 on the minimum mesh size in the codend of demersal trawl nets state that: "*The Members and Cooperating entities of GFCM shall adopt and implement, at latest by 31 January 2012, a minimum 40 mm square mesh codend or a diamond mesh size of at least 50 mm, of acknowledged equivalent or higher size selectivity, for all trawling activities exploiting demersal stocks when operating in the GFCM Area*".

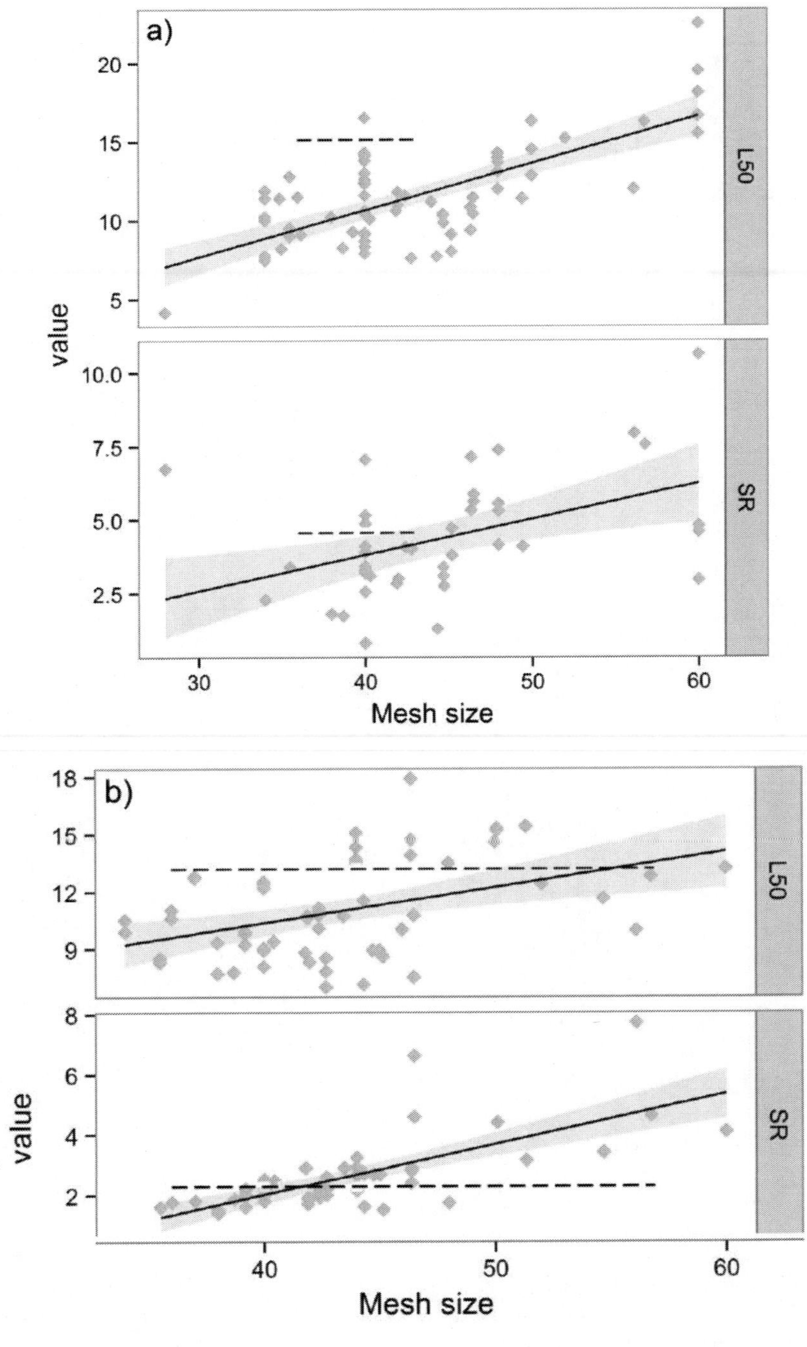

Figure 96. (Continued).

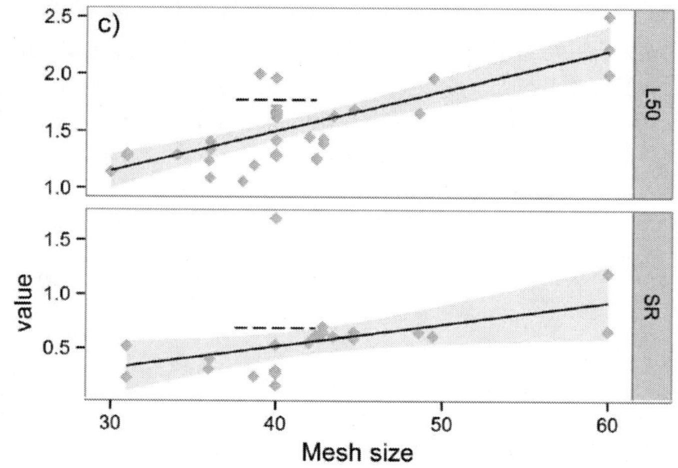

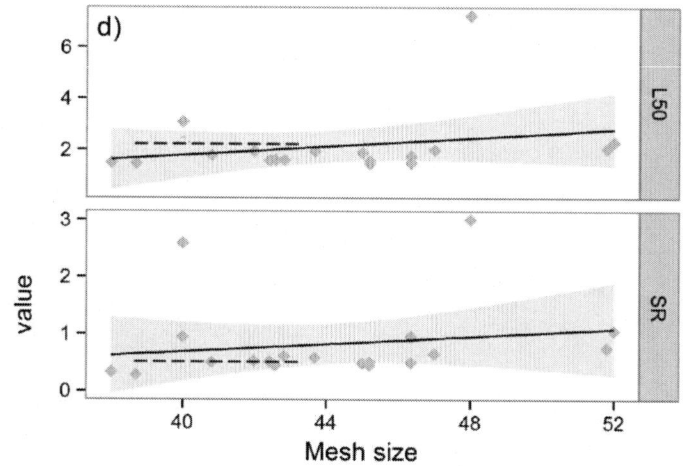

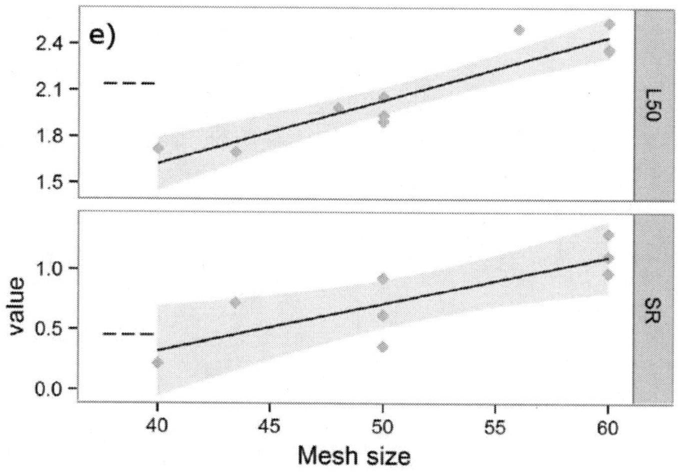

Figure 96. Relationship between selectivity parameters (L50% and SR) and mesh size for: a) Merluccius merluccius; b) Mullus barbatus; c) Parapenaeus longirostris; d) Nephrops norvegicus; e) Aristeus antennatus (from Sala et al., 2014, unpublished).

Square Mesh Rigging

In the Mediterranean bottom trawl, codend is usually made of the diamond 40 mm stretched minimum opening of meshes set by different legislations resulting in a non-particularly selective fishing gear. Therefore, the capture of unwanted species, immature and undersized fishes, is considerable. In the last twenty years, several studies have demonstrated that mesh configuration can greatly affect the selectivity performance of towed nets.

Some studies carried out in detail in the Mediterranean including Spain, Greece, Turkey and Italy showed that the use of a square mesh codend can reduce the catch of juvenile specimens and also the bycatch and discard of the bottom trawl net (Özbilgin et al., 2005a; Guijarro and Massutì, 2006; Ordines et al., 2006; Sardà et al. 2006; Lucchetti, 2008; Deval et al., 2009; Sala and Lucchetti; 2010).

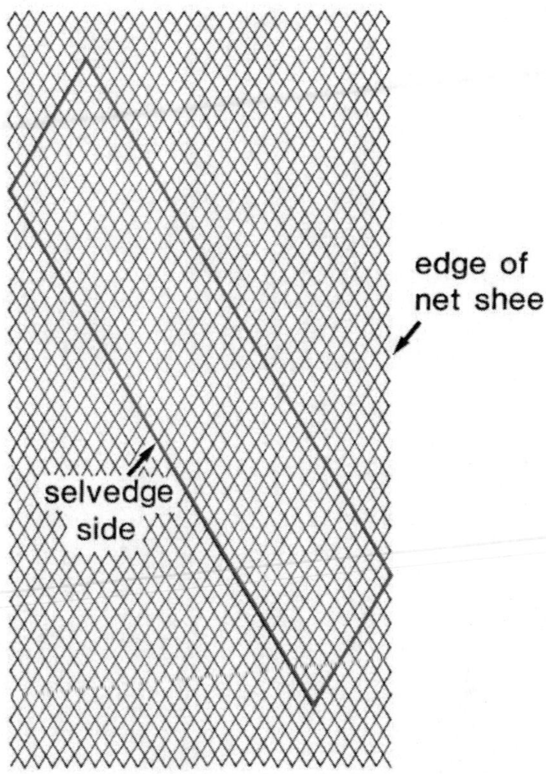

Figure 97. The method for cutting a square mesh codend panel from diamond mesh netting.

The square mesh shaped codend is usually obtained from the same PA netting panel used to prepare the traditional diamond codend by cutting the four sides of the panel on bar (Figure 97).

The square mesh codend is joined to the extension piece by turning the new panel of 45° (Figure 98, Figure 99 and Figure 100).

In the North Sea, the use of a new configuration called T90 in the codend meshes has proven to be effective in improving the selective performance of trawl nets. The construction of the turned meshes T90 is based on typical diamond netting that has to be turned through

90° (Figure 98). Therefore, the production of netting for codends with turned meshes does not require any changes in machinery or production technology in net factories.

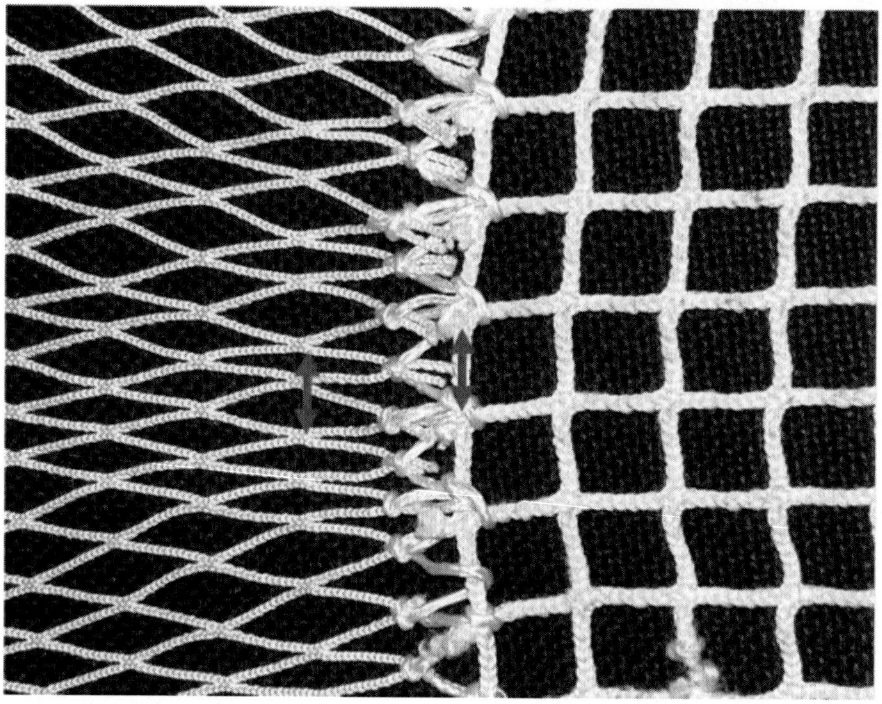

Figure 98. Joining of the square mesh codend to the extension piece.

Figure 99. Experiments with square (left) and T90 (right) codend in the Mediterranean.

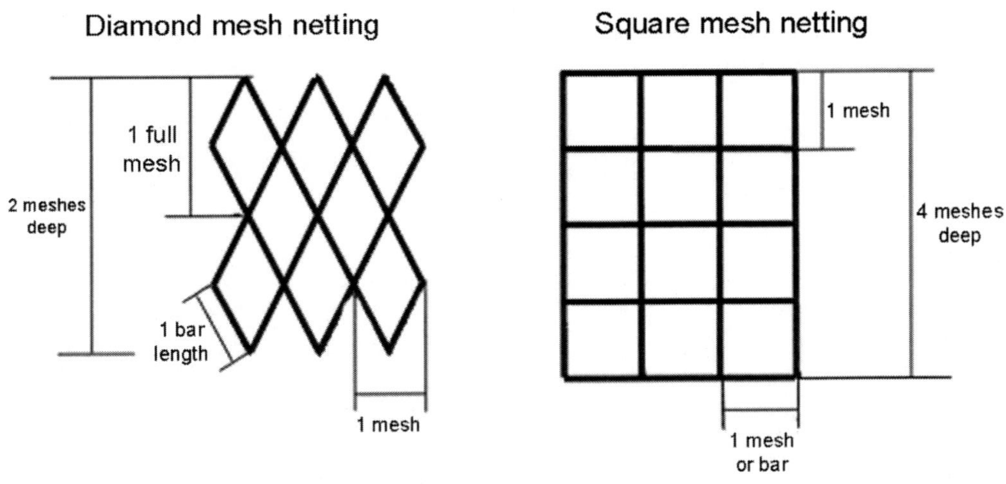

Figure 100. Nomenclature for diamond and square mesh nettings (by Robertson, 1986).

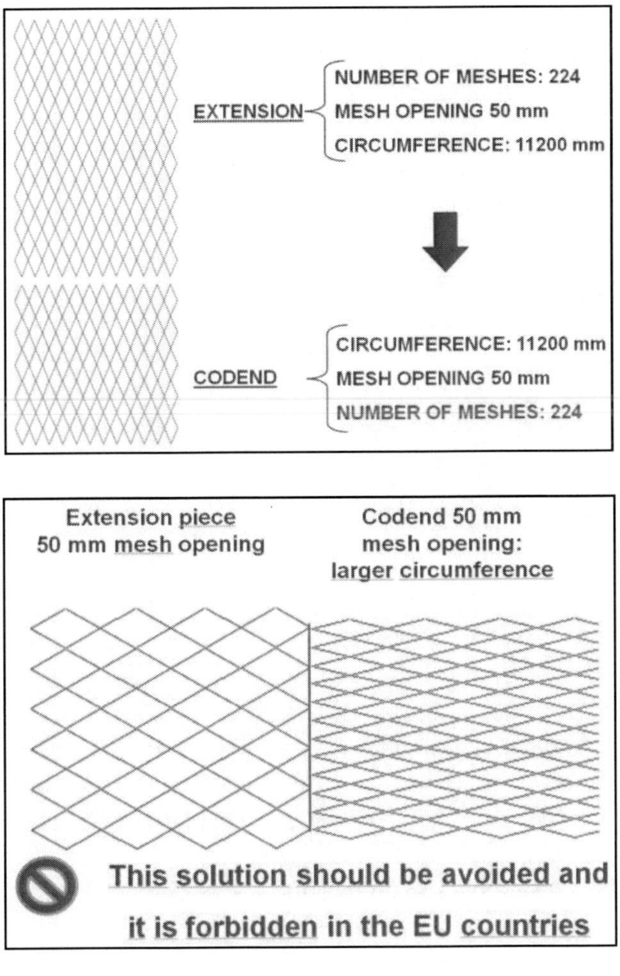

Figure 101. Rigging of diamond mesh codend with the extension piece.

Some examples of rigging can be used to join the extension piece (diamond mesh) and the codend (square and diamond mesh) are shown in Figure 101. It is noticeable that the use of a greater number of meshes in the codend circumference should be avoided. In fact, increasing the meshes at the circumference of the codend is a trick made by fishers to decrease the codend selectivity and catch smaller fish.

Square Mesh Selectivity

In the following section, the results of the selectivity study carried out by Sala et al., (2008) in the Mediterranean are reported (Figure 103). In this study, a traditional 40 mm diamond-mesh (38 mm mesh opening measured) codend was compared with an experimental 40 mm square-mesh codend in the Mediterranean demersal trawl fishery.

The results showed that square-meshes were unsuitable for flat and/or deep-bodied fish as they escape more readily than from diamond-meshes either because of their body shape or the forcing their body to penetrate through the mesh. In particular in this study were the better values of L50 found for all species except of scaldfish (*Arnoglossus laterna*).

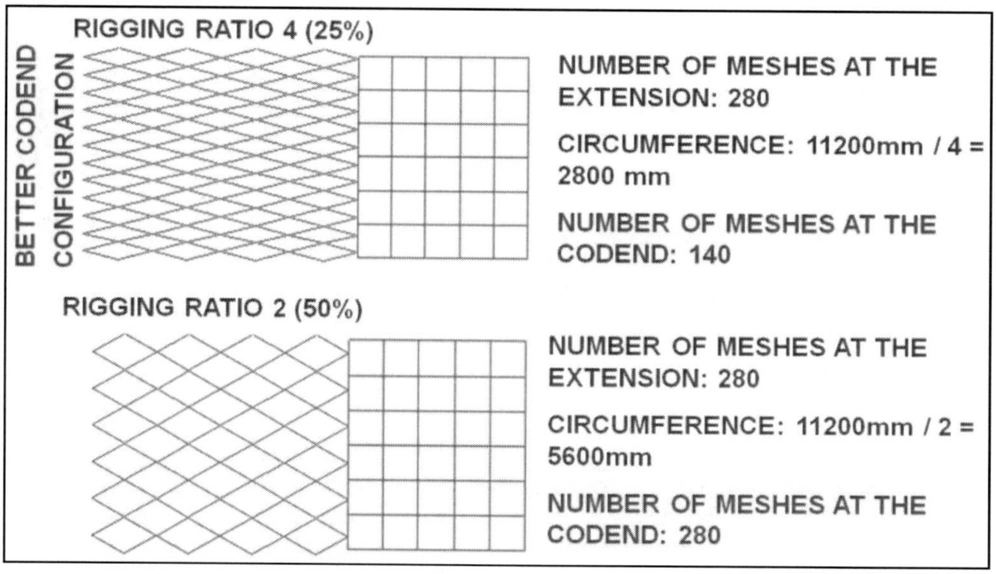

Figure 102. Rigging of square mesh codend with the extension piece.

It is noticed that the selectivity curves obtained with square mesh codend are shifted to right compared to curves of diamond mesh codend. This means that the L50 retention length improved with the use of square mesh codend.

Similar results were obtained in other research studies in the Mediterranean (Bahamon et al., 2006; Guijarro and Massutı, 2006; Ordines et al., 2006; Lucchetti, 2008; Sala and Lucchetti, 2011 etc.).

Some other studies showed that square mesh seems to protect the specimens of *Nephrops norvegicus* below the MLS in particular (Figure 104).

ARNOGLOSSUS LATERNA (SCALDFISH)

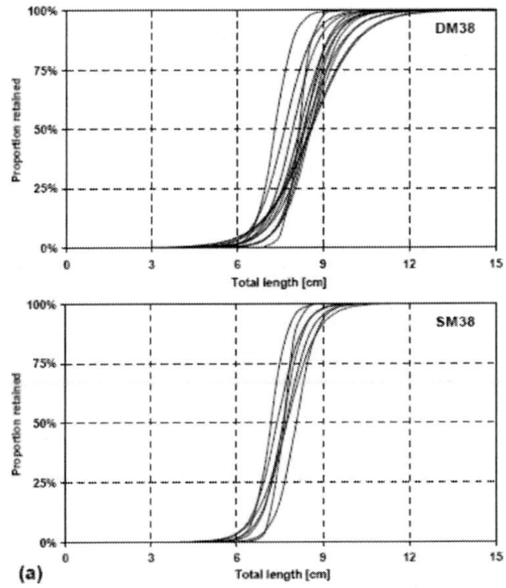

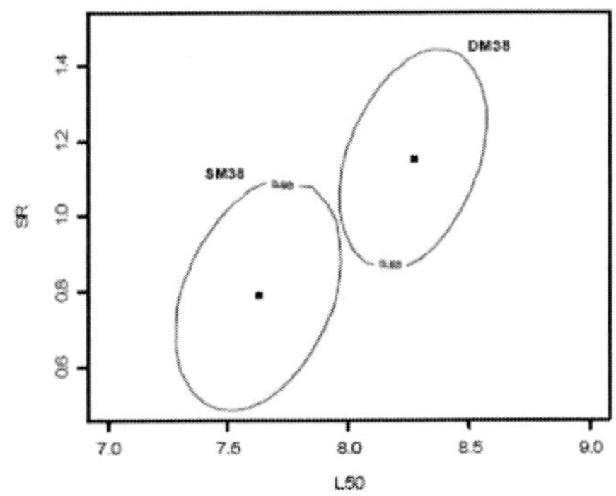

(b)

$$E\begin{pmatrix} L50 \\ SR \end{pmatrix} = E\begin{pmatrix} \alpha_1 + \alpha_3 m_i \\ \alpha_2 + \alpha_4 m_i \end{pmatrix}$$

Codend type	*pred.L50* [cm]	*pred.SR* [cm]	N.obs.	{D}		
				R_{i11}	R_{i12}	R_{i22}
DM38	8.27	1.15	18	0.0117	0.0041	0.0109
SM38	7.62	0.79		0.0151	0.0046	0.0119

ILLEX COINDETII (BROAD-TAIL SHORTFIN SQUID)

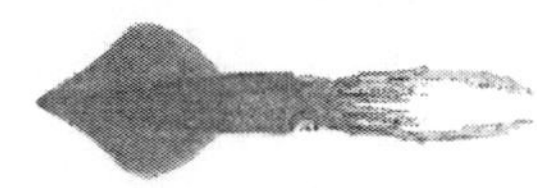

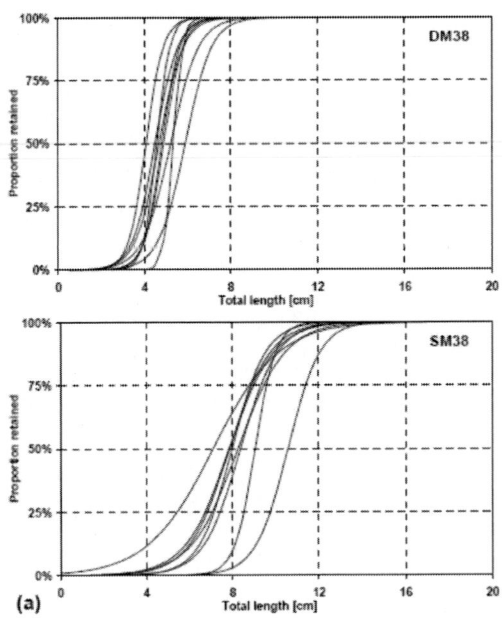

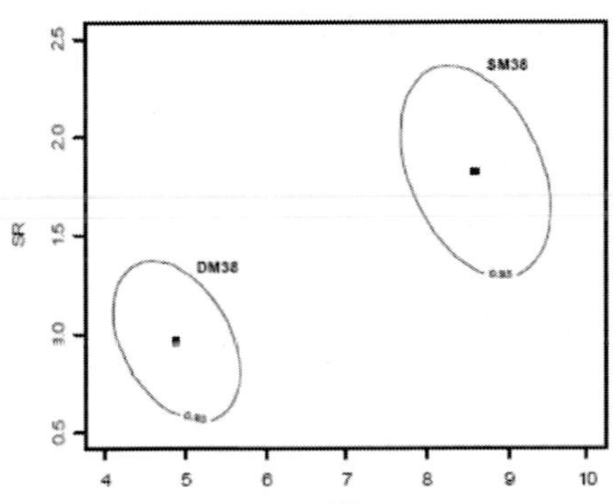

$$E\begin{pmatrix} L50 \\ SR \end{pmatrix} = E\begin{pmatrix} \alpha_1 + \alpha_3 m_i \\ \alpha_2 + \alpha_4 m_i \end{pmatrix}$$

(b)

Codend type	*pred.L50* [cm]	*pred.SR* [cm]	N.obs.	{D}		
				R_{i11}	R_{i12}	R_{i22}
DM38	4.89	0.97	14	0.0739	-0.0147	0.0194
SM38	8.60	1.82		0.1007	-0.0215	0.0337

Figure 103. (Continued).

MERLUCCIUS MERLUCCIUS (EUROPEAN HAKE)

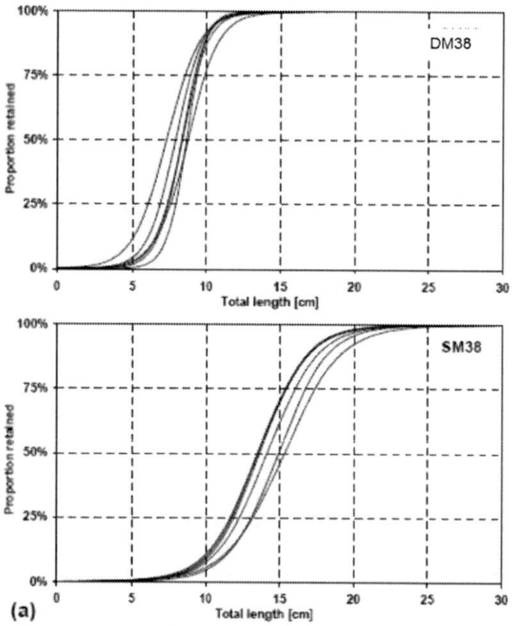

(a)

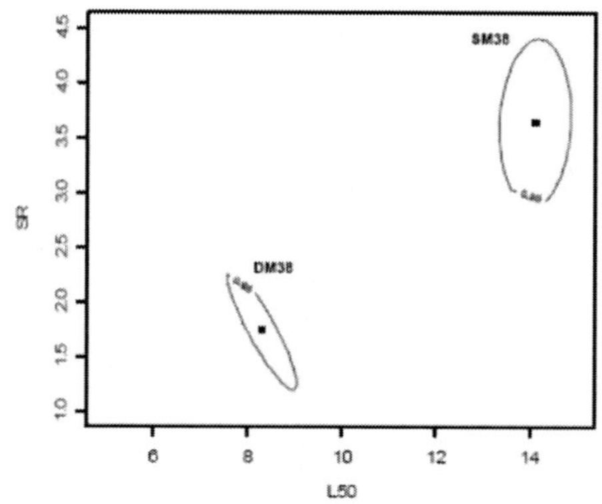

$$E\begin{pmatrix} L50 \\ SR \end{pmatrix} = E\begin{pmatrix} \alpha_1 + \alpha_3 m_i \\ \alpha_4 m_i + \alpha_8 d_i \end{pmatrix}$$

(b)

Codend type	*pred.L50* [cm]	*pred.SR* [cm]	N.obs.	{D} R_{i11}	R_{i12}	R_{i22}
DM38	8.30	1.75	10	0.0559	-0.0372	0.0302
SM38	14.11	3.64		0.0590	0.0052	0.0585

MULLUS BARBATUS (RED MULLET)

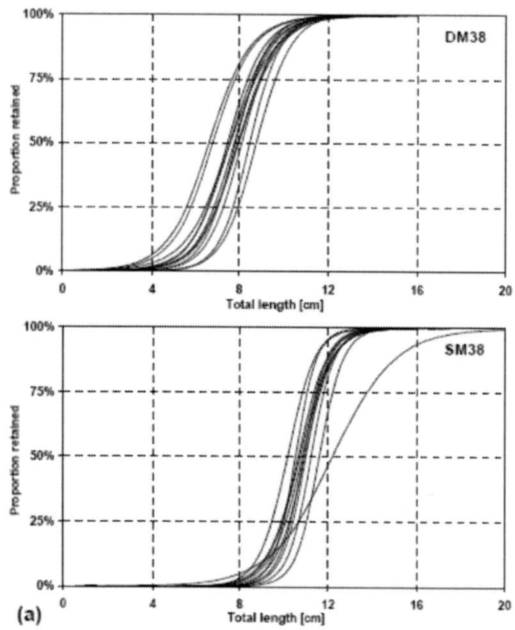

(a)

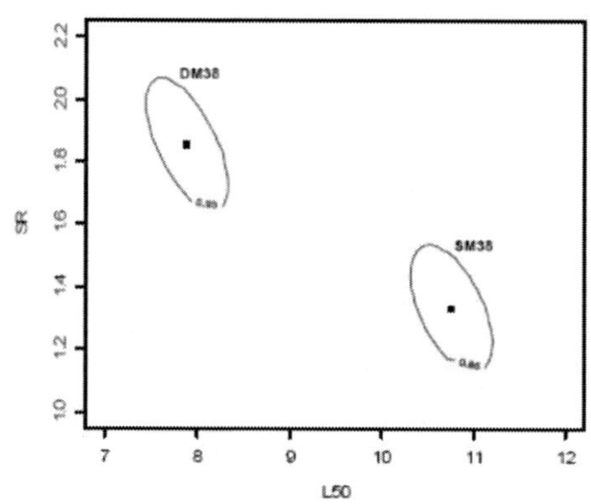

$$E\begin{pmatrix} L50 \\ SR \end{pmatrix} = E\begin{pmatrix} \alpha_1 + \alpha_3 m_i + \alpha_5 c_i \\ \alpha_2 + \alpha_4 m_i \end{pmatrix}$$

(b)

Codend type	*pred.L50* [cm]	*pred.SR* [cm]	N.obs.	{D} R_{i11}	R_{i12}	R_{i22}
DM38	7.89	1.85	21	0.0262	-0.0081	0.0060
SM38	10.75	1.33		0.0259	-0.0067	0.0055

Figure 103. (Continued).

NEPHROPS NORVEGICUS (NORWAY LOBSTER)

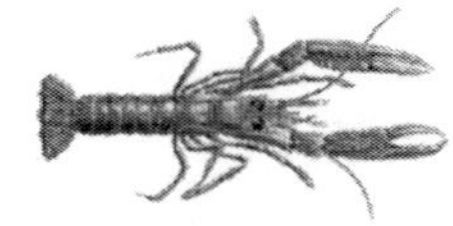

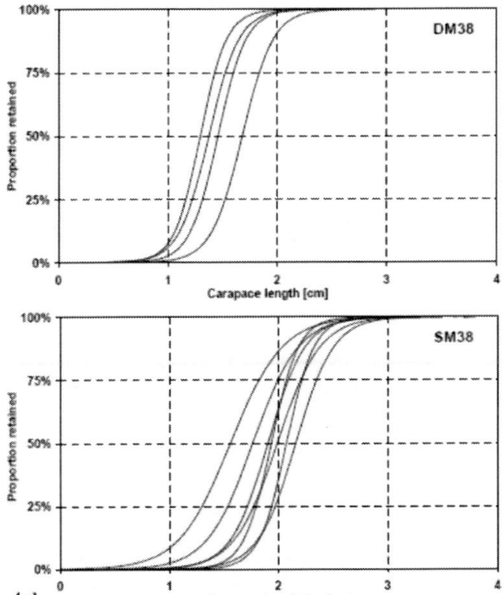

(a)

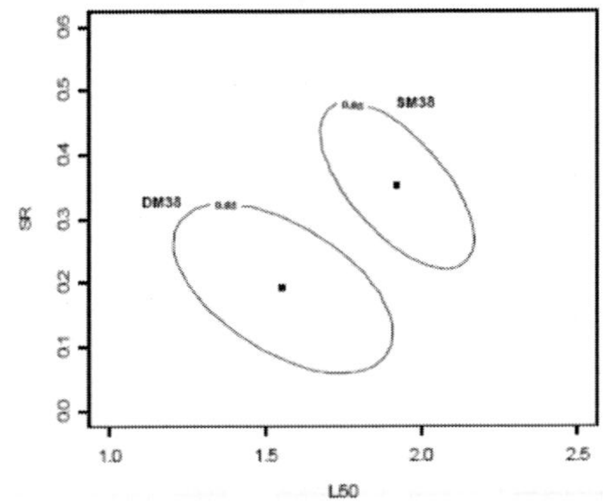

(b) $$E\begin{pmatrix} L50 \\ SR \end{pmatrix} = E\begin{pmatrix} \alpha_1 + \alpha_3 m_i + \alpha_5 c_i \\ \alpha_4 m_i + \alpha_6 c_i \end{pmatrix}$$

Species	Codend type	*pred.L50* [cm]	*pred.SR* [cm]	N.obs.	{D} R_{i11}	R_{i12}	R_{i22}
NN	DM38	1.56	0.19	10	0.0120	-0.0025	0.0017
	SM38	1.92	0.35		0.0059	-0.0020	0.0017

PAGELLUS ERYTHRINUS (COMMON PANDORA)

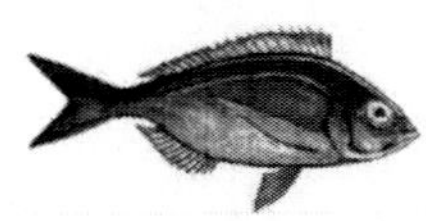

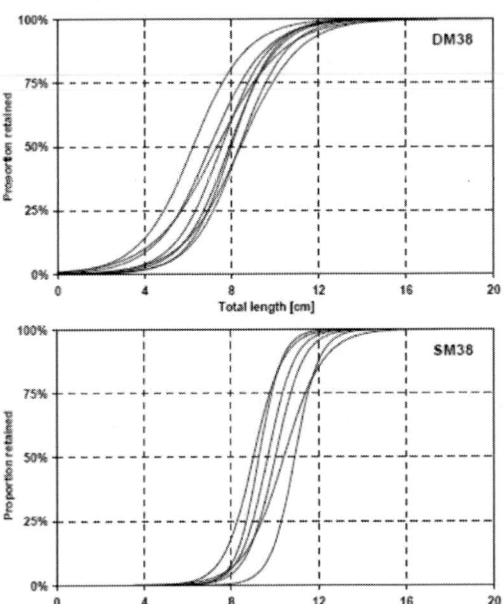

(a)

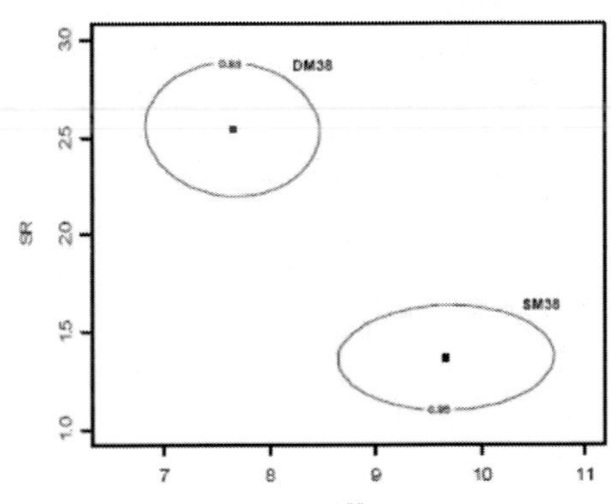

(b) $$E\begin{pmatrix} L50 \\ SR \end{pmatrix} = E\begin{pmatrix} \alpha_1 + \alpha_3 m_i \\ \alpha_2 + \alpha_4 m_i + \alpha_8 d_i \end{pmatrix}$$

Codend type	*pred.L50* [cm]	*pred.SR* [cm]	N.obs.	{D} R_{i11}	R_{i12}	R_{i22}
DM38	7.65	2.54	12	0.0744	-0.0014	0.0129
SM38	9.68	1.37		0.1142	0.0009	0.0080

Figure 103. (Continued).

PARAPENAEUS LONGIROSTRIS (PINK SHRIMP)

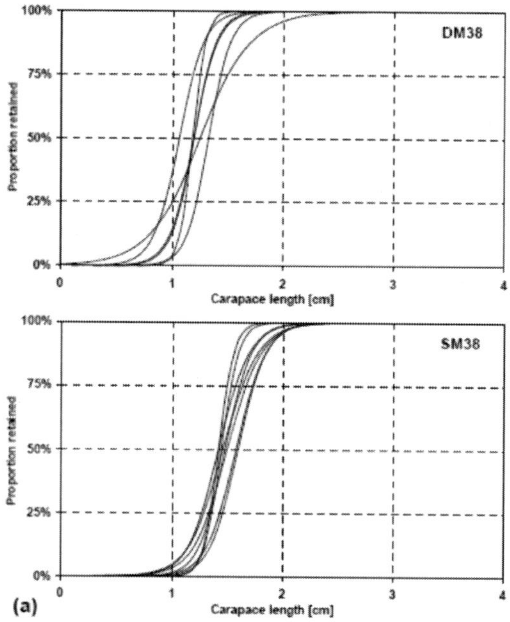

(a)

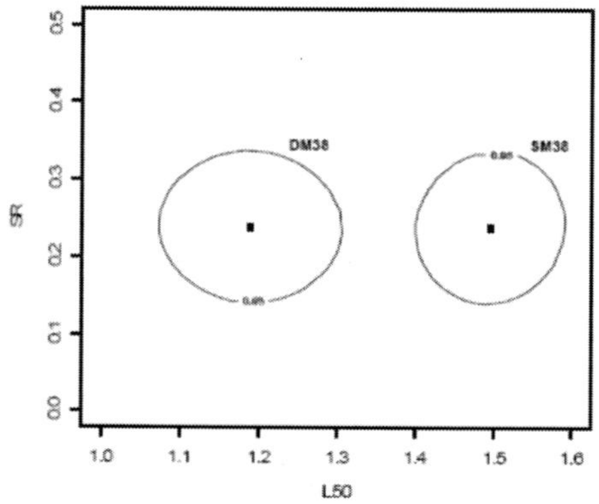

$$E\begin{pmatrix} L50 \\ SR \end{pmatrix} = E\begin{pmatrix} \alpha_1 + \alpha_3 m_i \\ \alpha_2 + \alpha_6 c_i + \alpha_8 d_i \end{pmatrix}$$

(b)

Codend type	pred.L50 [cm]	pred.SR [cm]	N.obs.	{D} R_{i11}	R_{i12}	R_{i22}
DM38	1.19	0.24	12	0.0015	0.0000	0.0011
SM38	1.50	0.24		0.0010	0.0001	0.0011

TRACHURUS SPP (MEDITERRANEAN HORSE MACKEREL)

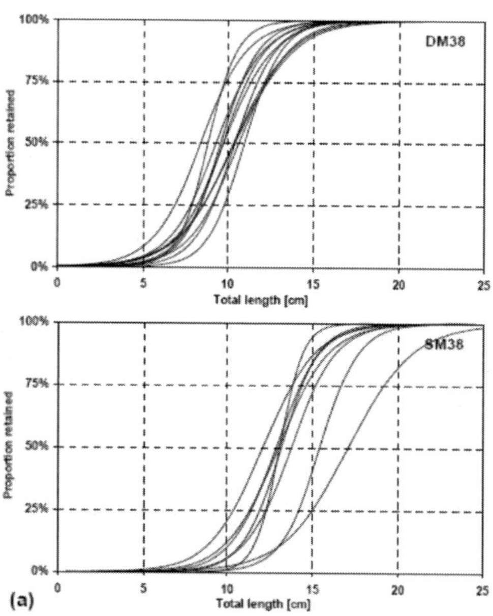

(a)

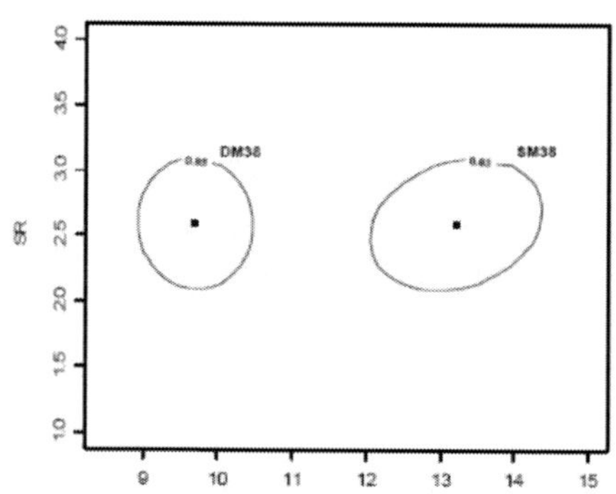

$$E\begin{pmatrix} L50 \\ SR \end{pmatrix} = E\begin{pmatrix} \alpha_1 + \alpha_3 m_i \\ \alpha_2 \end{pmatrix}$$

(b)

Codend type	pred.L50 [cm]	pred.SR [cm]	N.obs.	{D} R_{i11}	R_{i12}	R_{i22}
DM38	9.69	2.59	17	0.0761	-0.0017	0.0317
SM38	13.21	2.59		0.1705	0.0163	0.0317

Figure 103. (Continued).

TRISOPTERUS M. CAPELANUS (POOR COD)

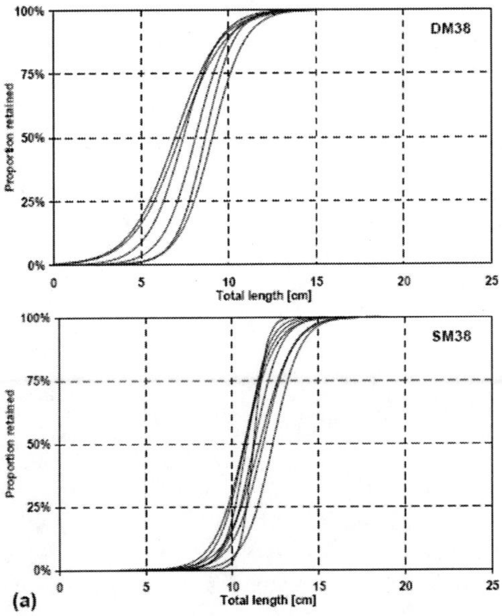

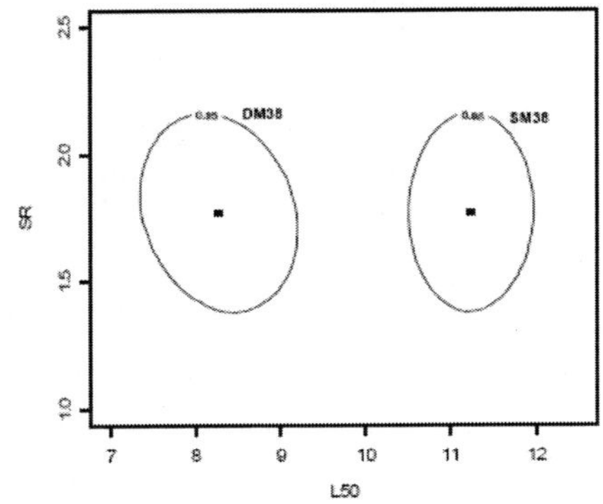

$$E\begin{pmatrix} L50 \\ SR \end{pmatrix} = E\begin{pmatrix} \alpha_1 + \alpha_2 m_i \\ \alpha_2 \end{pmatrix}$$

(b)

Codend type	pred.L50 [cm]	pred.SR [cm]	N.obs.	{D} R_{i11}	R_{i12}	R_{i22}
DM38	8.28	1.77	12	0.0933	-0.0076	0.0165
SM38	11.24	1.77		0.0584	0.0005	0.0165

Figure 103. (a) Selectivity curves for individual hauls (thin lines) and mean curve (thick line) Fryer (1991) method. (b) Confidence regions (95%) for expected mean REML values of selection range (pred. SR) plotted against the 50% retention length (pred. L50) calculated by the best model, based on the lowest value of Akaike's information criterion (AIC). DM38: diamond-mesh codend; SM-38: square-mesh codend.

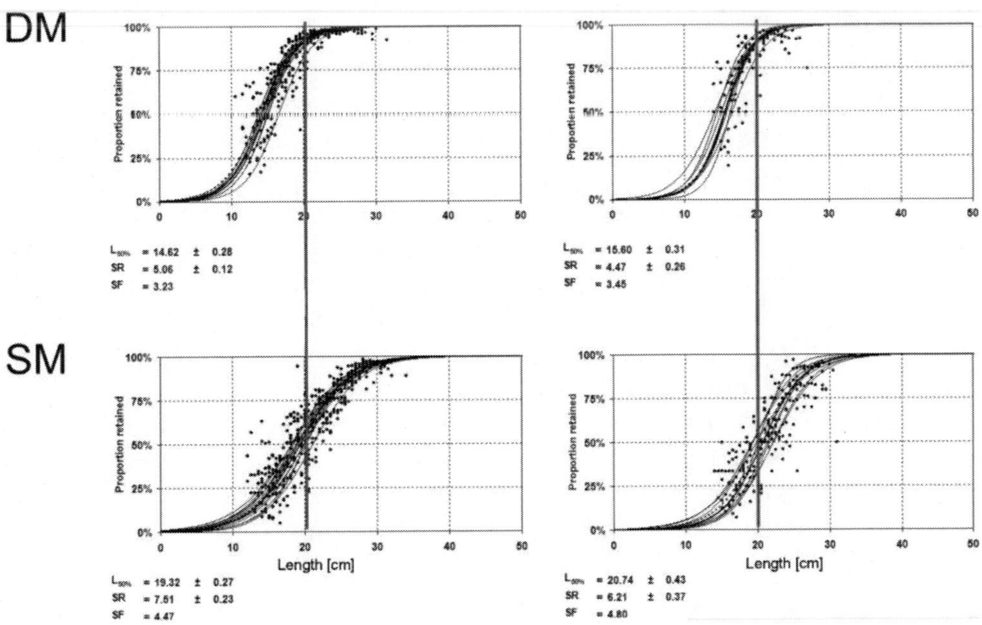

Figure 104. Selectivity curves obtained for Nephrops norvegicus in two different sample cruises.

7.2.3.3. Twine Diameter

Some Mediterranean countries have set the minimum mesh size for trawl codends at 40 mm stretched mesh resulting in the capture of large amounts of fish below the minimum landing size (MLS) with many of these being discarded. The effect of twine diameter on the selective performance of trawl net is poorly investigated in the Mediterranean.

The effects of PE netting and twine characteristics on selectivity in the North Sea have been reviewed by Ferro and O'Neill (1994). The mechanical effects affecting mesh shape and opening were found to be dependent on twine material as well as twine diameter and construction. In particular, the adverse effects of increasing twine diameter on PE codend selectivity were documented by Lowry (1995), Lowry and Robertson (1996), Kynoch et al. (1999), Ozbilgin and Tosunoglu (2003), Tosunoğlu et al., (2008). The reason of this is shown in Figure 105).

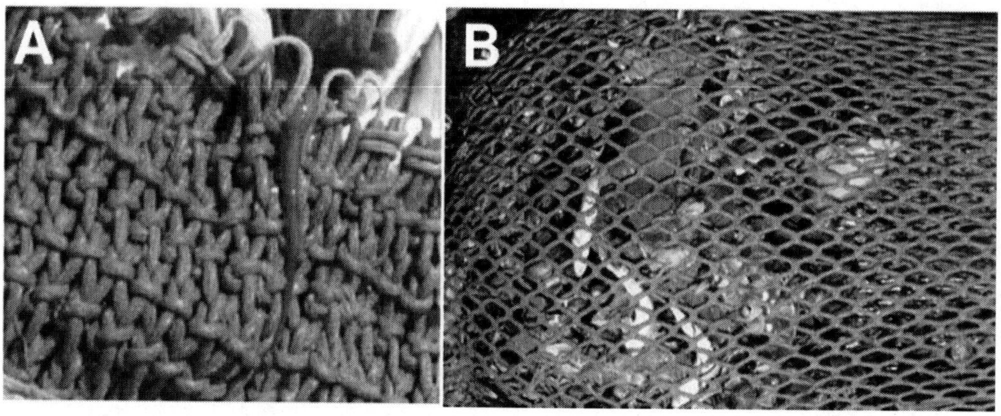

Figure 105. Two different codends made of PE (A) and PA (B) netting twine. It is possible to observe how the twine diameter affects the selectivity.

The evidence of these effects in the Mediterranean PA codends was assessed by Sala et al. (2007). This study examined the influence of twine thickness on the selectivity of the Mediterranean bottom trawl gear. The study showed that for all the species considered, an increase in twine thickness from 2.38 mm (Rtex 3644) to 2.89 mm (Rtex 5312) reduced selectivity by about 20–31% (Figure 106). For all the species, the 50% retention length showed a clear increase with the use of thin twine material.

The reasons for this difference are unclear, but changes in lateral mesh opening of netting are likely to play a major role. In practice, if heavier twine meshes were fitted to the same codend circumference, the greater amount of twine material would result in meshes being more closed than those made from lighter twine. The meshes might also be more visible deterring escape attempts. Therefore, some countries adopted rules defining the maximum twine diameter allowed. In EU countries, it is prohibited to carry on board or use any towed net constructed wholly or partly in the codend of single twine netting materials having a twine thickness of more than 3.0 millimetres as well as the use of multiple twines.

Mullus barbatus (red mullet)

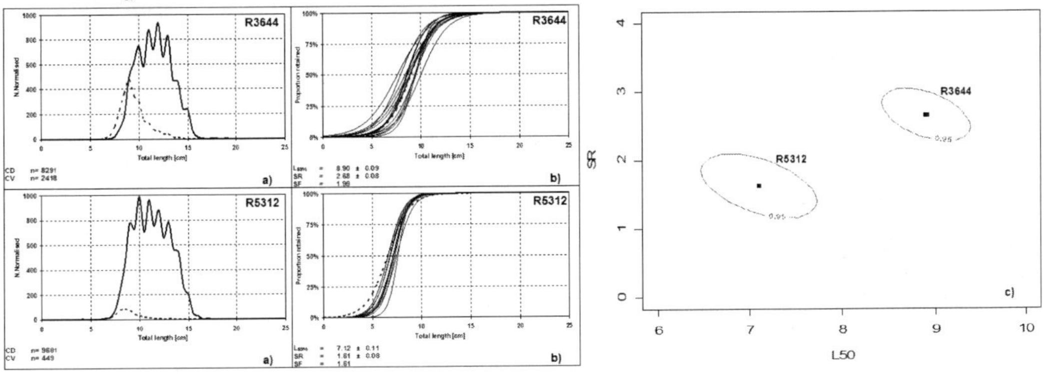

SPECIES	CODEND	MI	$L_{50\%}$ [cm]	CI [cm]	SR [cm]	CI [cm]	SF	υ_1	υ_2	{R} R_{i11}	{R} R_{i12}	{R} R_{i22}	{D} R_{i11}	{D} R_{i12}	{D} R_{i22}	LCD [cm]	LCV [cm]	LT [cm]
MB	R3644	F	**8.90**	*(8.80, 8.99)*	**2.68**	*(2.60, 2.76)*	1.99	-7.304	0.821	0.433	-0.156	0.280	0.039	-0.016	0.028	**11.61**	**9.47**	**11.13**
		P	**8.98**	*(8.95, 9.01)*	**2.59**	*(2.54, 2.64)*	2.01	-7.609	0.848	0.003	0.000	0.000	-	-	-	*(1.78)*	*(1.37)*	*(1.92)*
	R5312	F	**7.12**	*(7.02, 7.23)*	**1.61**	*(1.53, 1.69)*	1.61	-9.713	1.363	0.103	-0.039	0.053	0.020	-0.009	0.010	**11.28**	**8.86**	**11.18**
		P	**6.54**	*(6.48, 6.60)*	**2.43**	*(2.37, 2.48)*	1.48	-5.927	0.906	0.017	-0.002	0.000	-	-	-	*(1.88)*	*(1.58)*	*(1.94)*

Pagellus erythrinus (common pandora)

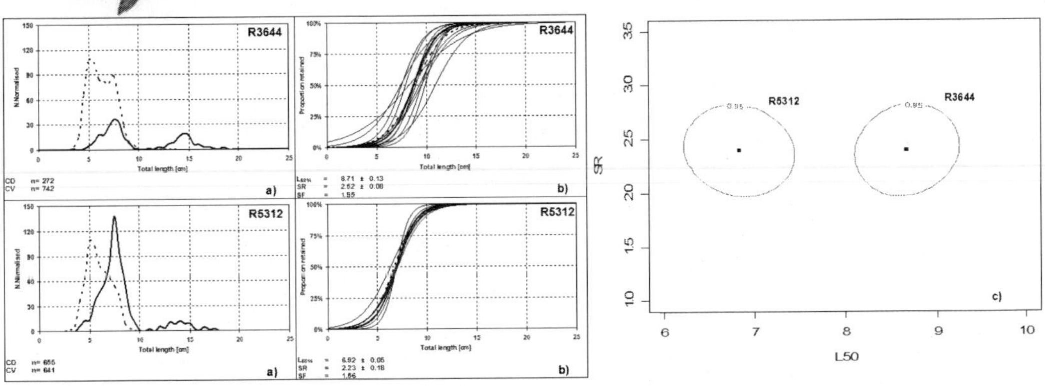

SPECIES	CODEND	MI	$L_{50\%}$ [cm]	CI [cm]	SR [cm]	CI [cm]	SF	υ_1	υ_2	{R} R_{i11}	{R} R_{i12}	{R} R_{i22}	{D} R_{i11}	{D} R_{i12}	{D} R_{i22}	LCD [cm]	LCV [cm]	LT [cm]
PE	R3644	F	**8.71**	*(8.58, 8.84)*	**2.52**	*(2.44, 2.60)*	1.95	-7.597	0.872	0.629	-0.037	0.002	0.071	0.007	0.028	**9.81**	**6.27**	**7.22**
		P	**8.87**	*(8.80, 8.94)*	**3.23**	*(3.13, 3.33)*	1.98	-6.033	0.680	0.016	-0.002	0.000	-	-	-	*(3.69)*	*(1.32)*	*(2.71)*
	R5312	F	**6.92**	*(6.87, 6.97)*	**2.23**	*(2.05, 2.41)*	1.56	-6.810	0.984	0.000	0.002	0.232	0.004	0.000	0.057	**7.96**	**5.92**	**6.95**
		P	**6.69**	*(6.64, 6.73)*	**2.51**	*(2.41, 2.60)*	1.51	-5.860	0.877	0.018	-0.003	0.000	-	-	-	*(2.45)*	*(1.22)*	*(2.20)*

Figure 106. (Continued).

Merlangus merlangius (whiting)

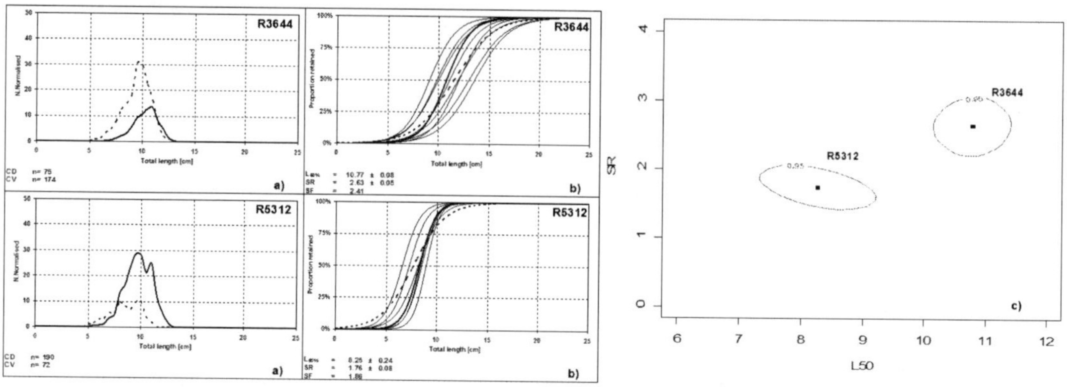

SPECIES	CODEND	MI	$L_{50\%}$ [cm]	CI [cm]	SR [cm]	CI [cm]	SF	υ_1	υ_2	{R} R_{i11}	{R} R_{i12}	{R} R_{i22}	{D} R_{i11}	{D} R_{i12}	{D} R_{i22}	LCD [cm]	LCV [cm]	LT [cm]
GM	R3644	F	**10.77**	*(10.69, 10.85)*	**2.63**	*(2.58, 2.68)*	2.41	-9.002	0.836	0.915	-0.056	0.003	0.055	0.006	0.024	**10.31**	**9.48**	**9.73**
		P	**11.60**	*(11.43, 11.57)*	**4.14**	*(3.97, 4.31)*	2.57	-6.104	0.531	0.065	-0.006	0.001	-	-	-	*(1.35)*	*(1.32)*	*(1.38)*
	R5312	F	**8.25**	*(8.00, 8.49)*	**1.76**	*(1.68, 1.85)*	1.86	-10.281	1.246	1.040	-0.173	0.030	0.136	-0.026	0.016	**9.74**	**8.52**	**9.41**
		P	**7.64**	*(7.49, 7.79)*	**3.47**	*(3.19, 3.75)*	1.72	-4.843	0.634	0.118	-0.013	0.001	-	-	-	*(1.30)*	*(1.46)*	*(1.45)*

Merluccius merluccius (European hake)

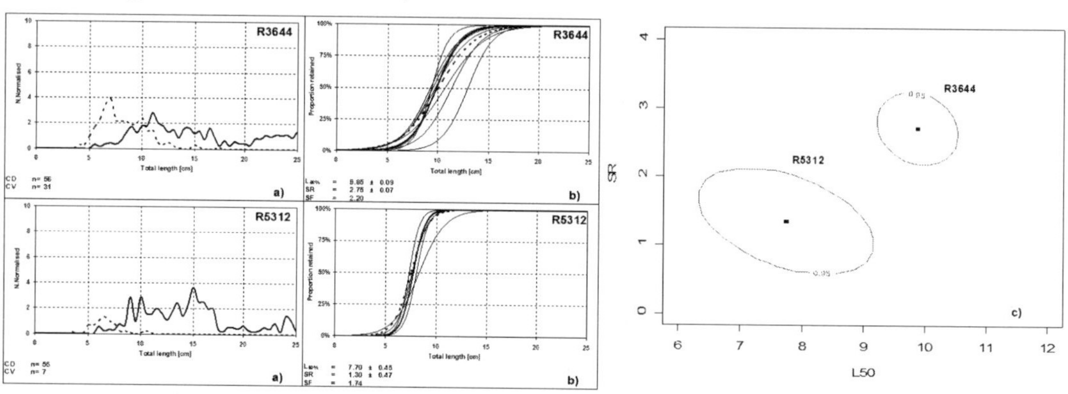

SPECIES	CODEND	MI	$L_{50\%}$ [cm]	CI [cm]	SR [cm]	CI [cm]	SF	υ_1	υ_2	{R} R_{i11}	{R} R_{i12}	{R} R_{i22}	{D} R_{i11}	{D} R_{i12}	{D} R_{i22}	LCD [cm]	LCV [cm]	LT [cm]
MM	R3644	F	**9.86**	*(9.76, 9.94)*	**2.76**	*(2.68, 2.82)*	2.20	7.881	0.800	0.950	0.236	0.099	0.067	-0.011	0.042	**18.47**	**8.25**	**14.78**
		P	**10.03**	*(10.00, 10.06)*	**3.91**	*(3.85, 3.97)*	2.24	-5.632	0.562	0.081	-0.008	0.001	-	-	-	*(8.01)*	*(2.28)*	*(8.19)*
	R5312	F	**7.70**	*(7.25, 8.15)*	**1.30**	*(0.83, 1.77)*	1.74	-12.979	1.686	0.013	0.006	0.003	0.060	-0.010	0.065	**16.59**	**6.76**	**15.53**
		P	**7.48**	*(7.41, 7.54)*	**1.85**	*(1.77, 1.93)*	1.69	-8.877	1.187	1.523	-0.188	0.024	-	-	-	*(7.05)*	*(1.44)*	*(7.34)*

Figure 106. (Continued).

Trisopterus minutus capelanus (poor-cod)

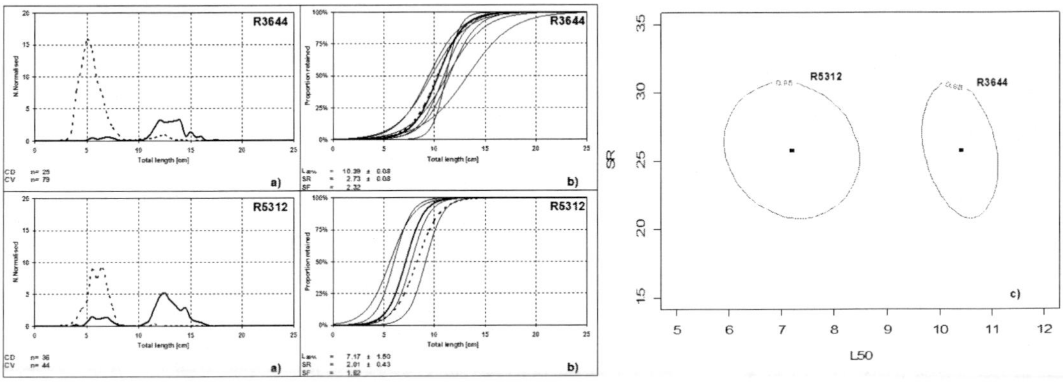

SPECIES	CODEND	MI	$L_{50\%}$ [cm]	CI [cm]	SR [cm]	CI [cm]	SF	υ_1	υ_2	{R} R_{i11}	R_{i12}	R_{i22}	{D} R_{i11}	R_{i12}	R_{i22}	LCD [cm]	LCV [cm]	LT [cm]
TMC	R3644	F	**10.39**	*(10.31, 10.48)*	**2.73**	*(2.64, 2.81)*	2.32	-8.375	0.806	0.319	-0.181	0.339	0.043	-0.009	0.044	**12.45**	**5.74**	**7.34**
		P	**10.32**	*(10.27, 10.38)*	**2.91**	*(2.86, 2.95)*	2.31	-7.805	0.756	0.076	-0.007	0.001	-	-	-	*(2.40)*	*(1.81)*	*(3.47)*
	R5312	F	**7.17**	*(5.66, 8.67)*	**2.01**	*(1.58, 2.44)*	1.62	-7.828	1.092	2.486	-0.170	0.012	0.669	-0.021	0.055	**11.76**	**6.04**	**8.63**
		P	**8.29**	*(8.19, 8.40)*	**2.49**	*(2.37, 2.61)*	1.87	-7.305	0.881	0.256	-0.034	0.005	-	-	-	*(2.91)*	*(1.06)*	*(3.55)*

Figure 106. (a) Selectivity curves for individual hauls (thin lines) and mean curve (thick line) according to Fryer (1991). (b) Confidence regions (95%) for expected mean REML values of selection range (pred.SR) plotted against the 50% retention length (pred.L50) calculated by the best model, based on the lowest value of Akaike's Information criterion (AIC). R3644: light codend; R5312: heavy codend.

7.2.3.4. Codend Circumference

Relatively little scientific work has been done to assess the effect of codend circumference on selectivity in variable multispecies conditions prevailing in the Mediterranean demersal trawl fisheries. The circumference of any cross section in a diamond-mesh netting of a trawl net shall be calculated as the number of meshes in that cross section multiplied by the stretched mesh size.

Recent studies carried out in the Mediterranean confirmed that the codend circumference can affect the selectivity of trawl codend (Özbilgin et al., 2005b; Sala and Lucchetti; 2011).

In Sala and Lucchetti (2011) study, four different codends were used in the trials. The nominal mesh sizes tested were 48 mm and 56 mm and two different nominal circumferences were associated with each mesh size; a standard one and a large circumference one. The two codends with standard circumference had the same circumference of the extension piece and, therefore, had rigging rules identical to those commonly used in commercial practice in the Mediterranean demersal trawl fisheries. The other two large codends tested had a 13–17% larger circumference than that of the extension piece.

The study found that an increase in mesh size from 48 mm to 56 mm could be made useless by an increase in codend circumference of only 13–17%. A simulation carried out by Sala et al. (2014, unpublished data), in agreement with Galbraith et al. (1994) and Tosunoglu et al. (2008), shows that selectivity increases with larger mesh size but decreases with wider codends.

The effect of circumference in the Mediterranean might be not evident because limited catch size does not allow meshes of the codend to stay open (Sala and Lucchetti, 2010).

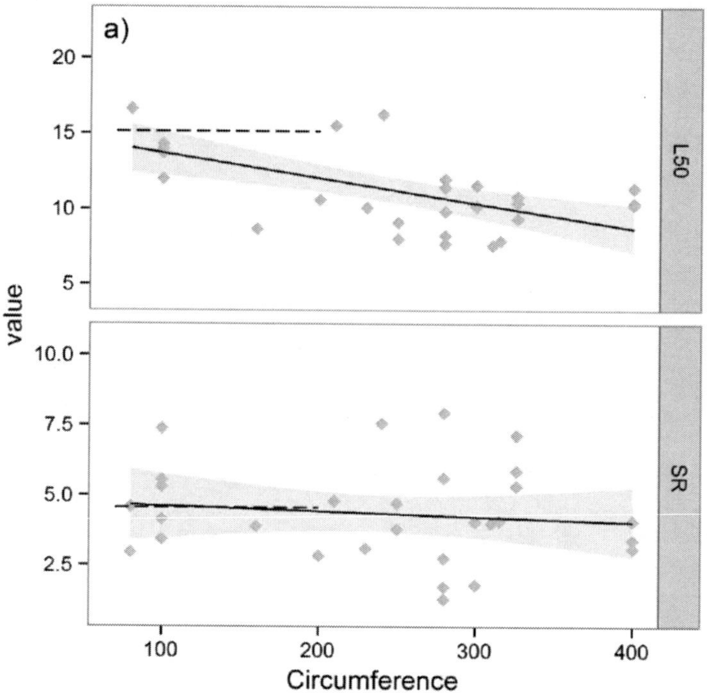

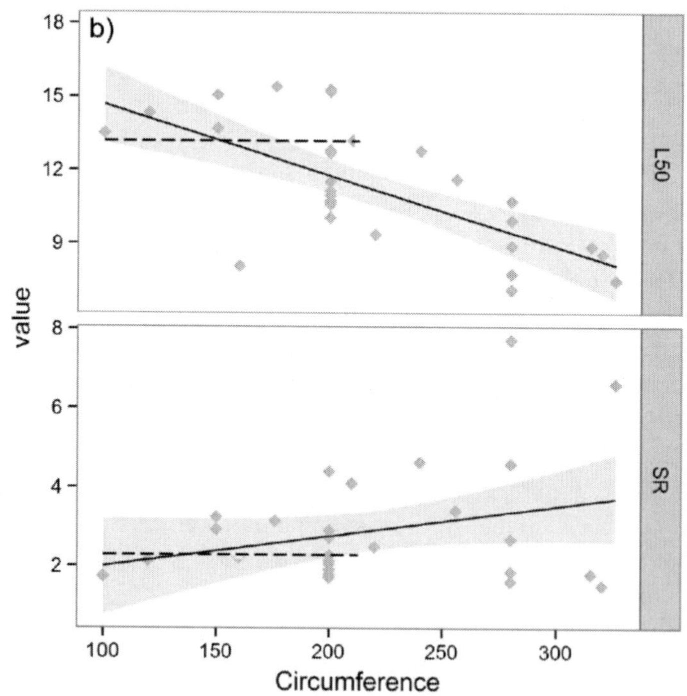

Figure 107. (Continued).

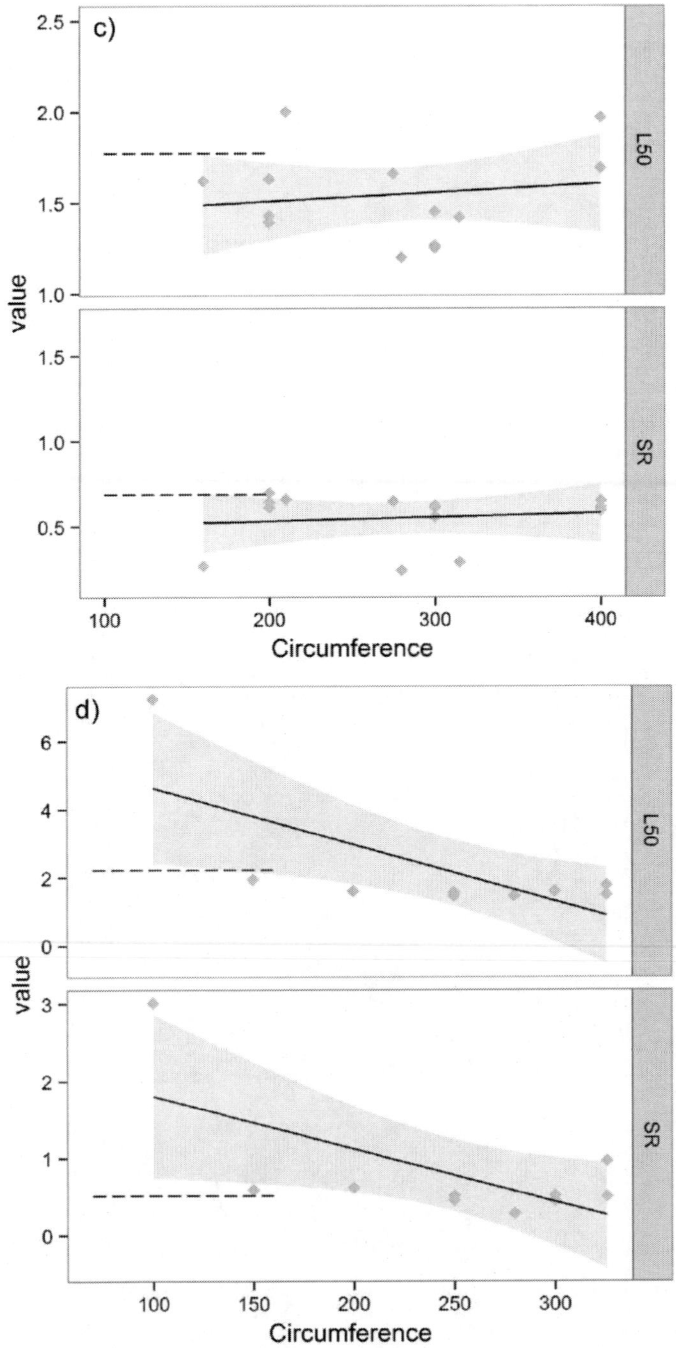

Figure 107. Relationship between selectivity parameters (L50% and SR) and codend circumference for DM codends for M. merluccius, M. barbatus; P. longirostris, N.norvegicus. Solid line represents DM codends, dashed line represents average value for SM codends.

However, the effect of different codend circumferences on selectivity it is not always clearly demonstrated. Increasing the number of meshes at the circumference of the codend is a trick made by fishers to decrease the codend selectivity and bypass legislation on mesh size.

Therefore, codend rigging is also regulated in EU countries. In particular, "*the circumference of the rearmost part of the trawl body (the tapered section) or of the extension piece (the non-tapered section) shall not be smaller than the circumference of the front end of the codend sensu stricto*".

7.2.3.5. Strengthening Bag

During trawling activities, a strengthening bag, which is a cylindrical piece of netting completely surrounding the codend of a trawl is commonly used (Figure 108). This additional and external bag is used to strengthen the codend during the hauling procedures in case of huge catch. The mesh opening and the dimension of this netting panel can strongly influence the selectivity of the codend. In fact the meshes of the strengthening bag can cover the meshes of the codend making it practically impossible for young fish to escape from the codend. Therefore, and for reasonable management of trawl gears, the strengthening bag should have a mesh opening and dimensions (length and circumference) greater than those of the codend. Thus, fisheries inspectors should also control the mesh opening as well as the dimensions of the strengthening bag provided that such technical measures are included in the legislation.

Figure 108. Details of a strengthening bag surrounding the codend of a bottom trawl.

7.2.4. How to Conduct a Selectivity Study

From the theoretical basis of selectivity curve, it can be assumed that the smallest and largest specimens are quite useless for a selectivity study. In fact, almost all the smallest specimens will be able to escape from the meshes while all the largest ones will be retained in the codend (100% of retention). Therefore, selectivity studies should be carried out in the period and area with specimens having length possibly within the Selection Range.

Therefore, and before starting work at sea, it is essential to collect preliminary information regarding the fishery especially in mixed fisheries like the Mediterranean. Information on the following should be collected:

- Fish populations (species composition, recruitment period, distribution and abundance, seasonal variations, bycatch);
- Fishing grounds (location, fishing yields, type of bottom, presence of other static gear in the area);
- Fish growth for the main species;
- Fish behaviour towards the gear; and
- Characterization of the area in terms of environmental conditions.

The main sources for this type of information are fishermen themselves, fish landings at local ports, and historic information from national databases.

Moreover, information on the following should be obtained to plan for sea cruise:

- Types of vessels used (Research vessel, commercial vessel, characteristics);
- On board facilities (space, working surface, net handling techniques, presence of refrigerator which can be very useful in case of subsampling); and
- Duration of fishing trips.

To investigate the selectivity of commercial fishing gears, it is highly recommended that the trials should be undertaken on board a commercial fishing vessel and commercial practices should be followed regarding fishing grounds, fishing depths, fishing season, towing time etc.

A crucial point in the set-up of a sea cruise for the study of selectivity is the number of hauls that is necessary to obtain reliable data. A power analysis (Fryer, 1996) before experiment can be very useful in determining how many hauls are likely to be needed to obtain significant results.

In the Mediterranean, the reliable number of hauls that is reasonable for a selectivity study is 8-10 hauls for each parameter to be studied.

In general, it is quite important to plan testing significant differences between a limited number of gear types or parameters. It is strongly recommended to test the effect of different parameters; e.g., the effect of mesh size or mesh configuration, individually rather than testing simultaneously a number of gears or parameters. Simultaneous testing of effects of several parameters on selectivity can lead to fail obtaining significant differences between any of them.

In selectivity experiments, it is quite common for catch to be so large that it is not feasible to take length measurements of each fish caught. In these cases, a sample of the target species catch or the whole catch if separation by species is impossible has to be taken from codend or cover and weighed. There is no standard procedure for subsampling catch and the procedure varies with circumstances. However, care is needed to avoid the normal tendency taking the sample from the top layer of catch or container under sea working conditions. It is essential to mix the bulk catch before taking the required sample. When the target species is abundant, the simplest procedure to subsample the catch is taking a uniform fraction from the entire catch. Most appropriately, a random selection of catch is made which might be difficult under normal working conditions at sea. When a small number of large fish is present in the catch, it is strongly recommended that these large fishes are measured individually. This is quite important particularly for cover catch where only few large specimens might be present. This procedure avoids the underestimation of large specimens by including them in subsample.

7.2.4.1. Data to Be Collected

Basically, two main categories of data have to be collected in selectivity trials; fishing gear, and vessel and haul data. This information is commonly included in the "*materials and methods*" sections in scientific papers.

For example, the cover used in the covered codend method can have a strong serious effect on the output of the study. Detailed information on net and cover designs is, therefore, of great importance for the evaluation of results validity.

The "haul data" include all information related to environmental parameters like area, season, time, depth, and type of bottom that can affect the selectivity of fishing gears. However, the most important data to be collected are the morphometric measures of fish including individual fish lengths in different codends and cover, weights of individuals, sample size measured etc. Selectivity can also be affected by the towing speed, haul duration, time of day (light intensity), weather conditions etc and, therefore, these data have to be registered also.

Obviously, not all data have the same importance. Some of them are indispensable for every selectivity experiment while others are useful but not essential. A checklist of the data that should be collected in the selectivity experiment might be useful when planning the experiment.

Here is a summarized example checklist:

- Selectivity method applied: covered codend, twin trawl, trouser trawl etc.;
- Vessel properties: Overall length, power, Year of construction;
- Area: Position (Latitude, Longitude), depth, type of bottom;
- Time: Season, month, hour of the day;
- Sea conditions;
- Gear properties: Gear design would be essential if possible, net dimensions of each netting section such as body, extension, codend;
- Codend properties: Mesh opening measured or the nominal value should be reported, mesh rigging (square or diamond usually), twine thickness, codend length, codend circumference;

- Gear performances (only with acoustic sensors): Horizontal net opening, door spread, vertical net opening;
- Haul properties: Haul duration (start & end of haul), vessel speed, vessel course;
- Total weight of catch in codend and cover separately;
- Eventual subsample rate;
- Total weight of each species after sorting; and
- Individual length of each species: Method adopted should be noted on the length record form. Overall length, fork length or carapace length for fish; carapace length, total length or tail length for crustaceans (carapace length is commonly used); mantle length, total length for cephalopod. The unit of measurement is usually 1 cm for species that grow larger than 30 cm, and 0.5 cm for species that do not reach 30 cm.

7.2.4.2. Software for the Study of Trawl Selectivity

There are software packages that provide tools for a fast and methodic assessment of trawl selectivity. In this section a summary of the FISHSELECT, SELNET and PRESEMO, which are most representative and useful tools for trawl selectivity are reported. Each tool focus on a particular selectivity aspect: FISHSELECT focus on interaction between selectivity fish morphology; SELNET provides many selectivity models and methods and PRESEMO models the selectivity simulating the fish escapement on the basis of the codend geometry. The combined use of these software allows for a whole coverage of the selectivity analisys in trawl fisheries.

FISHSELECT

FISHSELECT (Herrmann et al., 2009) is a fish morphology data- and simulation-based methodology that can be applied to investigate the basic size selective properties of meshes of different shape and size for individual fish species (Figure 109).

The FISHSELECT methodology, which includes procedures, tools and software, determines whether a fish is capable of passing through a certain mesh based on the morphology and compressibility of the fish, and shape and size of the mesh. By means of computer simulation, the method is ultimately used to predict the selective properties of diverse fishing gears.

The fish morphopology is evaluated through a morphometer by measuring different cross sections of fish. The scanned sections are modelled and each section shape is related to the fish length. On the basis of the relationship among sections shape and length, a virtual population is created and the interaction between fishes and different mesh size and configuration is estimated on the basis of the fish and mesh morphology.

The method has already been applied in a number of studies to estimate the selective properties of towed fishing gears for both round-fish (Herrmann et al., 2009; Krag et al., 2011; Sistiaga et al., 2011) and crayfish (Frandsen et al., 2010).

Predictions from FISHSELECT can contribute to improving the management of a fishery because they can help identify the gear changes that would be necessary to achieve different exploitation patterns. Studies conducted using FISHSELECT, for example, are globally applicable to fisheries, meaning that they can be used to predict selection properties of not only trawl gear but of gillnets and pots as well.

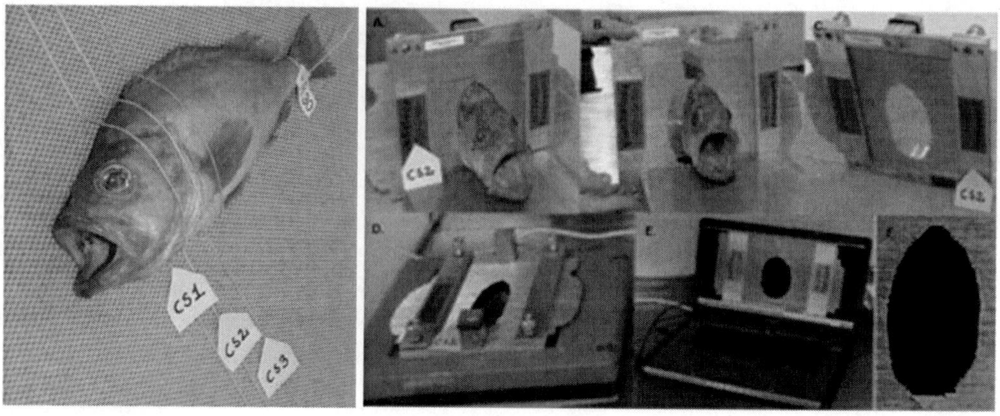

Figure 109. Fishselect: morphology measurements of a redfish by a morphometer and evaluation of the cross section area.

SELNET

SELNET (Size Selection in trawl NETting) is a flexible software tool to acquire and view catch, to analyse catch comparison and size selectivity data for towed fishing gears, to provide analysis of the data, to view resultrs from analysis, to simulate cath and discard dased on the single and multi species fishery and make simple assessments on economical consequences (Sistiaga et al., 2010).

Furhter to standard selectivity studies, based on the relationship between length of species and mesh size and configuration, SELNET enables the analysis of data for experimental designs involving multiple compartments by means of complex selection models.

PRESEMO

PRESEMO (PREdictive SElective MOdel) is based on an individual-based structural model of the selection process in the cod-end of a trawl fishing gear. It models different populations of fish entering the cod-end during a tow. Each fish is assigned a weight and a maximum width and height dependent on its length, and is assumed to be of elliptical cross-section. It is also allocated to each fish a travel time down the cod-end, a time it can swim in the cod-end without being exhausted, a time between escape attempts and a packing density for swimming in front of the catch. An escape attempt is deemed successful, if the fish can pass through the mesh opening at the point of the codend, where the attempt takes place. The openness of a mesh is a function of the codend geometry, calculated external to PRESEMO and imported into PRESEMO. Fish that do not escape fall back and become part of the catch when their exhaustion time is reached (Figure 110). The codend shape is continually updated, as the catch builds up during the tow. At the end of a simulation, a logistic function is automatically fitted to the simulated selection data to obtain estimates of the 50% retention length (L50) and selection range (SR). The model is thus based on information about the fundamental mechanical, hydrodynamic and biological processes that govern cod-end selection.

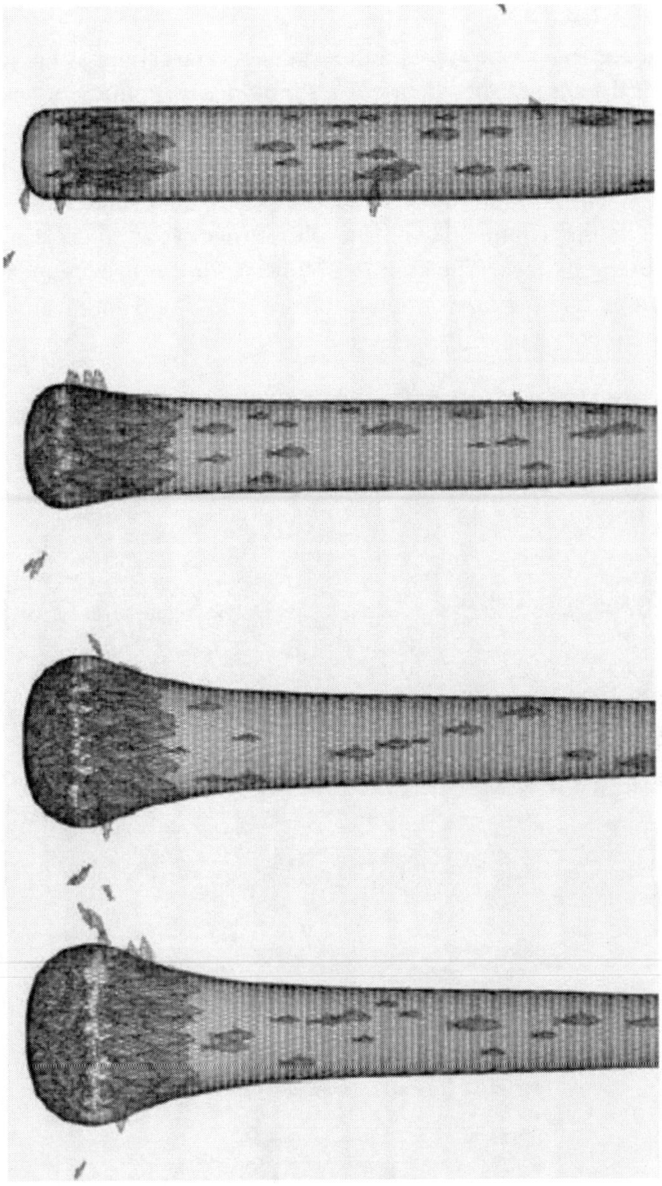

Figure 110. Screen dumps from PRESEMO when simulating a haul (from Priour et al., 2006).

Dredge Selectivity

The selectivity of dredges is usually the sum of two selective processes: the selectivity of the main gear (the dredge or cage) and the selectivity of the sieve. Therefore, the selective process in the hydraulic dredge fisheries has two stages: the first stage takes place at sea while fishing, the second occur on board the vessel.

Many factors are known to affect dredge selectivity. These include the technical properties of the dredge, such as dredge setup and design (teeth length and angle, blade length and angle, space between rods, dredge weight) and operational factors (towing speed, tow duration) and environmental conditions (sea state, type of sediment etc.).

Selectivity studies have been carried out in the Mediterranean Sea to investigate the selectivity performance of striped hydraulic dredgers targeting venus clam (*Chamelea gallina;* Sala et al., 2014 unpublished data). The simplest way to measure dredge selectivity is to use the same approach followed in a classic covered codend method for the estimation of trawl selectivity. The population exploited by the dredge during towing operation can be obtained by using a small mesh bag (small frame armed with a small mesh netting) mounted on the mouth of the dredge (Figure 111). The role of this bag is to collect all the organisms encountered in the dredging tack. The comparison between the size frequency distribution of clams inside the dredge and the size structure of the whole population investigated with the cover, enabled to obtain a classic ogive selection curve.

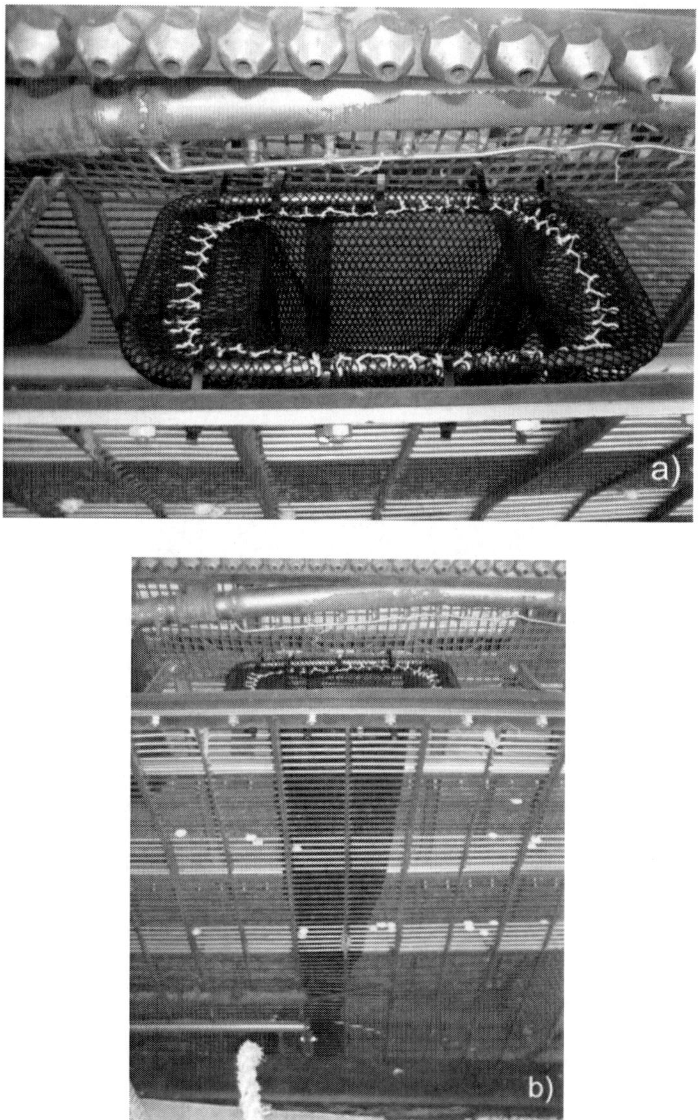

Figure 111. a) Detail of the frame mounted on the dredge mouth; b) detail of the small mesh bag inside the dredge.

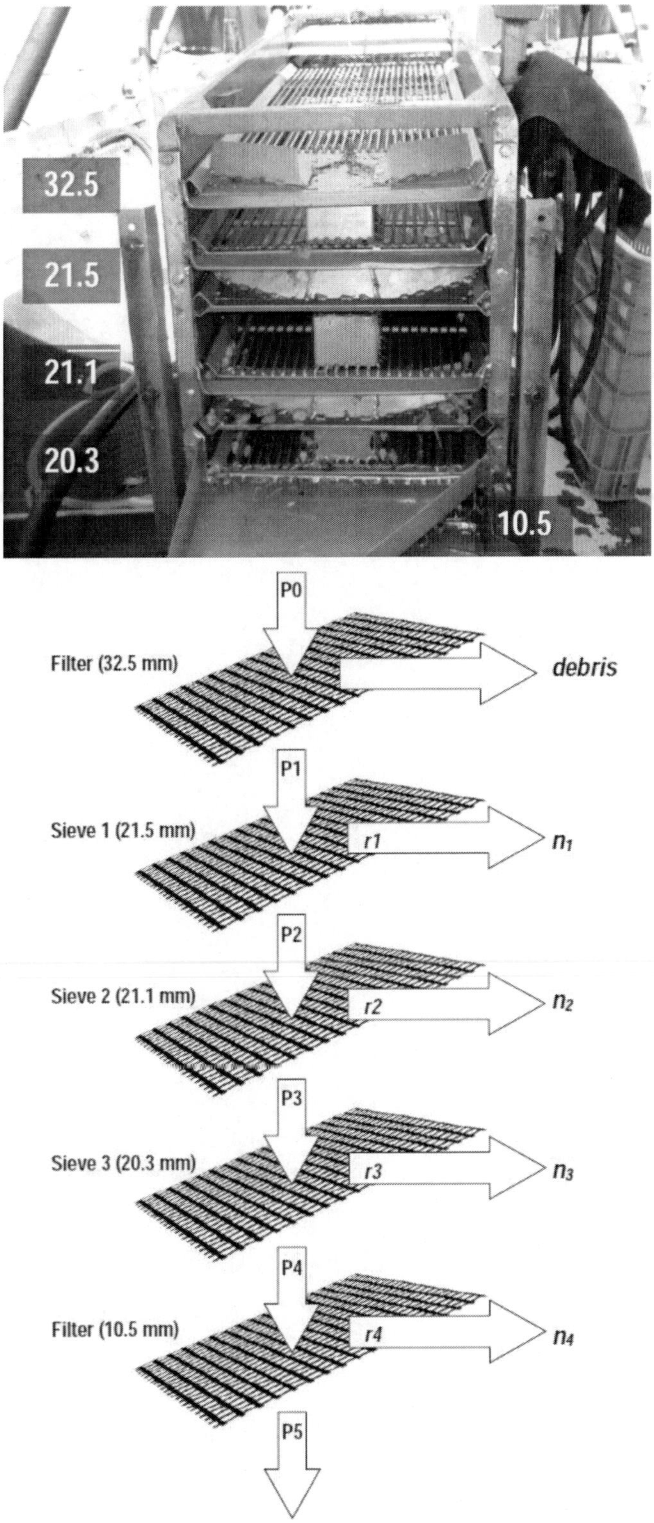

Figure 112. Method used for the study of sieve selectivity; sieves with different hole diameter are tested.

In the same way it is possible to calculate the selectivity of the different sieves (Figure 112).

It is possible to observe that "drawers" with decreasing diameter of holes are used to select the product in different commercial classes. Each drawer or sieve allows to retain a portion of the total catch. P_0 is the initial population. The first sieve (32.5 mm hole diameter) only retains debris and marine litter. P_1 (the fraction passed through the sieve 32.5 mm) is the population considered for the calculation of selectivity of the sieve 1 (21.5 mm). P_2 (the fraction passed through the sieve 21.5 mm) is the population considered for the calculation of selectivity of the sieve 2 (21.1 mm) and so on.

For each sieve the selection process can be summarized as follows. Sieve 1 allows to divide initial population n in two fractions: the retained portion of the catch ($n_1 = r_1 \cdot n$) and the escaped fraction (n-n_1), where r_1 is the retained percentage. The fraction n-n_1 is than considered as the initial population for the calculation of sieve 2 selectivity. The retained portion of sieve 2 (n_2) can be obtained by: $n_2 = r_2 \cdot (n-n_1)$ where r_2 is the retained percentage of sieve 2, and so on.

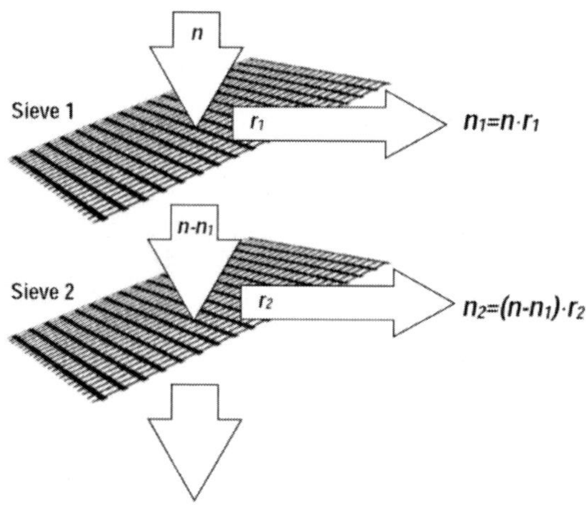

Figure 113. Details of the selectivity calculation.

Therefore, the retention probability of each sieve (r_i) can be obtained by the ratio between the retained fraction and the total population entering that sieve (the sum of retained and escaped fraction).

$$\begin{cases} r_{sieve1} = \dfrac{r1}{r1+r2+r3+r4} \\[2ex] r_{sieve2} = \dfrac{r2}{r2+r3+r4} \\[2ex] r_{sieve3} = \dfrac{r3}{r3+r4} \end{cases}$$

and the selectivity model is the same used for bottom trawl selectivity:

$$r(\ell) = \frac{e^{\upsilon_1 + \upsilon_2 \ell}}{1 + e^{\upsilon_1 + \upsilon_2 \ell}}$$

Where $r(\ell)$ is the retention probability of a clam having length ℓ.

7.4. Passive Nets Selectivity

The catch process and also the selective properties of a passive net depend on three distinct conditions:

- The probability that the use of a certain fishing gear coincides in time and space with the presence of a single or different species;
- The probability that fish either a single or multiple species encounters the fishing gear; i.e., fish are accessible to the gear; and
- The probability that the passive net retains fish provided they have encountered it; i.e., that fish are vulnerable to the gear.

The latter condition is strongly correlated with the technical characteristics of the fishing gear.

When several species interact with a passive net, the catch process mainly depends on the behavior displayed by each species towards the fishing gear. When a single species encounters the net, the retention of a fish is driven by its specific characteristics such as age, length or girth, shape, presence of spines, teeth etc. In this case, the selection is often expressed as a function of fish length though the escape-capture events are related mainly to the girth of a fish. This is mainly because it is easier to measure the length of fish rather than its girth.

The weakness in gillnet selectivity studies is the lack of knowledge on the population encountering the gear, while this information is collected in trawl selectivity studies using a cover or an unselective gear. Thus, one of the main solutions to study the selectivity of a gillnet is to compare the size frequency distribution of the specimens caught with gillnets having the same technical properties (drop, length hanging ratio etc.) but different mesh sizes, the so-called indirect technique. Moreover, some basic assumptions are essential for this type of investigation. The most important is the Baranov's "Principle of Geometric Similarity". This principle states that if the selection depends only on the relative geometry of the fish; i.e., fish girth and indirectly fish length and meshes, then all selection curves are similar. Therefore, the shape of selectivity curves is the same for any combination of fish length and mesh size for which their ratio is constant (Hamley, 1975). Therefore, catch should be thought of as the product of encounter probability and retention probability even though when indirect methods are applied, the encounter probability cannot be estimated.

Generally, it is assumed that all length classes have the same probability of encountering the nets even if larger fishes are more active than smaller ones in real situation and, thus, having a higher probability of encountering the nets.

7.4.1. Passive Net Selection Curve

The selectivity curve of gillnet has a bell shape with a modal length L_m (*L50*) which is the optimal length at which the fish is caught (Figure 114). Noticeably in this figure, the maximum girth (Figure 115) of small specimens in yellow enables them to pass through the meshes. For fish caught in red, the mesh perimeter is greater than the gill girth but smaller than the maximum girth. For fish with gill girth greater than mesh perimeter (in green), head dimension hampers the fish to be gilled or enmeshed. In general, there is no standardized method of conducting girth measurements. The main problem is girth round soft parts of the fish. In fact differences in the degree of maturity or in stomach content of fish caught can account for these effects, as well as differences in body compressibility.

In case of gilling or wedging catches, the catching process mostly depends on the relative geometry of the mesh and the fish. Therefore, the fish measurements should also include girth perimeter, which is in principle more related to mesh size than length values. Therefore, some authors have used fish girths in evaluating the selectivity, even if the fish length is easier to measure and it is a more precise method.

In gillnets, the fish modal lengths increase with increase in mesh sizes in general. This is mainly explained by the principle of geometrical similarity of Baranov (1948). However, the morphology of certain species can violate the principle and, hence, there are different ways of capturing the fish. Therefore, it is possible to observe a distortion in the shape of the selection curves in some cases, particularly in the right arm. This can be caused by an increase in mesh size which leads to a higher capture of entangled fish belonging to the higher length classes. The asymmetry of the curve is more frequent in the trammel nets. For example, an increase in the vertical slack will modify the shape of the right arm of the curve due to an increased percentage of larger individuals that become entangled. Selectivity curve can also be non-unimodal. In these cases, it is likely that the fish are caught in different ways. Therefore, it is important to record the way of enmeshment and proportion of the catch enmeshed in the different ways.

The bell shaped selection curve can be expressed as:

$$S_L = \exp\left[-\frac{(L-L_m)^2}{2\cdot s^2}\right]$$

Where:

- S_L is the selectivity curve which is the probability that a fish of length L is caught between 0 and 100%;
- L_m (*L50*) is the optimal length at which the fish is caught which is proportional to the mesh opening; and
- s is the standard deviation of the normal distribution.

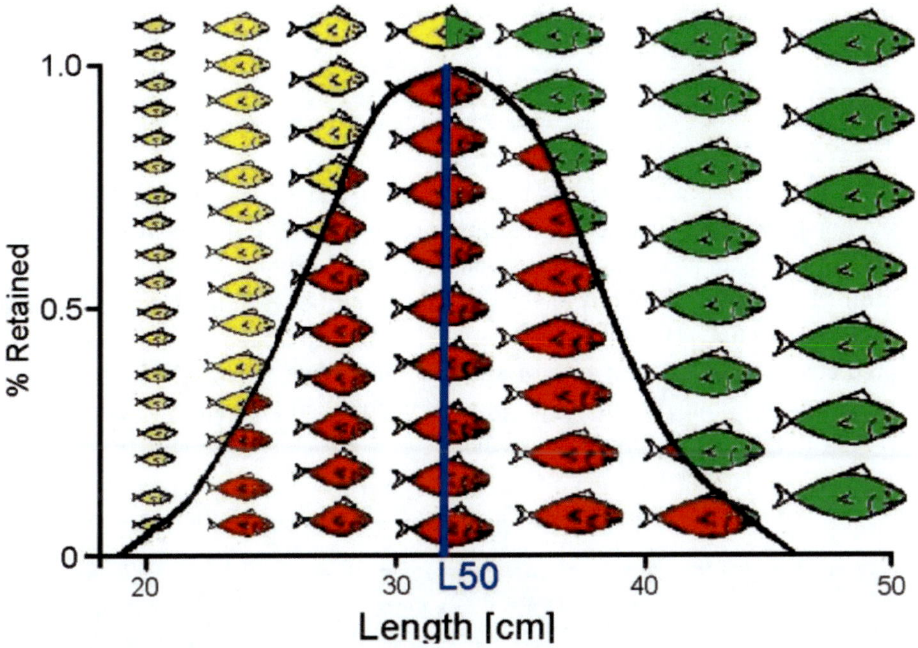

Figure 114. The bell shaped selection curve for a gillnet. In yellow: small specimens escaped; in red: captured; in green: large specimens escaped.

Figure 115. Maximum girth and gill girth measurement. It is possible to use a cotton thread or a tape measure.

The bell-shaped selection curve for gillnet can be explained as follows:

- Small fish can easily escape from the meshes so that the retained percentage is really low. Head and body girth are smaller than the mesh perimeter which is double the mesh size;
- When fish size increases, the retained percentage increases also up to a maximum of L50 (head girth is smaller than the mesh perimeter but the body girth is equal to the mesh perimeter);

- When fish size increases further, its dimensions fish start to prevent it from entering the meshes so that the retained percentage decreases down to zero (head girth is bigger than mesh perimeter and fish is too big to be caught).

Two different methods can be used for the estimation of passive net selectivity:

- Direct estimation which requires knowledge of real size structure of the population by comparing the frequency distributions obtained with passive nets and those obtained with an unselective gear such as a bottom trawl); and
- Indirect estimation which is based on the relation between the girth of fish and mesh opening. Sechin (1969) and Kawamura (1972) approach allows selectivity to be predicted from measurements of species' maximum and head girths provided that wedging and gilling are the main ways of capture, and the analysis is independent of size distribution data for gillnet catches.

Table 4. Data analysis for gillnet selectivity calculation

Length	Retained (in relation with the maximum girth)	Escaped (in relation with the gill girth)	Product retained-escaped
20	0.005	1.000	0.005
21	0.017	1.000	0.017
22	0.048	1.000	0.048
23	0.120	1.000	0.120
24	0.215	1.000	0.215
25	0.366	0.990	0.362
26	0.581	0.970	0.564
27	0.740	0.940	0.696
28	0.844	0.890	0.751
29	0.900	0.820	0.738
30	0.940	0.740	0.696
31	0.960	0.640	0.614
32	0.974	0.460	0.448
33	0.990	0.280	0.277
34	1.000	0.160	0.160
35	1.000	0.090	0.090
36	1.000	0.040	0.040
37	1.000	0.013	0.013
38	1.000	0.006	0.006
39	1.000	0.001	0.001
40	1.000	0.000	0.000

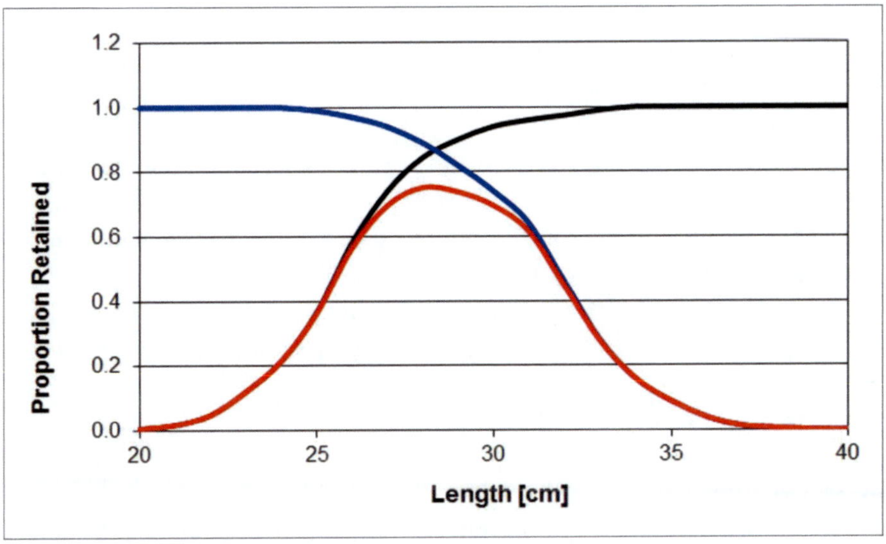

Figure 116. Selectivity graph obtained from Table 4 data.

7.4.2. Factor Affecting Passive Net Selectivity

Several parameters related to fish, gear, boat and fishing operations can affect the catching process and the selectivity of passive nets.

Passive net technical properties:

- Net length and net drop;
- Mesh size;
- Horizontal and vertical hanging ratio. Vertical slack is defined as the ratio between the stretched length of inner and outer nettings in trammels nets. Increase in horizontal hanging ratio and vertical slack will lead to an increase in the catch of larger individuals that become entangled; and
- Twine characteristics (material, thickness, colour, flexibility). A thinner twine normally has a better catch efficiency because it's probably less visible than thicker one but mechanical breaking point decreases with decreasing diameter also.

Parameters related to fish and environment:

- Fish size;
- Fish shape (girth at different body points);
- Presence of spines, teeth etc.;
- Fish abundance;
- Fish availability to the net;
- Fish behaviour towards the net; and
- Presence of predators which can reduce the soaking time.

Parameters related to fishing operations:

- Space on board;
- Net handling techniques;
- Soaking time;
- Environmental parameters which can affect the swimming performances of fish);
- Seawater turbidity and light level which can affect the visibility of the net);
- Sea state and currents which can affect the movement of the net on the bottom);
- Seabed type (it can affect the turbidity of water);
- Amount of catch; the presence of other specimens caught can increase the visibility of the net; the catching efficiency can be affected by net saturation;
- Depth;
- Presence of debris.

7.4.2.1. Mesh Opening

Passive net selectivity is mostly a girth/mesh-opening related process. Mesh opening is the most important parameter affecting gear selectivity. Fish modal lengths normally increase with an increase in mesh sizes used. The shape of the selection curves, particularly concerning the right arm, can be affected by an increase in mesh size, since the use of higher mesh sizes can lead to a higher capture of entangled fish belonging to the higher length classes.

It is well known that large mesh size nets have a greater efficiency for large fish. This theory can explain why gillnets are species-specific. Fishermen are used to make gillnet with a mesh opening decided on the bases of target species and size).

7.4.2.2. Hanging Ratio

One of the main parameters affecting the catching process of a passive net is the hanging ratio. The hanging ratio (E) is commonly defined as

$$E = L/L_0$$

Where:

- L is the length of rope on which a net panel is mounted;
- Lo is the l*ength* of stretched netting hung on the rope.

It is possible to define two different hanging ratios;: the primary or horizontal hanging ratio and the secondary or vertical hanging ratio.

Primary or horizontal hanging ratio (Figure 117) is defined as

$$E1=L/L_0$$

Where:

- L is the hung length of the netting or the mounted length of the main mounting rope;
- L_0 is the length of the same netting when fully extended (L_0= *number of meshes in horizontal x mesh length*).

Secondary or vertical hanging ratio (Figure 117) is defined as

$$E2=H/H_0$$

Where:

- *H* is the hung height or depth of the netting panel, or the mounted length of the side hanging line;
- H_0 is the fully extended height of the netting (*number of meshes in vertical x mesh length*).

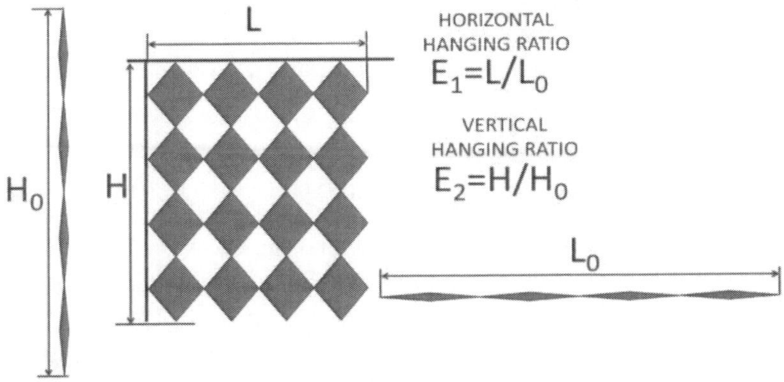

Figure 117. Details of the horizontal and vertical hanging ratio.

Passive nets with high hanging ratios (>0.5) will have a stretched netting panel with meshes clearly open. In these cases, the catching of fish will happen mostly by gilling or enmeshing.

In nets with low hanging ratios (<0.4), netting will present a marked "slack" so that catching of fish will happen mostly through entanglement. Obviously, the effect of using a low hanging ratio depends on the size and the shape, the presence of spines or cutaneous expansions as in anglerfish. Therefore, passive nets with low hanging ratios are less selective than those with high hanging ratios.

The vertical slack is known to affect selectivity of passive nets. The selection curves obtained for trammels with vertical slack of 1:1 have the same shape as those for gill nets with the same mesh size. An increase in the vertical slack will mainly modify the shape of the right arm of the curve, due to an increased percentage of larger individuals that become entangled, with a lower influence in the selection of smaller individuals. Usually an increase in the vertical slack is used to increase the net efficiency, even if this is valid only within a certain range of values for this parameter.

7.4.2.3. Floatation

The net floatation and its weight affect the vertical slack of passive nets and the drifting of bottom nets with the tide. The increase of vertical net slack is obtained by decreasing the number of floats and this solution is commonly used by fishermen as a mean to enhance the fish entangled (mostly large fish or marine organisms with spines, teeth etc.). The net floatation can also be adjusted in order to improve the drift over the ground and increase the catch (especially with flatfish such as Common sole) but can also cause a lot of trouble on hard bottom.

7.4.2.4. Net Colour

Passive nets can be produced in various netting colours (red, brown, transparent, green, white etc.) depending on regional preferences and traditions. The efficiency of each net strongly depends on water turbidity. Many observations, however, show that nets of clear colours are often more powerful than dark ones though if great seasonal variation was even observed.

Andreev (1955) recommended the use of dark nets in good light conditions, and clear water and lighter coloured ones in turbid conditions. Therefore, the efficiency of a net depends on its contrast with the background and particularly on colours of water and of bottom.

7.4.2.5. Twine Material

The technical characteristics of netting materials that affect the selectivity properties of passive nets are those that can affect mesh opening and slackness of the panel.

Twine diameter, elasticity and flexibility, and manufacturing processes (monofilament, twisted, multi-monofilament) are the main twine parameters. Usually, the elasticity of material enhances the efficiency of enmeshing method. When fish passes through the mesh, the elasticity of the twine tightens around fish body preventing its escape.

7.4.3. How to Conduct a Selectivity Study with Passive Nets

Mesh size is the most important Parameters affecting selectivity in passive nets, therefore most of studies are focused in this parameter. Generally selectivity studies account for variations in mesh size while keeping constant the other factors that are known to affect selectivity. However data on technical properties of passive nets except that of mesh opening are usually missing in published studies.

Furthermore, taking into account that twine characteristics (material, thickness, colour, flexibility) strongly affect the reaction of fish that is primarily determined by visual stimuli, it is essential that a selectivity experiment should not accidentally introduce additional visual, olfactory or auditory variables. In fact, most of selectivity studies aim at evaluating the variations in mesh size with other technical factors that are known to affect selectivity kept constant. Therefore, it is essential to collect preliminary information on the characteristics of nets used in commercial fishing in order to test the target parameter under commercial conditions before starting work at sea.

In selectivity studies the number of mesh sizes tested can vary, with at least one of them in commercial size. Usually, a difference of 10 mm in mesh opening between different meshes can be considered as a good compromise in the Mediterranean. Therefore, the different mesh sizes must be carefully checked out with official mesh gauges during net construction to avoid errors due to differences between actual and nominal mesh size.

The most important issue to take into consideration in selectivity studies with passive nets is that it is essential to have a good degree of overlap between size distributions of fish caught by different mesh sizes. Therefore, a plausible scenario is to start with a certain number of mesh sizes (6 or 7) and delete some of them after the previous analysis. Alternatively, it might be possible to start with few mesh sizes and incorporate intermediate ones if needed. In any case, the use of at least one mesh opening that's used in commercial fishing in strongly recommended.

In selectivity tests, the gang consists of one netting sheet from each mesh size joined together. In some cases, it is possible to observe that larger fish can be guided along small mesh sizes till they are captured by a larger one. Therefore, a random distribution of different sheets along the gang would be appropriate (Figure 118).

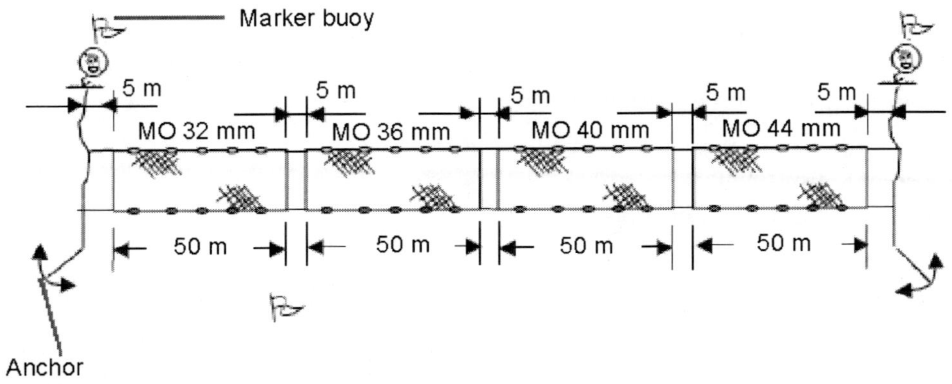

Figure 118. Passive net selectivity study: netting panels with different mesh opening (MO) distributed in the same gang.

7.4.3.1. Data to Be Collected

Net type

- Bottom set gillnets
- Trammels
- Combined nets (gill nets/trammel nets) or
- Driftnets

Mesh sizes

- Nominal mesh size; and
- Mesh size measured with an official mesh gauge

Net dimensions

- For each sheet, the number of meshes in the drop and width.

Twine characteristics

- Netting material (PA, PET);
- Yarn type (monofilament, multifilament, multi-monofilament etc.);
- Knotted/knotless
- Twine colour; and
- Linear density

In case of trammels and mixed nets (gill nets/trammels), data on mesh opening, dimension and twine material must be discriminated for inner and outer walls of netting.

Floatline and ground rope

- Length (metres);
- Material (PA, PE, PP);
- Diameter;
- Construction (twisted, braided, floats incorporated);
- Linear density;
- Number of floats;
- Dimensions of floats;
- Floats material (Plastic, Rubber, other...);
- Floats buoyancy
- Number of weights; and
- Material of weights (Fe, Pb,...).

Hanging ratio

- Horizontal hanging ratio
- Vertical hanging ratio
- Hanging ratio of the inner panel in trammel nets

Vessel data

- Research vessel/commercial vessel;
- Vessel name/registration;
- Length overall;
- GRT;
- Engine power;
- Hauling equipment; and
- Hauling speed

Haul data

- Date;
- Haul number;
- Shooting time for each gang;
- Hauling time for each gang;
- Coordinates for each gang;
- Start shooting;
- End shooting;
- Depth; and
- Sea state

Catch data for each mesh size

- For each species
- Scientific name;

- Common name; and
- Weight by species.
- For each target species
- Sample size measured;
- Length of individuals (Total length and/or fork length, carapace length, mantle length);
- Girth perimeter;
- Weight of individuals;
- Sex; and
- Maturation state

Regarding girth measurement, it should be taken into consideration that this measurement is highly time consuming and length (either fork- or total length) is measured quicker. However, girth measurement (measurement of maximum girth) remains really important. For example, it is important in preliminary design of an experiment to establish the correct mesh sizes to be used. As already seen, there is no standard method for girth measurement. In the case of flatfish it might be preferable to measure width instead of girth and converting to girth by multiplying by two.

7.5. Hook Selectivity

The catching performance of a longline is directly correlated with the technical properties of hooks, bait types and shapes, and longline properties and setting.

Three main processes should be taken into account about catching and selectivity of a hook:

- Attraction of fish by bait towards hook;
- Hooking when fish bites the bait; and
- Catch or escape during hauling of the gear.

Longline is considered a size-selective fishing gear (Bjordal, 1988; Bertrand, 1988) but few studies regarding selectivity of hook and line fishing have been published. Species selectivity of longlining is clearly affected by fishing strategy regarding horizontal and vertical distribution of fish (Lokkeborg and Bjordal, 1992). Fishermen use their experience to catch the desired target species by setting the longlines in selected areas and substrates, or at specific depths.

7.5.1. Hook Properties

Hook is the core of longline system. The technical properties of hook (Figure 119a) include:

- Eye (flat or loop),
- Shank,
- Length,
- Width,
- Point (without barb or not),
- Gap,
- Throat,
- Material (steel, inox), and
- Flat or twisted hook.

In longline, hooks size and shape are normally used according to target species and its behaviour. The name of the hook (J hook, Circle hook etc.) indicates its basic shape, while the quality number identifies the varieties within this hook group. Circle hook are generally circular in shape with the hook point bent back at the hook shaft (Figure 119b).

Fish caught by longlines are generally hooked in the mouth, particularly in the jaw, or in the alimentary tract if the hook is swallowed. Since hook shape is designed to increase the probability of penetrating the inside of the mouth for a particular species of fish (Huse and Fernø, 1990), hook design is also a parameter that affects the species selectivity of longlining.

The size of a hook is normally identified by the length, width and gap between point and shank. However, there is no standard system of hook measurements and dimensions can vary greatly from one pattern to the other.

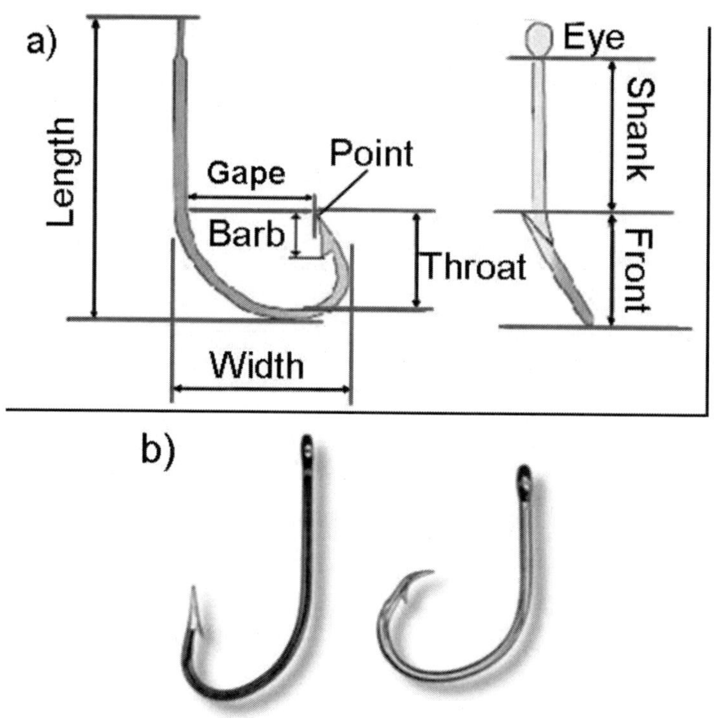

Figure 119. a) Technical properties of a hook affecting selectivity; b) Different hook shapes: J shape on the left, circle hook on the right.

In hooking process, the most important dimensions of hook are width and the gap which can be correlated with mouth size of the fish. Throat depth finally ensures deeper penetration of the point and better holding power of the fish.

Hook size is among the main parameters affecting selectivity: in theory, the larger the hook the bigger the fish caught since this is correlated with mouth of the fish. A large hook is less readily broken or straightened and its wider gap may allow the hook point to engage more deeply in the mouth cavity. Large hooks may therefore be more effective in preventing hooked fish from escaping (Lokkeborg and Bjordal, 1992). However, the mouth cavity of small fish has thinner skin that is more easily penetrated by the hook point. In general, several authors found no clear evidence for hook size selectivity.

7.5.2. Bait Properties

Bait type, size and orientation can be considered as the main factors affecting the first steps of catching process in longline; i.e., attraction of fish to the hook, since this fishing method is highly dependent on the feeding behaviour of the target species.

Several studies showed the influence of bait on catch efficiency of longline. Once the fish has come to baited hook, not only size but also shape of the bait may affect the likelihood of it attacking and ingesting the bait. Several studies have also shown the species-selective effects of baits so that each species is mostly attracted by certain bait type. For instance, squid bait seems to be more attractive for Bluefin tuna than the mackerel one. Moreover, bait that is effective in one fishery might not be effective in other seasons or areas.

As for hook size, bait size strongly affects the size of fish that can be caught. Bait size is considered as the most important factor affecting both size selectivity and catching efficiency (Løkkeborg and Bjordal, 1992). Smaller baits are usually more efficient than large baits since even small fish can bite the hooks.

Bait efficiency is based on an odour source that releases feeding attractants that trigger food search behaviour in the target species and it must release these attractants over a considerable period of time. Moreover, taste, texture, shape and size are also important factors that encourage fish to approach and swallow baited hooks.

The olfactory and gustatory systems represent the major chemosensory stimulus in search for bait. In fact, visibility in water is limited and underwater vision has a relatively short range. For this reason, the odours released by baits are therefore important for food search behaviour.

Bait orientation on the hook can be extremely important for some species also. For example the efficiency of mackerel as bait increases when it is hooked horizontally (Broadhust et al., 2001).

7.5.3. Hook Selection Curve

Estimates of size selectivity of longline and hook type give important information for conservation and optimum exploitation of fisheries resources. Through reasonable manipulation of hook size and bait type, the longline can be considered highly species-selective given the small proportion of non-target species usually caught.

The choice of a size selectivity model for longlines is less obvious than for gillnets and trawl experiments. Both logistic type models typically used to describe the selectivity of trawls, and the unimodal models used in gillnet selectivity studies have been used in hook selectivity studies (Czerwinski et al., 2009; 2010). If the hook size is the sole factor determining selectivity, the selection curve should take some sigmoid shape. However, we have stressed that several different parameters can affect the selection process. Therefore, the model that best suits most of experimental catch data is the logistic selection curves:

$$S_{ij} = \frac{1}{1 + e^{-b_i(l_j - L50_i)}}$$

Where
S_{ij} is the size selectivity for hook size *i* and size class *j*,
b_i is a parameter determining the slope of the selection curve for hook size *i*,
l_j is the midpoint of the size class *j*, and
$L50_i$ is the length at 50% selection.

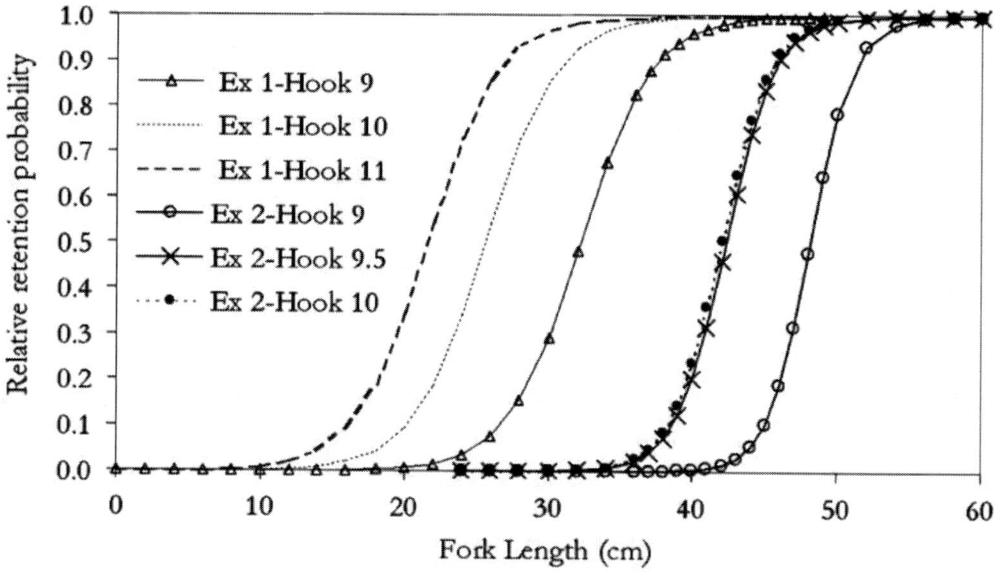

Figure 120. Logistic curves obtained with different hook sizes for the spot seabream (Pagellus bogaraveo) in the Strait of Gibraltar (by Czerwinski et al., 2009).

To estimate the parameters of the selection curves, it is normally assumed that the parameters of the selection curve are a function of hook size. Usually the mean length, depth, width and overall hook size are used to estimate the parameters of the logistic curves as linear functions of hook size:

$$b_i = AH_i + B, \text{ and } L50_i = CH_i + D$$

Where

A, B, C and *D* are parameters of the linear functions, and

H_i is the mean of each dimension like width, length and overall hook size for hook size i.

The use of a logistic curve for the explanation of longline selectivity can be justified the fact that very small mouth of young fish doesn't allow them to bite the hook making the retention potential near zero. Fish mouth dimension increase as they grow so that some specimens start to be caught. As the size of fish increases, the percentage of retained specimens increases. Over a certain size where mouth dimension is large enough, most of fish are able to bite and swallow the hooks so that the retained percentage is about 100%.

Chapter 8

BYCATCH MITIGATION MEASURES

Bycatch is a major issue mainly in multi-species fisheries like those in the Mediterranean where several species coexist together at the same time and in the same place. In the Mediterranean, bottom trawl may simultaneously catch target species like Norway lobster (*Nephrops norvegicus*), hake (*Merluccius merluccius*), squid (*Loligo vulgaris*), short-finned squid (*Illex coindetii*), monkfish (*Lophius* spp.), red mullet (*Mullus barbatus*), cuttlefish (*Sepia officinalis*) or other species. Most of these species attain different sizes when fully grown, have different shapes and behaviours and finally have different Minimum Landing Sizes (MLS). In addition to the target species, the codend retains large quantities of non-target organisms. Besides, concerns about the impacts of towed fishing gears such as trawls and dredges on benthic habitats and organisms have increased over the last two decades as already explained earlier. Benthic habitats provide shelter and refuge for juveniles and their fauna provide important food sources for demersal fish etc. Therefore, mitigation measures to reduce the environmental impacts on seabed as well to improve the selectivity of current fishing gears are essential to guarantee a reasonable management of fisheries resources.

8.1. TRAWL

8.1.1. Improving Selective Properties: Bycatch Reducer Devices (BRDs)

Bycatch in fishery is practically unavoidable. The capture of undesirable species is a recognized problem with all fishing methods. Bycatch can include species that may be targeted in other fisheries, undersized fish in the target fishery as well as accidentally caught endangered or protected species. In all cases, these fish and shellfish are part of a species population and an ecosystem. The wide range of developments in fishing gear technology continue to have a significant impact on bycatch, and consequently on discarding. Much of this bycatch consists of juvenile and low-value fish that are often discarded, usually dead. Therefore, removing them affects the food chain and ultimately the economic and social aspects of the fishery in many ways. For example, good results happen when fishermen release bycatch, retain it for consumption or sale, or when weakened or dead bycatch becomes food for scavengers in the food chain. Bad results occur when an endangered or protected species is removed from its already small population. Fishermen primarily aim to catch species of high commercial value, and consequently bottom trawl fishing operations are usually characterized by frequent and quick changes of fishing ground that cause substantial differences in species composition of catches. In such highly efficient gear, the amount of bycatch caught is sometimes so large that it causes serious problems in the management of fisheries and resources. The main problem with the bottom trawl bycatch is that most of this portion of the catch might not survive because it is damaged in the net, brought up from depths too quickly, or thrown back too late.

The problem of bycatch is particularly important in multispecies fisheries like the Mediterranean where different species of animals are found together in the same time and place. The management of a multispecies fishery is really difficult since most of fishes and Invertebrates caught attain different sizes when fully grown, have different shapes and behaviours, and finally have different Minimum Landing Size (MLS) making it hard to target only one of them in their shared habitat. At present, the management of fishing stocks in Mediterranean is mainly based on defining closed areas and seasons, minimum landing sizes, minimum mesh sizes, and limiting fishing effort. In the meantime, it is very difficult to define a minimum codend mesh size for towed nets in a multispecies fishery where several species are caught simultaneously because a mesh size appropriate for one species could be unsuitable for many others (Stewart, 2001). In recent years, there have been many initiatives to improve selectivity of fishing nets or more correctly to reduce the capture and discard of non-target fish, but it has also become clear that the natural behaviour patterns of many species prevent effective selection. For a trawl gear to be truly selective, fish entering the net should be tested to insure that those that are small enough to pass through the meshes escape and those that are above the minimum landing size are retained (Glass and Wardle, 1995).

In the last twenty years, several studies showed that technical modification of traditional fishing gears might improve the release of undersized fish and unwanted bycatch (Kennelly, 1995; Wileman et al., 1996; Broadhurst, 2000; Walsh et al., 2002; Valdemarsen and Suuronen, 2003; Graham and Ferro, 2004; Suuronen, 2005). Improved selectivity can be achieved by modifying gear design and/or operation, and by using alternative fishing gears. The changes usually involve modifying the meshes of the codend only; the twine thickness etc. (see 8.2.3 for more details); or small parts of the gear in use such as devices inserted either in the codend or in the aft part of the extension piece like grid devices, square mesh windows etc: for the successful separation of targets and non-targets species.

In general, these devices aim at reducing the catch of juveniles as well the incidental catch of unwanted species (bycatch). In the latter case, they are known as Bycatch Reducer Devices – BRDs. BRDs are commonly used in trawl fisheries to allow fish that are not targeted by the fishers to escape from the net before it is hauled back into the boat. However, most of BRDs (Bycatch Reducer Devices) experimented around the world to facilitate the escape of unwanted fish or to reduce the catch of small size specimens such as sorting grids are difficult to apply in the Mediterranean where the medium size of fishes is very small. To minimize the biological impacts on bycatch and to promote ecologically sustainable fisheries in the Mediterranean, bottom trawls should be modified to address both the size and species selectivity issues.

The commercial mesh size codend traditionally used in the Mediterranean is shown to be too small to allow the escape of immature fish. The economic consequences of introducing gear modifications are possibly the single most important constraint. In many cases, bycatch costs fishers time and money. This further emphasizes the need for a close partnership with industry in the introduction of BRDs and more selective gears in a gradual and adaptive manner. Therefore, fishermen must be a part of the research process for the rapid development of effective BRDs and to successfully reduce bycatch. This will result in a higher compliance with the rules and regulations.

By avoiding the catch of rocks and debris, BRDs might also improve trawl performance and procedures on board through:

- Reducing damage to the codend,
- Saving time usually spent in shooting and hauling operations given that the catch is less the trawl duration can be longer,
- Reducing time spent for sorting of catch on board,
- Improved the quality of catch, and
- Reducing injuries to crew from dangerous animals.

On the other hand, experimental trials are essential in order to find out the right setup and to reduce short-term economic loss. In fact, the risk of losing the large specimens is really high especially in mixed fisheries.

Windows, grids etc. are often designed to take advantage of the general upwards escape behaviour of fish and are, therefore, normally positioned in the upper half of the codend or extension piece. Some BRDs are described below:

8.1.1.1. Fisheye

Fisheye is a small opening held by a rigid frame that is attached to the top or sides of the codend through which fish can escape (Figure 121). Fisheye is designed to allow fish to voluntarily swim through an escape opening in the codend (Crespi and Prado, 2002). In its simple design, fisheye is a triangular or rectangular shaped-cut made in the netting. An aluminium or steel frame is more commonly used to provide an elliptical or "eye shaped" rigid escape opening. Fisheyes are generally used in shrimp trawling and allow for the escape of swimming species (fish).

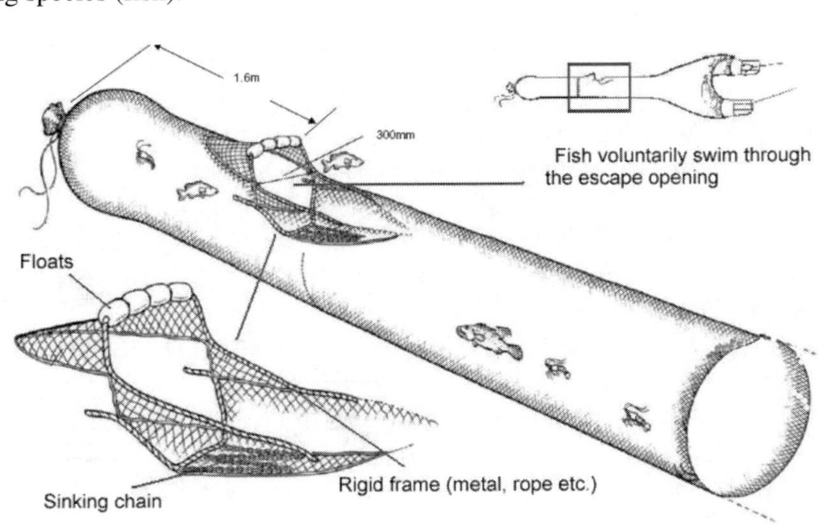

Figure 121. (Continued).

Figure 121. Fisheye operation and models (adaptd from Eayrs 2007).

In shrimp trawl fisheries, the poor swimming ability of prawns and the location of the fisheye at the top of the codend reduce their chances of escape. To escape through the hole, fish must swim towards the mouth of the net and out of the escape opening in the fisheye. Generally, the fisheye should be located as close as possible to the codend drawstrings to enable small fish to escape but care is required to ensure that prawns cannot escape also.

8.1.1.2. Sorting Grids

Sorting grids are generally rigid panels of spaced bars made of aluminium, steel or plastic fitted to the extension or tunnel of the trawl, immediately ahead of the codend. Sorting grids are used to physically block the passage of bycatch into the codend and guide it toward an escape opening while small animals (including shrimp) pass through the bars of the grid and enter the codend. As animals pass through the net, they are eventually deflected by the grid and guided towards the escape opening. The grid can be oriented to guide animals through an escape in the upper or lower panel of the codend. Sometimes, a non-rigid inclined panel of netting is used to drive bycatch towards the escape opening at the top of the trawl. Thus, depending on material used to construct the inclined grid or netting panel, it is possible to identify either "hard" or "soft" grids. The escape opening is a hole cut in the codend and is usually covered with a flap of netting or other material to prevent the escape of marketable catch, shrimp included. The cover must be constructed with a mesh size small enough so that large animals do not become caught as they are trying to escape. The cover must also act as a trapdoor, allowing unhindered escape of these animals and quickly returning to close the escape opening.

The grid is sewn into the net at an angle normally between 40 and 60 degrees and immediately behind an escape opening cut into the netting. Depending on the positioning and bar spacing of the grid, these devices can be tailored to reduce the capture of either large or small unwanted animals (Eayrs, 2007).

Sorting grids are mandatory in certain fisheries, and are widespread in others. In most fisheries where the grids have been introduced, fishermen have initially complained that they are unwieldy, unpractical and cause the loss of commercial species. However, the use of the grids has subsequently become a marked success in reducing bycatch, although the loss of marketable catch remains a problem. There is little evidence that organisms are injured as they are deflected off the grid.

Different types of grid devices are available for the reduction of unwanted bycatch. One of the most commonly used grids in the Southeast Asia (Thailand, Brunei Darussalam, Vietnam, Indonesia and Malaysia) is the Juvenile and Trash Excluder Device (JTED) that aims at excluding juvenile- and trash-fish (small, undersized fish and other animals that were traditionally discarded overboard because they had no economic value in bottom trawl shrimp fisheries). In recent years, this part of the catch (juvenile- and trash-fish) has become an important source of income for many small-scale fishermen given that this part of catch can be sold as fish meal or food for cultured fish or shrimp. Shrimp trawling is generally regarded as one of the least selective fishing methods because the bycatch may consist of over several hundred teleost species. Thus, the development of technical solutions designed to exclude small fish from the trawl and maintain the catch of shrimp and large fish is essential.

The JTED is a device composed by three sections hinged together (Figure 122), The first two sections are metal grids and the third section is a metal frame supporting a panel of fine-mesh netting. Small fish swim between the bars of the grids and escape. The netting panel in the third section helps maintaining the orientation of the device and hampering shrimp returning forward in the codend and escaping, and prevents small fish from re-entering the codend.

For instance, Juvenile and Trash Excluder Devices (JTEDs) typically having a very small bar spacing of 50 or 55 mm is used in Norwegian waters and Barents Sea to exclude non-target fish species and juveniles, while JTEDs in southeast Asia have been tested with spacing ranging from 12 – 25 mm.

A similar solution called Nordmore grid is also applied in other regions (Norway, North Atlantic etc.). Nordmore grid is an aluminium grid mainly designed to exclude large animals from shrimp trawls. In this type of grid, the space between bars is typically 100 mm. A funnel of netting guides all animals to the bottom of the grid. Large animals are guided by the grid towards the escape opening in the top of the codend, while shrimps and other small animals pass through the grid and enter the codend. The setting of the grid angle is essential to avoid the loss of shrimp. This angle is usually managed by means of floats.

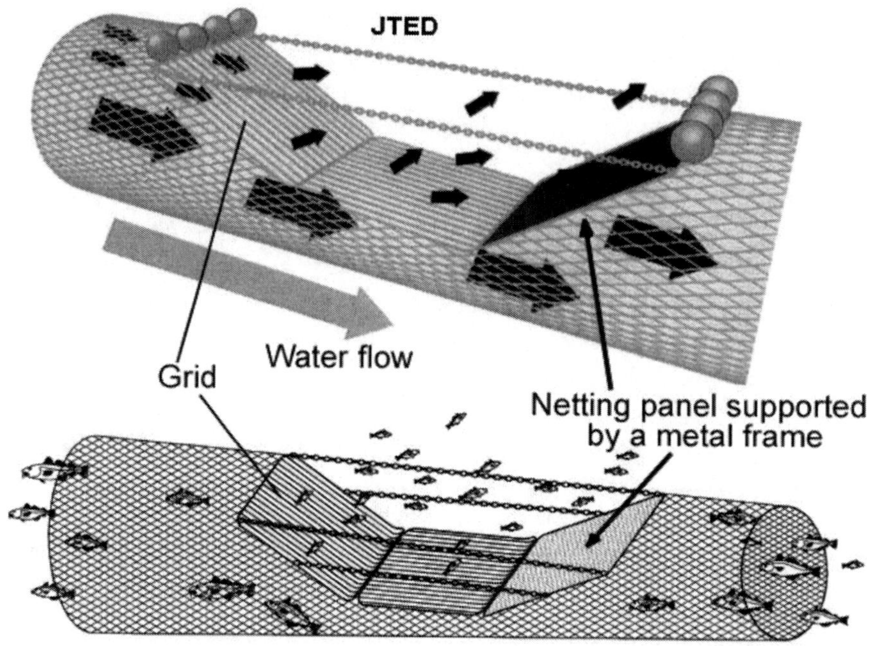

Figure 122. Juvenile and Trash Excluder Device: juveniles are able to escape from the net.

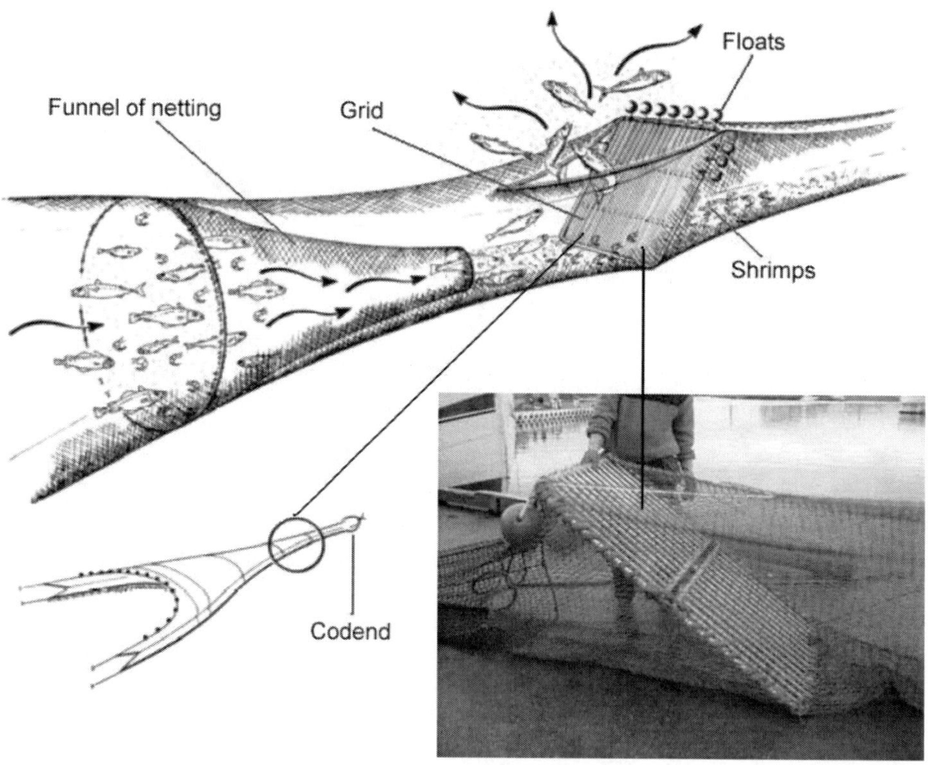

Figure 123. Nordmore grid used in prawn trawling to reduce fish bycatch.

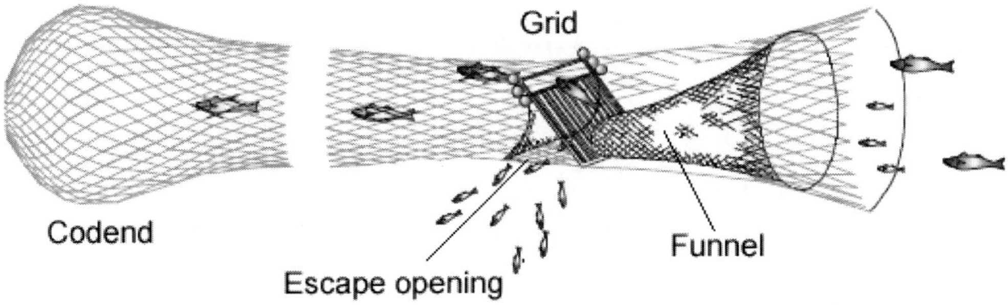

Figure 124. Grid used to reduce the catch of small fish and shrimps (adapted by Sacchi, 2008).

The grids can also be used to target large specimens while avoiding the catch of juveniles (Figure 124). In this case, the escape opening is usually positioned in the lower section of the net.

8.1.1.3. Separator Panels

This is a horizontal panel of net that divides the trawl into two compartments each leading to separate codends or escape opening to allow unwanted fish or unwanted catch to escape (Figure 125). In this case, the codend were the unwanted catch is retained should be kept open.

The general aim of separating panels as a selectivity tool is to take advantage of behavioural differences between fish and other animals to separate the catch into species groups (theoretically target and non-target species). During the fishing operation, many animals enter the trawl and may exhibit species-specific behaviour. The segregation of different species in the two compartments can be caused by morphological differences (size and/or shape), or by different reactions to fishing gears.

Fisheries that have benefited from the separator principle are shrimp fisheries (for Panaeid, Pandalid and Crangonid species), prawn fisheries (for *Nephrops* spp.) which sometimes have high levels of finfish bycatches associated with them. The careful selection of mesh size in each codend; e.g., square-mesh codend could allow the size selection of fish to be optimized.

In the Mediterranean where several species are marketable, the separator panel also allows rocks, shell, crabs and other bottom-dwelling animals to be kept separated from the commercial part of the catch (Figure 126). This usually increases the quality of fish that are guided in the upper codend.

The main goal of a separator panel is to split catch on the basis of inter-species variations in natural behaviour and/or the reactions to fishing gear. For example, flatfish species and crustaceans generally live very close to the seabed and, hence, they enter the trawl gear at a low level. On the other hand, species like hakes, sparids, red mullets tend to rise upwards in the mouth of the trawl and consequently enter the net at a higher level.

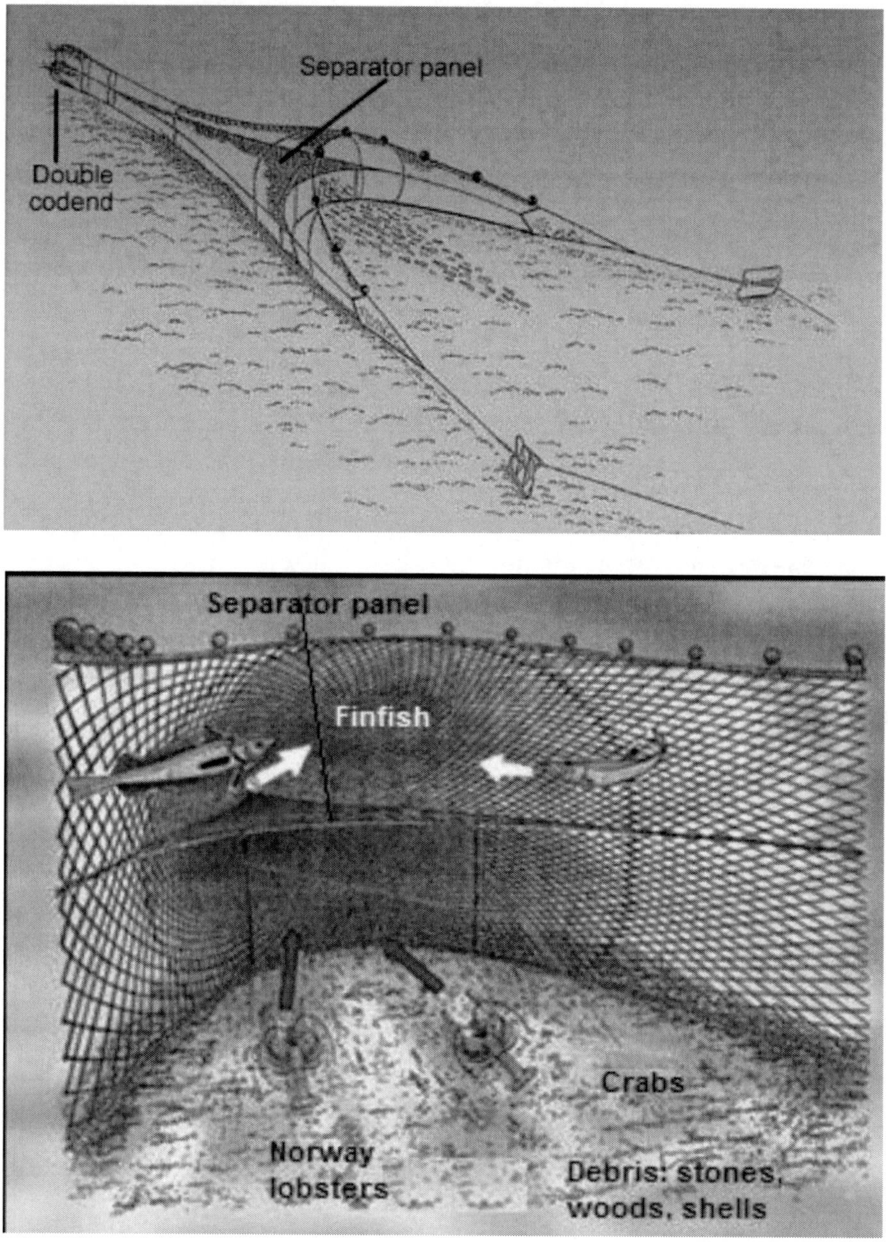

Figure 125. Separator panel that divides the trawl in two compartments.

8.1.1.4. Square-mesh Windows

As already observed, the most obvious and widely used technique to change the selectivity of a trawl gear is to increase the mesh size of the main catching region, the codend. Alternatively, it's possible to replace the traditional diamond meshes with a square-mesh codend. The advantage of square-mesh is that it remains open under load unlike diamond meshes which close, thus hampering the escape of small animals. The effect of square-mesh codend has been already described. In some countries, the use of square-mesh windows has been widely investigated and they have been included in the legislation.

Figure 126. Bottom trawl catch segregated into two compartments.

The size of square-mesh used in the panel directly determines the size of fish that are allowed to escape.

Square-mesh netting panels can be simply constructed from diamond mesh material by turning the netting panel 45 degrees. However, these panels may suffer from slippage of the knots used in the construction of the mesh and each mesh may become rectangular in shape. As a result, square-mesh codends are usually constructed from knotless material. Repairing damaged square-mesh netting is a more difficult task, but they have been demonstrated to substantially reduce bycatch in some fisheries. This device can be used in shrimp and fish trawl fisheries.

The use of square-mesh windows has been extensively studied in a variety of locations in the upper panel of the trawl, including the codend extension piece and near the bosom of the headrope (Figure 127). They have been shown to be particularly effective in fisheries where particular species (such as haddock) exhibit distinctive upward escape reactions.

A particular type of square-mesh window is the Bacoma panel. The Bacoma-selectivity window is a relatively recent development, emanating from the Baltic Sea area. The BACOMA codend is a codend with the upper panel consisting mainly of one large square-mesh window that allows juvenile fish to escape, an idea dating back to the mid-1990s (Madsen et a., 1999). Square-mesh panels inserted in the top of the net have been shown to allow smaller fish to escape, where otherwise they would have been trapped in the codend. However, square-mesh panels have several problems: they are harder to insert and repair for fishermen and their effectiveness depends on their position in the net, visibility and ambient light levels. The BACOMA escape window became part of technical regulations in the IBCFC (International Baltic Sea Fisheries Commission) in 2001.

Studies on the effect of square mesh windows in the Mediterranean are scarce (Figure 128).

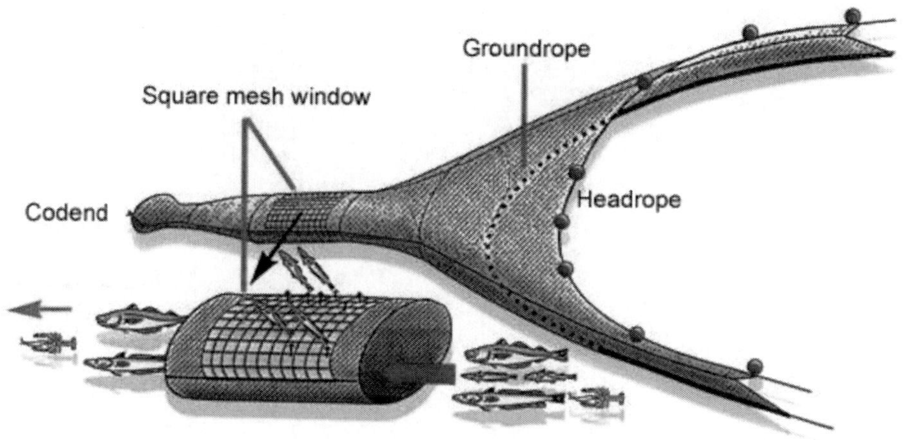

Figure 127. Scheme of a square mesh window mounted on the extension piece.

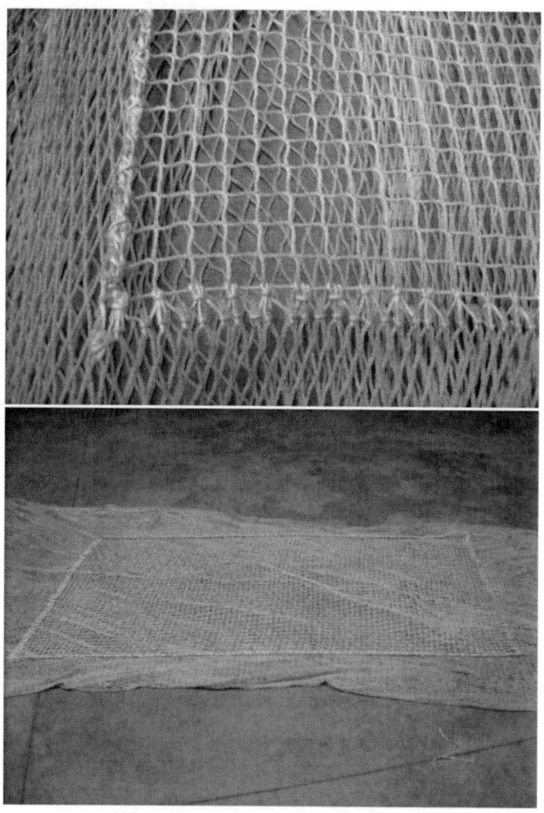

Figure 128. Square mesh panel tested in the Mediterranean Sea.

8.1.1.5. T90 Codend

This is the name given to diamond mesh that has been turned 90 degrees. Traditionally, the orientation of diamond mesh netting is determined by the orientation of the knot used in construction of the netting. By turning the netting 90 degrees, the meshes do not close as tightly under load and small fish and other animals can escape.

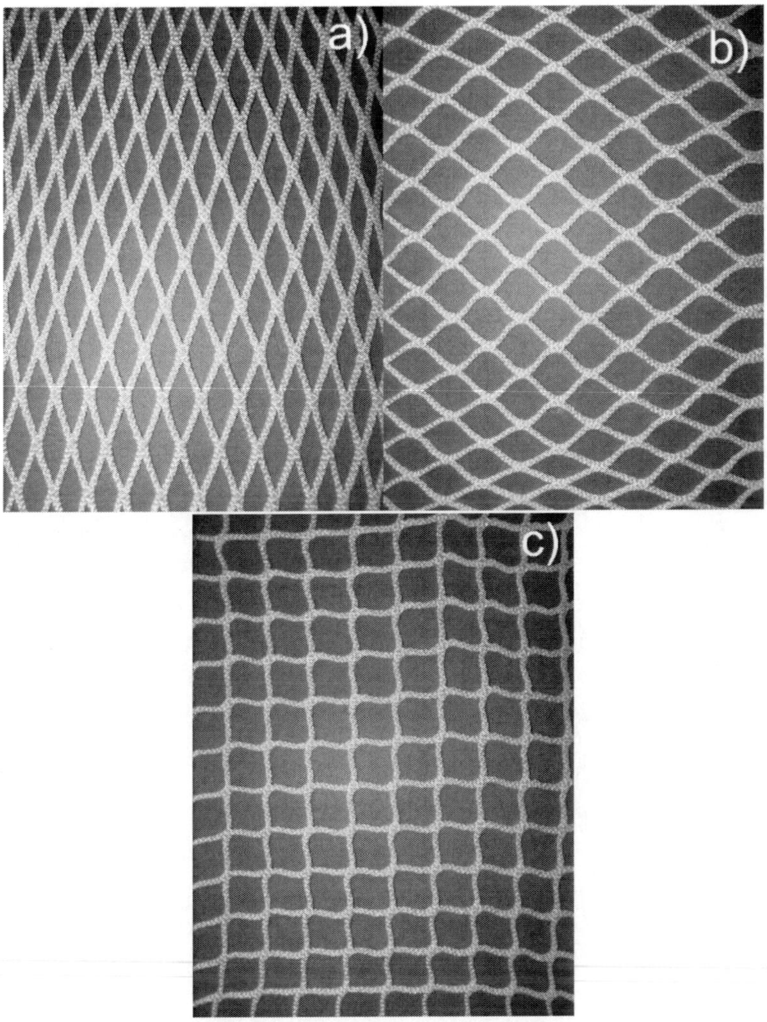

Figure 129. Mesh configuration tested in the Mediterranean Sea: a) Diamond; b) T90; c) Square.

The use of T90 codend has shown that the selectivity of the codend is higher letting more small fish escape the trawl. It's worth noticing also that the escapees will be of better quality meaning a survival rate.

The T90 codend has the advantage that it is easier to mount on a net than a square mesh and there is no loss of net in the construction of T90 panels. However it is evident that the square mesh stays more open and selectivity is also affected by this (Figure 130).

8.1.1.6. Turtle Excluder Devices - TEDs

In addition to the target species and bycatch of fish and fishlike species, other animals might incidentally be captured with fishing gears. This latter incidental catch might include several species of marine mammals, turtles, sharks and seabirds. Although bycatch of such organisms is seldom a threat to their population size, public concerns make it necessary to reduce such bycatch.

Figure 130. Different codend mesh configuration: a) Diamond meshes; b) T90 meshes; c) Square meshes.

In the Mediterranean, interactions of sea turtles with fishing gears mainly involve the longlines. Nevertheless, a growing number of sea turtles were accidentally caught by bottom trawlers in the last few years. Trawling has a serious impact on sea turtles. In some shrimp-trawl fisheries, several thousand turtles are caught and drowned each year.

Bottom trawl activities mainly impact turtles in the demersal phase since they prefer the shallow waters of North Adriatic Sea, South Turkey, Tunisia and Egypt. The Central Mediterranean provides important neritic habitats for loggerhead turtles (*Caretta caretta*), but Mediterranean bottom trawlers catch an estimated 30000 turtles a year with 25% mortality.

Incidental catch of *Caretta caretta* probably occurs during towing operations when turtles are foraging on the bottom. When captured by a trawl net, loggerhead turtles may drown becoming first comatose and eventually dying. Therefore, towing time is one of the main factors affecting the mortality rate (Henwood and Stuntz, 1987) but additional factors in bottom trawl are probably involved also (Stabenau et al., 1991).

One of the most important management measures to protect sea turtles, especially the juveniles and sub-adult size class, is the use of Turtle Excluded Devices (TEDs). TEDs seem to be very effective mainly in prawn trawl fisheries. Several countries in southeast Asia and South America adopted TEDs as mandatory management measure in order to export their product to USA (Laurent and Lescure, 1994).

Only few studies are available until now on the TEDs in the Mediterranean (Atabey and Taskavak, 2001; Sala et al., 2011). They obtained very good results as both *Caretta caretta* and *Chelonia mydas* were excluded by the Supershooter TED and unwanted incidental catches such as jellyfish, sharks, and rays were also excluded. The final convlusion of the authors was that modified Supershooter greatly assists fishermen in reducing catches of turtles and unwanted bycatch but the loss of commercial species should be considered.

TED Design

A variety of hard TED designs is available for fisherman but it is generally very difficult to introduce new technical solutions taking into consideration that innovations can be easily accepted only if economic losses are insignificant. Thus, there is a need to minimize the losses of target species and providing benefits to fishermen to effectively implement TEDs,

Since the turtle excluder device (TED) was first introduced to the U.S. shrimp fishery in the late 1980's, research and development to improve its performance has continued using SCUBA divers and video cameras attached to shrimp trawls under actual working conditions.

The available TED designs can be summarized as follows (Figure 131):

- The simplest of TED designs is the oval grid type, commonly called the Georgia-Jumper style. It is usually constructed from steel rods and features a horizontal cross brace for added strength.
- Hooped TEDs have a circular or oval hoop on the front and or rear of the TED. It has the advantages of sturdier construction for fishing in rugged conditions and the fixed non-changeable angle of deflector bars as the trawl webbing stretches.
- Fixed angle TED features a single hoop used to strengthen the TED frame and to maintain TED angle. The hoop and the deflector grid are sewn to the trawl extension in order to "fix" the angle of the TED in the trawl.
- The Super Shooter (Reg. trademark) and Anthony Weedless (Patented) style TEDs are specially designed to reduce the accumulation of debris such as seagrass, which can prevent shrimp from passing through the TED and into the codend, on the TED deflector bars. These TEDs are constructed of solid aluminium rods or pipes and require little or no horizontal cross bracing. Each of these TED designs features modified deflector bars to eliminate seagrass and other debris from the TED.
- The flounder (flat fish) TED has been developed for use only in areas where flounder is the targeted catch. The 4-inch wide horizontal slots located at the bottom of the TED frame allow flounder and other fish to pass through the TED and into the codend.

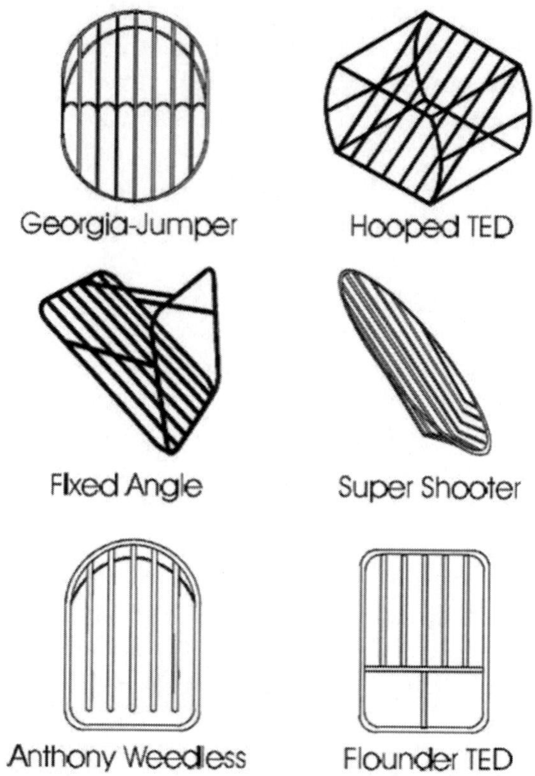

Figure 131. Basic hard TED designs (by Mitchell et al., 1995).

According to the "Guide to By-catch Reduction in Tropical Shrimp-Trawl Fisheries" (Eayrs, 2007), the circumference of a grid should be about 60% of the stretched-mesh circumference of the codend. The same author reported that TEDs have been used effectively measuring between 52 - 75% of codend circumference.

TED Materials

TED regulations specify that hard TEDs should be constructed of solid steel rod, fiberglass rod, aluminium rod or heavy gauge steel or aluminium tubing. Regardless of material, the TED should be built to withstand the rough conditions of seagoing work. A hard TED must be constructed of one or a combination of the following materials with minimum dimensions as follows:

MATERIAL	MINIMUM OUTSIDE DIAMETER
Solid steel Rod	0.64 cm
Fiberglass or Aluminium Rod	1.27 cm
Steel or Aluminium Tubing	1.27 cm (schedule 40 tubing)

Exit Hole Position

Any hard TED can be installed with the turtle escape hole positioned at either the top or bottom of the TED frame (Figure 132). Both configurations expel sea turtles but there are advantages and disadvantages of each depending on fishing conditions.

Bottom-opening TEDs can exclude debris such as grass, sticks, shell and sponge from their catch with water flow and gravity assisting the TED sliding debris down the grid face and out of the exit hole. In addition, unwanted bycatch like jelly balls, sharks, and rays can also be excluded. Excluding debris and bycatch from the trawl would result in less damage to shrimp and faster sorting time on deck.

When trawling over an area that is relatively clean of debris, it might be better using a top-opening TED in view of the tendency of shrimp to accumulate near the bottom of the net as the water flow carries them to the codend making top-opening TEDs more efficient in retaining shrimp than bottom-opening ones. This has been documented through comparative tows of top-opening versus bottom-opening TED on board commercial shrimp trawlers (Mitchell et al., 1995).

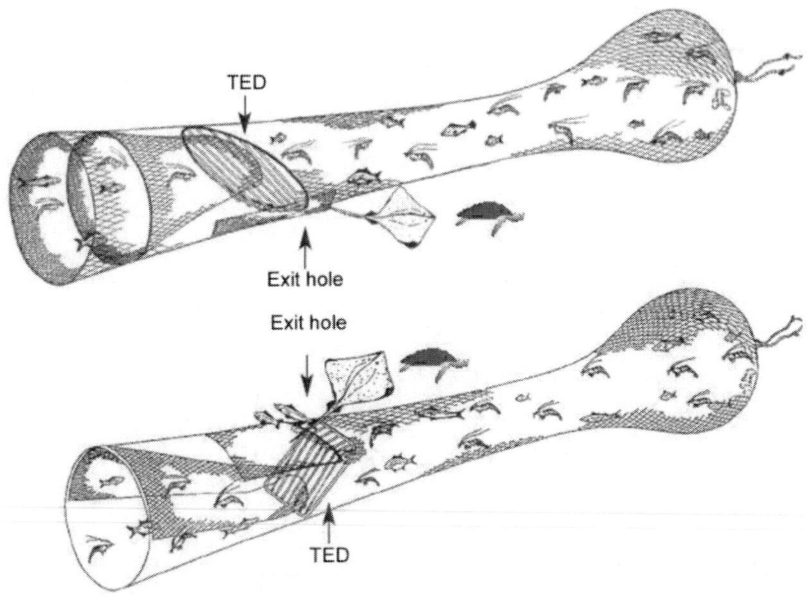

Figure 132. Top and bottom exiting TEDS (by Mitchell et al., 1995).

TED Angle

The angle at which the TED operates during a tow is an important factor in preventing shrimp loss. TED regulations specify that all hard TEDs must be installed at angles between 30° to 55° from the horizontal (Figure 133).

Grid TEDs work best for turtle exclusion and shrimp retention when operating at a 45° angle. However, the TED angle can change over time due to stretching of the webbing extension around it resulting in an angle less than 45°. A reduction in angle can occur with single grid hard TEDs when the TEDs are installed in new webbing and the webbing knots tighten due to heavy catches or "muddying up".

If a TED is operating at an angle of less than 40°, shrimp loss can occur due to diversion of water through the exit hole (Figure 134). TEDs operating at angles greater than 55° can prevent turtles from escape and trash will not slide down the deflector bars resulting in clogging of the grid. Shrimp collect with the accumulated trash and get discharged through the exit hole during haulback. When using a grid TED that has been recently installed in new webbing, it is a good practice to check the grid angle after several days of fishing to ensure that it is between 40° and 55°. Whether the TED is new or old, the grid angle should be checked and, if necessary, readjusted to the proper angle before fishing operations continue.

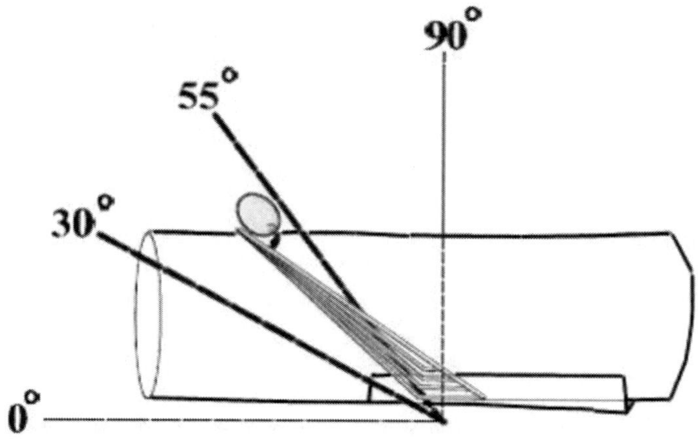

Figure 133. Allowable range for TED angle (by Mitchell et al., 1995).

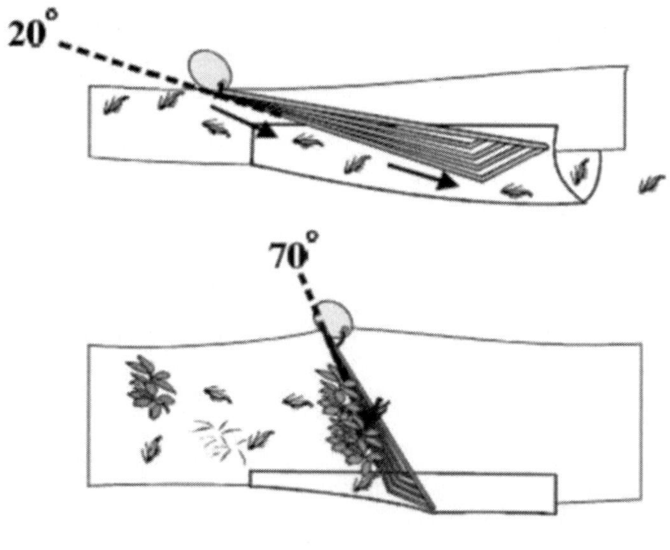

Figure 134. Effect of angle change on TED operation (by Mitchell et al., 1995).

Floating

Floats are an important component of any hard TED. Floats help stabilize the TED in the water and prevent it from rolling over during deployment or retrieval. More importantly, flotation insures that the TED will not chafe against seafloor during operation, thus preventing the need for mending and additional maintenance. When a TED rides along the seafloor during a tow, debris wouldn't discharge out the exit hole. Debris gets trapped in the TED obstructing passage of shrimp. A properly floated TED should operate 46-51 cm off the sea floor (Figure 135).

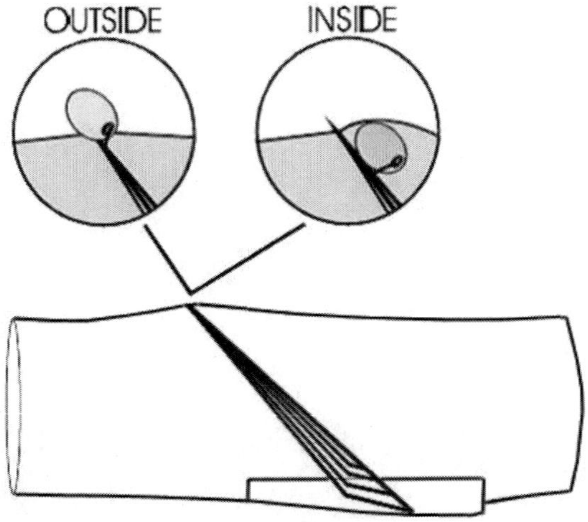

Figure 135. Allowable positions for floats on bottom opening TEDS (by Mitchell et al., 1995).

Size of Escape Opening

TED Regulations (Mitchell et al., 1995) specify the size and position of the cut in the trawl webbing which allows a turtle to exit the trawl. The cut must be centred on the top or bottom of the trawl (depending on whether your TED is a top or bottom excluder).

The minimum size of the cut or opening in the trawl webbing is dependent on the maximum width of the TED frame.

8.1.1.7. TED Experience in the Mediterranean

In the Adriatic sea, Sala et al. (2011) investigated the use of a typical Supershooter and three new types of low-cost TEDs made using different designs and materials. The performance of the TEDs was investigated under commercial fishing conditions in diverse trawl grounds.

Four different types of TEDs were designed, manufactured and tested at sea (Figure 136). The tested TEDs differed in their material, shape, size and bar spacing and would, therefore, be expected to behave differently, which justifies the testing and performance comparison that are the main objectives of the study.

The first grid type (TED1) was a light and rigid grid made of aluminium. The bars of TED1 were removable so that the space between them could be adjusted. Unfortunately, TED1 broke down after the first haul due to the large quantity of debris that was not discharged from the exit hole and, thus, got trapped in the TED causing rupture.

The second grid type (TED2) was a flexible grid made of mixed cable (steel inside and polyethylene outside). This flexible grid was designed so that the TED position could adjust according to net movements during tow.

The third grid type (TED3) was a semi-rigid and resistant grid. This grid was made of steel and rubber to combine the flexibility of rubber and the resistance of steel.

TED4 was a classic aluminium Supershooter grid commonly used in shrimp fisheries in several countries (Figure 136 d). The Supershooter is usually designed to reduce the accumulation of debris on the TED deflector bars, which can prevent fish from passing through the TED and into the codend. Taking into account the complex nature of Adriatic fisheries (crustaceans, molluscs and fishes together), the space between reflector bars was greater than that of standard models.

During sea trials, the angle of the grid was measured by means of a small sensor mounted on the side bar.

An escape gap in the bottom of the net was covered by a clear, weighted sheet of small-mesh netting. This escape gap cover was designed to allow the exclusion of large objects, while preventing the loss of smaller commercial species; i.e., hake, red mullet, shrimp, etc.). A bottom-opening TED was used as it is important to exclude debris, such as grass, sticks, shell and sponge from commercial catch in Adriatic demersal fisheries. A TED cover was attached under the TED opening to collect any escapees (Figure 137).

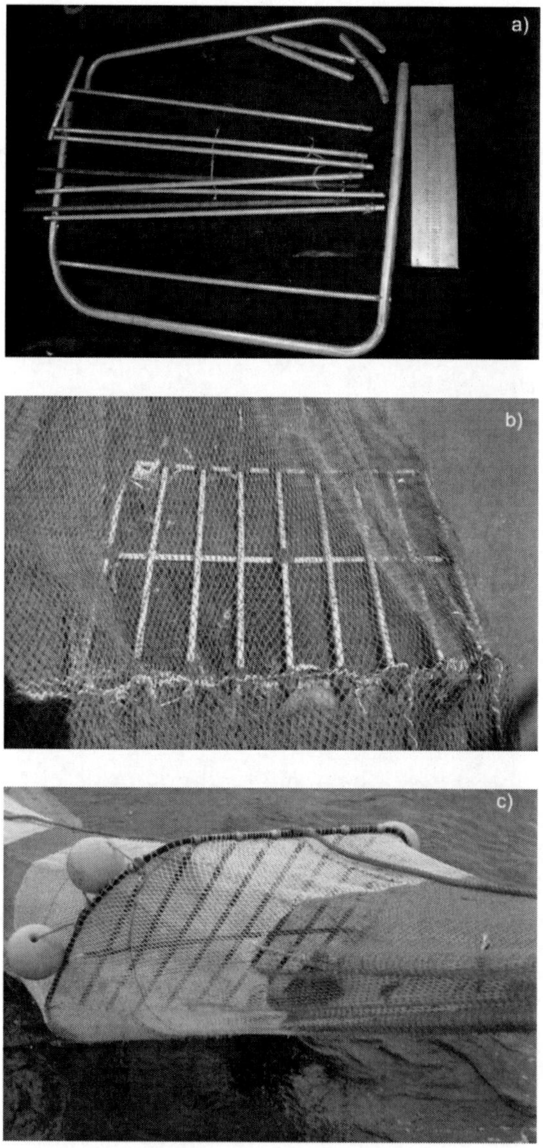

Figure 136. (Continued).

Figure 136. Dimensions (mm) of the Turtle Excluder Devices (TED): (a) TED1, (b) TED2, (c) TED3, (d) TED4.

Figure 137. Supershooter TED tested in the Adriatic Sea (Sala et al., 2011).

The TEDs reduced anthropogenic debris and, consequently, sorting operations on board. Among the four TEDs tested, both the semi-rigid TED and the Supershooter performed in accordance with the design objectives; i.e., total discards were reduced but total commercial catches were not significantly reduced. With the Supershooter, all European hake (*Merluccius merluccius*) individuals equal to or above 16 cm were found in the codend and 10–15% of those between 5.0 and 15.5 cm were released. In general, the total discard rate of the TED-equipped nets was reduced to around 20–60%. However, the reduced catches of commercial species might certainly deter some fishers from using trawls equipped with them. The Supershooter TED was finally able to expel the turtle accidentally caught in the net (Figure 138).

Figure 138. Turtle (Caretta caretta) expelled from the TED towards the escape cover.

Recently a new type of TED was tested in the Adriatic Sea in the framework of TartaLife project with promising results. The flexible grid that we tested had the following characteristics (Figure 145):

- Height: 1130 mm
- Width: 845 mm
- Circumference: 3110 mm

Flexible grid was chosen in order to not affect net handling on board; TED was able to keep its rigid position while trawling, and to ride along on the net drum, at the end of each haul Preliminary tests showed that the commercial loss is negligible and the TED is effective in reducing marine litter in the catch (Figure 140 and Figure 141).

In the future, the TED should be tested in the prawn fisheries in order to evaluate its efficacy.

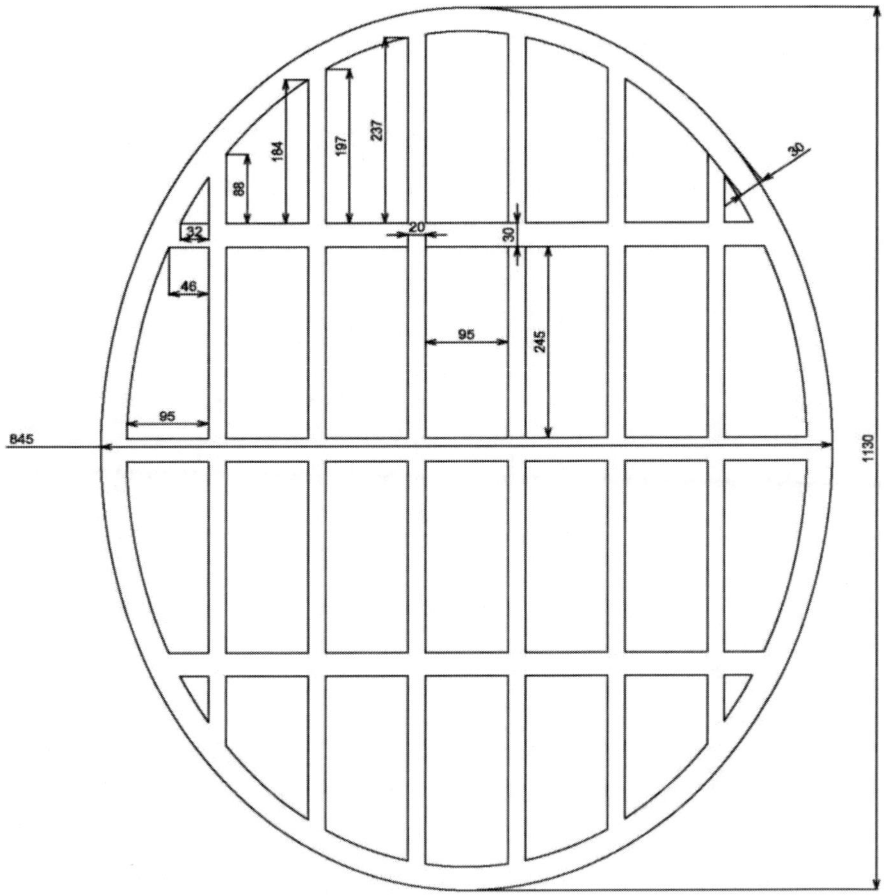

Figure 139. Technical draw of the tested flexible grid.

Figure 140. Flexible grid used in the Mediterranean tests.

Figure 141. Underwater image showing the right setting of the grid.

8.1.2. Reducing Bottom Impacts

Bottom-tending mobile gears have an effect on the seabed and the impacts are poorly understood and documented in many fisheries. A number of technical measures are available or in development such as semi-pelagic trawls, drop-out windows and electrical stimuli in beam trawls, and ground gear modifications. Many of these offer potentially powerful means of mitigating seabed impact.

Reduced impact of bottom otter trawling on the seabed can basically be achieved by minimizing the impacted area and by reduction of the pressure of gear components on the bottom.

8.1.2.1. Otterboards

Otterboards are considered as the most significant source of bottom impact of the entire trawl system. This is why recently a growing interest has been addressed in the development of more hydrodynamic efficient doors, maximizing spread forces while minimizing resistance. There are basically two methods to reduce the physical bottom impact of otterboards:

1. The use of otterboards with a lower shoe angle relative to the towing direction;
2. The reduction of bottom pressure by shortening wire ropes;
3. The use of otterboards with a higher aspect (height/length) ratio.

Traditional otterboards were designed in a way that they relied on the ground shear to spread the trawl (rectangular and V-doors). Such doors loose much of their spreading force when lifted off the bottom. For a huge net spreading they need to be rigged with a high angle of attack, thus creating a bigger shear on the bottom. Recent development has been towards more efficient otterboards, maximizing spread forces and minimizing resistance. Another recent trend is to increase the size and weight of otterboards. In conventional rectangular or V-door designs, the shoe is parallel to the trawl door surface. In modern bottom trawl door design, however, the flat surface has been replaced with one or several curved surfaces. In these new curved designs, the angle of attack is reduced resulting in narrower furrows in the bottom (Figure 142). Another practical solution is to reduce the angle of attack relative to the towing direction as described by Sterling and Eayrs (2006).

In simplified terms the pressure of a trawl door against the bottom substrate (BP) can be expressed as a function of the doors' weight (W), the warp tension (T) and the warp slope angle (α): BP= W-T$\cdot$sin(α). By shortening the warp length, the slope angle increases, resulting in an increased lift force T$\cdot$sin(α) (Figure 143). For a certain warp tension there is an angle when the lift force exceeds the weight of the door, resulting in a situation wherein the trawl door will lose bottom contact. In normal bottom trawling situations proper bottom contact of the door is regarded as important for efficient capture. Therefore, longer warp than required to keep bottom contact is often used in commercial fishing situations.

High-aspect ratio doors are more efficient in terms of hydrodynamic performances than low-aspect ratio ones (Figure 144). Such doors are commonly used as pelagic otterboards. The absence of bottom contact implies a reduction in gear drag and bottom impact, while a hydrodynamic profile allows for great horizontal door spreading and vertical lifting. To compensate such lifting force weights or compensation chains are added to the rigging, taking into account that the effect of the chain is less impacting than the entire otterboard. The use of this type of otterboards implies the need of instrumentation for the control of door performance. Therefore, the recent development of gear instruments that monitor the vertical roll and pitch angels of otterboards during towing may facilitate the use of such otterboards in bottom trawling (Figure 145).

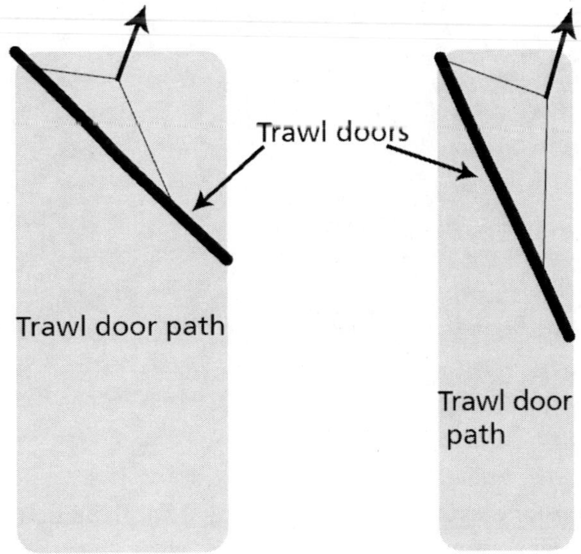

Figure 142. Impacted widths of two equal trawl doors when door angles are 45° (on the left) and 25° (on the right; Valdemarsen et al., 2007).

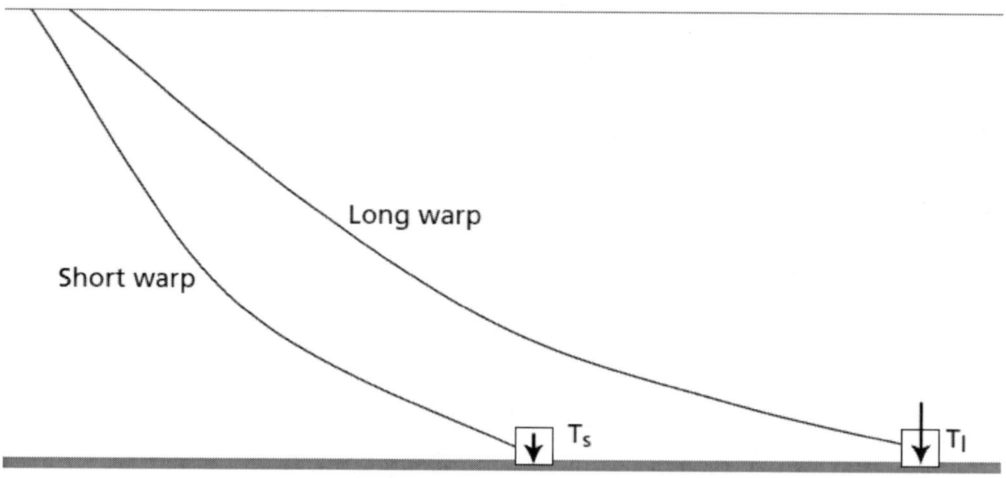

Figure 143. Pressure (Ts and Tl) against the bottom floor when using short and long warps in relation to the fishing depth (Valdemarsen et al., 2007).

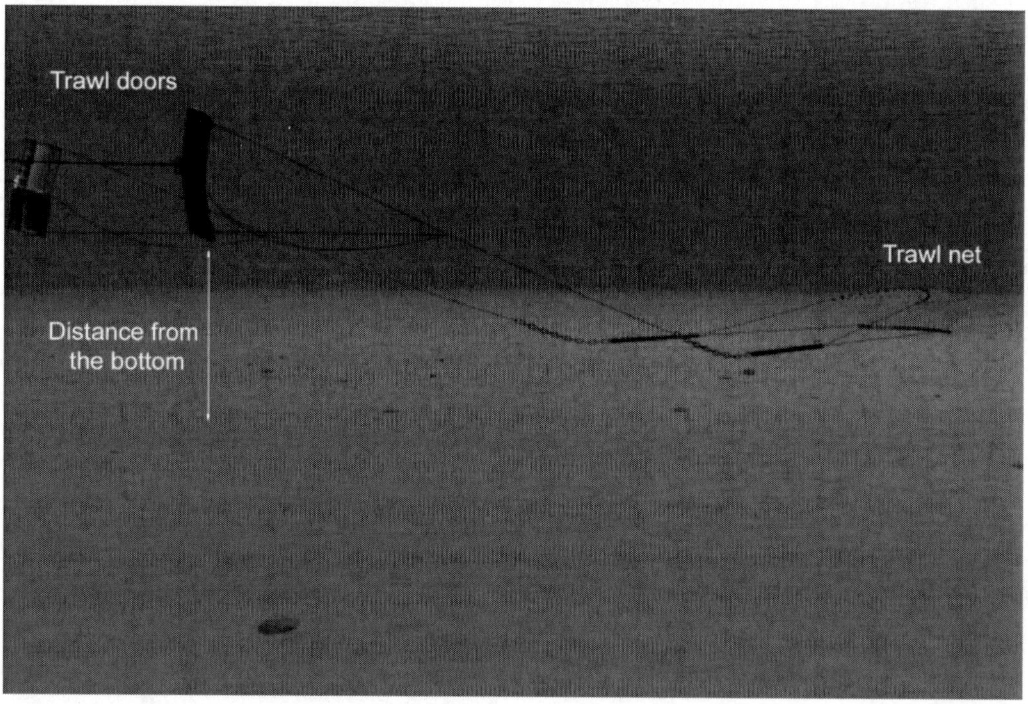

Figure 144. Rigging and behaviour of a high profile door: the impact on the bottom can be strongly reduced.

8.1.2.2. Sweep/Bridles

The impact of sweeps/bridles in several bottom trawl configurations needs to be clearly investigated because sweeps seem to have the largest contact area with the bottom. This impact is poorly understood but certainly affects sea bottom by scrapping and flattening the bottom surface protruding objects on the bottom.

Figure 145. High profile otterboards with low bottom impact. Pitch, roll and depth sensors are mounted on the otterboards for the real-time monitoring of the performances.

The impact of sweeps/bridles can be reduced with the use of discs or bobbins along the cable to raise the sweeps off the bottom and by off-bottom rigging of the lower bridle. Another option is to reduce the sweep/bridle length or angles of attack. Implementation of the above options may result in reduced capture efficiency for target species and acceptance of the technology by the fishing industry in their commercial fishing activity might therefore be difficult.

8.1.2.3. Modified Groundgears

Existing groundgear Designs have been developed with the aim of ensuring that the trawl maintains seabed contact to increase catch effciency and can safely pass over obstacles without sustaining damage.

Bottom trawls frequently have the bottom of the net fitted with a ground rope weighted with lead rings and chains that drag along the bottom. This ground gear is designed to maintain bottom contact and stimulate fish and other animals to enter the trawl. A consequence of this groundgear is that bycatch living on the bottom, such as fish, sponges, soft corals and benthic invertebrates may also be caught. Modification to the groundgear can go some way to reducing this bycatch and options include increased dropper-chain length, removal of tickler chains or specific sections of the groundgear, or ground gear replacement.

In some fisheries such as beam trawling and shrimp fisheries, tickler chains are used to increase catches of the target species (flat fish). However, they often impact the seabed harder and result in greater catches of bottom dwelling animals, as well as rocks, shell and other debris. Removal of these equipments will reduce this bycatch but clearly will not be favoured by fishermen.

Raised-Footrope

This was designed to provide vessels with means of continued fishing without catching groundfish and benthic organisms. In this type of trawl, the footrope is about 0.45 to 0.6 m above the bottom (Carr and Milliken, 1998). The raised footrope and net allows complete flatfish escape and has much less contact with the seafloor than does the traditional groundgear.

This can be achieved by replacing the groundgear with several weights, each attached to the footrope at a pre-determined position (Figure 146a). As only a few weights are used (five or six) and replace the entire groundgear, animals on the seabed have a good opportunity of avoiding contact with the trawl and subsequent capture. Damage to seabed animals is also reduced dramatically.

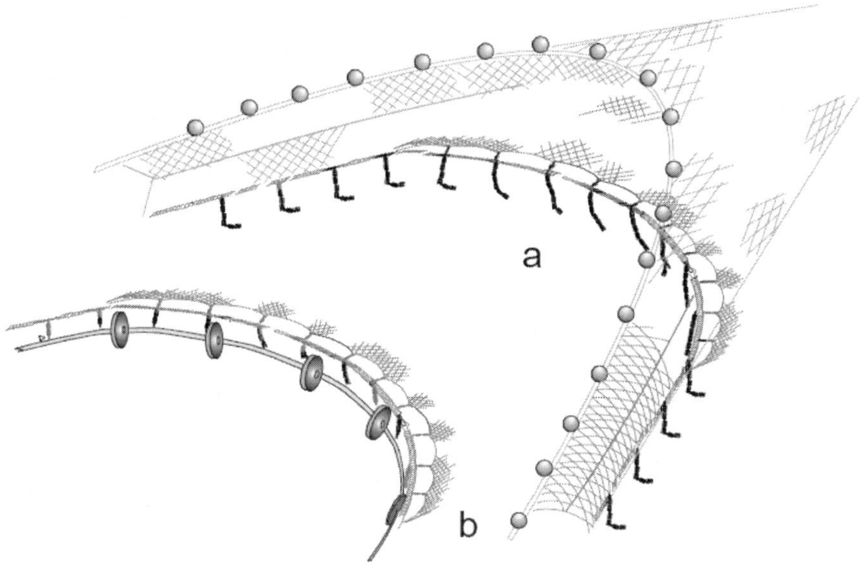

Figure 146. Scheme of a Raised-footrope (a), and Roller Gear (b) (by Sacchi, 2008).

Tickler Brushes

Tickler chains are commonly used in bottom trawl fisheries. A project was developed to reduce sea bottom impact of the tickler chain by replacing it with tickler brushes (K. Zachariassen, unpublished). Tickler brushes were made of nylon, cylindrical in shape and 15 cm in diameter. Underwater video observations showed that tickler brushes can drastically reduce suspension of sediments while maintaining catches of target species.

Roller Gear

To reduce the bottom impact of heavy groundgears, several configurations of wheels, or rolling gears were tested starting with pairs of rubber disks mounted on an axle and attached to the fishing line at a fixed angle (Figure 146b). Using larger rubber bobbins will increase the height of the groundrope above the seabed and reduce the amount of bottom-dwelling animals and debris caught.

Figure 147. Tickler chains used in beam trawling (North Europe) having high bottom impact.

In this best configuration, each wheel can rotate independently and maintain orientation in the towing direction. Video recordings of this roller gear showed a reduction in target species but showed also clear reductions in sand clouds.

Some experiments were conducted to replace the rubber groundgear or bobbins of otter trawl by a series of weighted rollers to allow the trawl moving over seabed rather than ploughing through it (Ball et al., 1999). The action of rollers appears to stimulate fish to rise off the seabed, eliminating the need for tickler chains. Preliminary fishing trials indicated no reduction in commercial catch. Due to reduced ground fiction, there is also potential of a reduction in fuel consumption during trawling.

8.1.3. Alternative Gears

The bottom impact exerted by bottom trawling as well as the overexploitation of demersal resources in addition to poor selectivity all encourage fishing technologist to propose and develop alternative gears as possible mitigating solutions. However, it has to be kept in mind that any local fishery is developed through "*trial and error*" for generations to find the most efficient way of fishing. So, it is not an easy task to introduce alternative solutions in a fishery. Moreover, what's an effective and responsible fishing method in one area does not necessarily work well in another one since several factors can affect the efficiency of a fishing method such as species behaviour, bottom conditions, temperature, fishing season etc..

One of the possible alternatives for exploiting near bottom organisms is to increase the use of stationary fishing gears that have much less impact on bottom habitats. However, the shift from trawl fishing to another technique with less impact must take into account the risk of profitability loss due to lower fishing efficiency or the lack of know-how at least during the first years of exploitation.

For some target species, bottom set gillnets and longlines might be both practical and economically viable alternatives to bottom trawling but other species like shrimp and small non-shoaling fish species cannot be caught economically with such gears.

Some marine organisms targeted with demersal gear have diurnal and/or seasonal vertical migrations. For these targets, an off-bottom harvesting fishing technique might be a bottom-friendly alternative.

The use of pots is usually considered as a valid alternative to high impacting gears mainly because pots can be used on all types of bottom. To accommodate more pots in the limited space available on deck of a vessel, certain models of pots need to be either collapsible or designed so that they can be piled one upon each other.

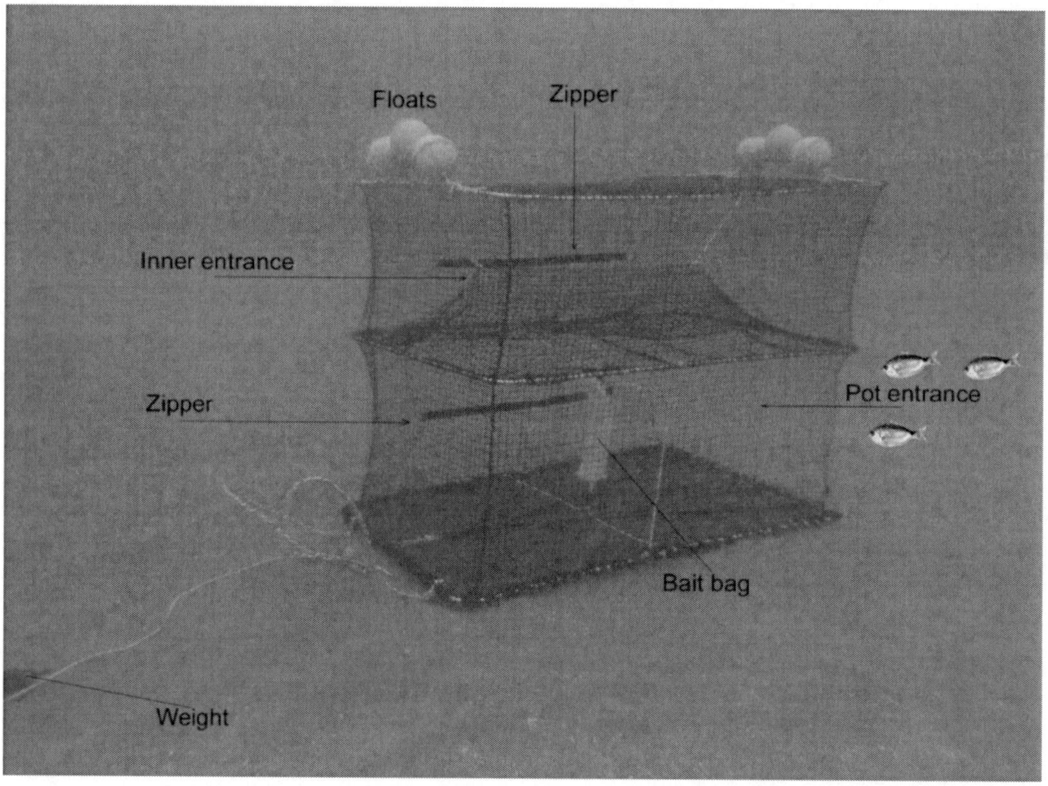

Figure 148. Collapsible fish pots as a possible solutions for the reduction of trawl impact.

An advantage of using traps is that catches are normally still alive and uninjured when the gear is retrieved so that unwanted bycatch organisms can be released with a good chance of survival in most cases. However, factors such as on deck injury and exposure, traumas due to pressure shock, or thermal shock may affect the survival of released organisms. Therefore, the use of traps offers the possibility of low bycatch mortality in comparison with many other fishing activities in general. The bycatch obtained in traps can also be further minimized by modifying pot design including choosing the correct size, shape and material; right pot entrance and escape openings etc.; mesh sizes; and the right location, season and time settings. The use of certain types of bait also has the potential attracting the target species and/or repelling unwanted species.

The main concern with traps as with other passive gears is the loss of gear at sea during fishing operations with trap continuing to catch fish or other organisms (ghost fishing). The use of biodegradable materials, and various escape vents have successfully been implemented in traps to reduce their ghost-fishing capacity. However, the loss of pots in seabeds with low natural structure or complexity such as sandy or muddy habitats may sometimes diversify the monotonous sea bottom by adding habitat complexity that offers refuges and shelters for various species and so acting in the same manner as artificial reefs.

The information to be collected when an alternative gear needs to be tested can be summarized as follows:

- State of the art: gear(s) used including net-gear drawings, dimensions, mesh opening, netting materials, floats etc., target species, catch per day, vessel types and dimensions, type of bottom, depth, seasonality;
- Descriptions of the behaviour of the target species including typical swimming speeds and depth range, variations in behaviour associated with diurnal/lunar/ seasonal cycles and changes in environmental conditions;
- Detailed description of the fishing grounds and bottom habitats which strongly affect the kind of alternative gear possible to use;
- Detailed descriptions of the alternative gear: net-gear drawings, dimensions, mesh size, netting materials, floats etc., target species; and
- Modifications to the traditional gear like reducing damage from ground contact, costs for modification.

Chapter 9

SEA TURTLE MORTALITY AND MITIGATION MEASURES

Incidental catch and rarely intentional killing for blood and meat represent the major threats to the survival and the general decline of marine turtle populations in the Mediterranean (Margaritoulis et al. 2003). Currently, knowledge about the interaction of sea turtles with fishing gears and the possible mitigation measures are still insufficiently studied (Casale et al. 2004).

The leatherback- (*Dermochelys coriacea*), the green- (*Chelonia midas*) and the loggerhead-turtle (*Caretta caretta*) are regularly found in the Mediterranean basin (Tudela 2004). The latter two species are listed as "Endangered" while the leatherback turtle is listed as "Critically Endangered" on the Red List of Threatened Species of the International Union for Conservation of Nature and Natural Resources (IUCN; Hilton-Taylor 2000).

Loggerhead is the most common species of marine turtles in the Mediterranean and the knowledge of its biology is a key factor to evaluate the impact of different fishing activity in different areas.

The loggerhead and the green turtles are known to nest in the Mediterranean particularly on the beaches of the eastern part in Cyprus, Egypt, Greece, Israel, Italy, Lebanon, Libya, Syria, Turkey and Tunisia (Figure 149). Broderick et al. (2002) estimated that 2,280-2,787 loggerhead and only 300-400 green female turtles nest annually in the Mediterranean, while. Margaritoulis et al. (2003) have estimated 3,375-7,085 nests per season. The estimates of Groombridge (2,000 loggerhead turtles) in 1990 are similar considering an average of three clutches per season.

The number of adults should be even higher considering that most of the individuals do not breed every year. Moreover, additional numbers migrate from the Atlantic population to Mediterranean through the strait of Gibraltar during the first half of the year (Camiñas 1997a; 1997b), Figure 149).

Three main ecological phases are known for the loggerhead turtle. The pelagic phase in which loggerheads feed on pelagic preys; the demersal phase in which they swim near the bottom to feed on benthic species; and finally the intermediate neritic phase in which loggerheads shift from pelagic–oceanic to benthic–neritic foraging habitats (Tomas et al. 2001), Figure 149). The highest density of individuals in the demersal phase is found in shallow waters (<100 m). Loggerheads do not dive deep and the maximum recorded depth is 110 m (Lutcavage and Lutz 1997). The bathymetry and environmental characteristics of different areas determine the distribution and abundance of marine turtles. Loggerhead turtles exhibit adheres highly to migratory routes, foraging areas and wintering sites between and within years and after successive breeding migrations.

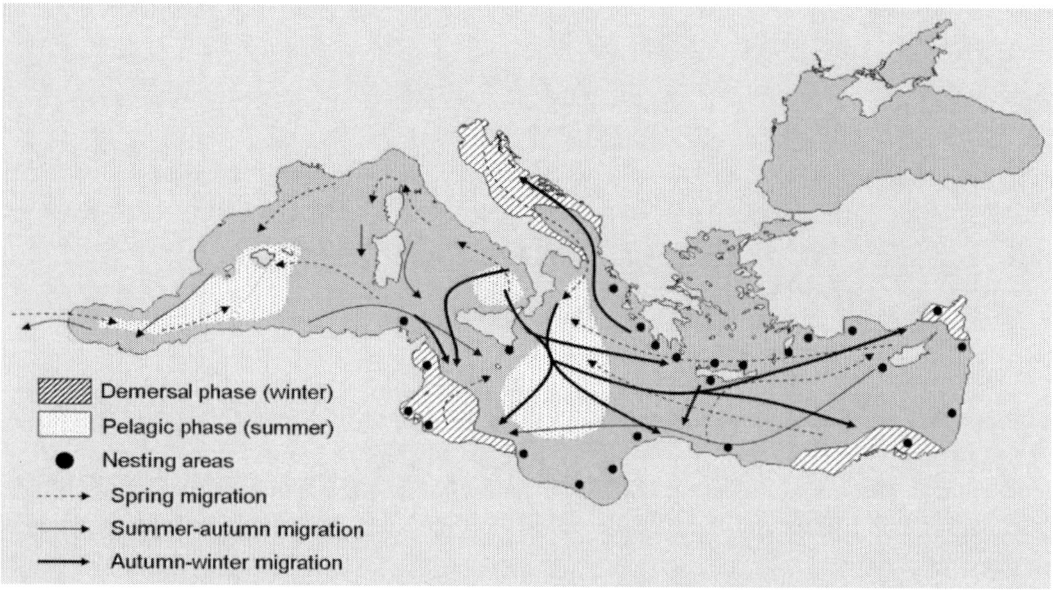

Figure 149. Loggerhead migrations in the Mediterranean Sea; the main routes, nesting beaches, pelagic and demersal areas are shown (Bentivegna 2002; Broderick et al. 2007; Camiñas 2004; Maffucci et al., 2006; Lucchetti and Sala, 2012).

Thus, the different types of fishing gears (towed or passive, on the bottom or in the water column) can result in different captures and mortality rates (Gerosa and Casale 1999) and may affect different ecological phases (Figure 150).

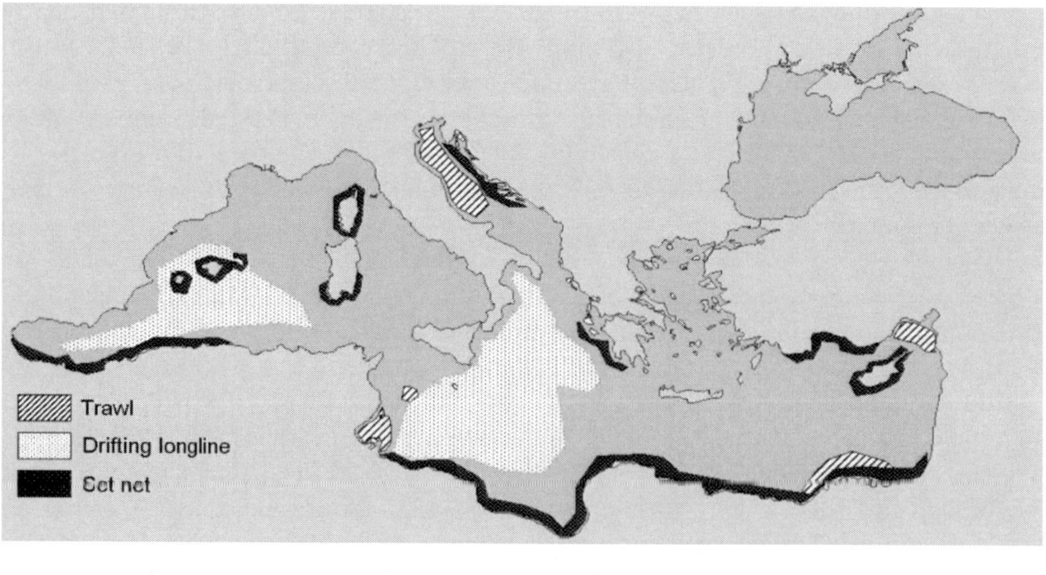

Figure 150. Location of turtle bycatch in Mediterranean by gear type and area.

9.1. Assessment of Loggerhead Sea Turtle Bycatch

More than 60,000 turtles are estimated to be accidentally caught per year as a result of fishing practices in the Mediterranean (Lee and Poland 1998). The direct mortality rates range from 10% to 50% and even reach 100% of individuals caught. Moreover, the delayed mortality after release at sea is mostly unknown. Casale (2008) highlighted an even worse conditions suggesting that over 150,000 captures per year may be taking place in the Mediterranean by trawlers, longliners and set netters with possibly over 50,000 mortalities per year by interaction alone. A more pessimistic scenario is defined considering that official fleet statistics do not include all the existent fishing vessels and that the number of small boats is underestimated. This leads to an estimate of more than 200,000 capture events per year.

9.1.1. Drifting Longline

Drifting longline targeting swordfish (*Xiphias gladius*), albacore (*Thunnus alalunga*) and Bluefin tunas (*Thunnus thynnus*) deployed over the continental shelf and in offshore waters is considered as the main threat to the marine turtles in the Mediterranean in terms of catch per year (Gerosa and Casale 1999; De Florio et al. 2005; Margaritoulis et al. 2003), Table 5). The incidental capture of loggerhead turtles in pelagic phase with longline extends mainly from spring to late autumn with most captures occurring in the summer.

Casale (2008) estimated that over 50,000 individual are being caught with pelagic longlines with a mortality of 40% mainly in Spain, Morocco, Italy, Greece, Malta, Libya, Table 5). Lewison et al. (2004) estimated at least 60,000-80,000 captures per year in the Mediterranean, while Panou et al. (1992) found an annual catch of about 35,000 individuals for the Central-Western Mediterranean only which is one of the main concentration areas of adult loggerheads and for targeted species (swordfish) as well. In fact, Spain is the country with the highest number of turtle captures per year followed by Morocco and Italy (Table 5).

Bycatch due to foreign industrial longline fleets operating in the Mediterranean waters (Japanese, flag of convenience) could lead to an even worse situation. Data on annual catches are available for other non-EU countries but in some cases there is a doubt in their validity.

9.1.2. Bottom Trawl

Bottom trawl has been considered as the second most impacting fishing gears for sea turtle populations in terms of number of catch per year (Table 5). Bottom trawl mainly impacts turtles in demersal phase since they prefer coastal shallow waters less than 50 m or even 20 m deep for feeding (Hare 1991; Caillouet et al. 1991; Bradai 1994; Epperly et al. 1995) with most of them being juveniles and sub-adults usually far from their hatching ground (Bolten et al. 1998). The Mediterranean continental shelf, which is the feeding habitats for several turtles, is usually narrow in most areas. Nevertheless, in some places (Gulf of Gabés, Northern Adriatic Sea, Southern Turkey and Egypt), the continental shelf is expansive and turtles in demersal phase are used to spend winter time in these areas. Mediterranean bottom trawlers are estimated to catch around 30,000 individuals per year altogether (Table 5) with a mortality of 25% mainly in Italy, Tunisia, Croatia, Greece, Turkey, Egypt, and Libya (Casale et al. 2004; Casale 2008; Laurent et al. 1996; Lazar and Tvrtkovic 1995; Oruc 2001), Figure (2). Moreover, it is well known that the same turtle can be caught more than one time. For this reason, Casale (2008) has estimated more than 40,000 “capture events” for the Italian waters alone.

9.1.3. Other Gears

Few official and published data are available for drift nets targeting swordfish, which are illegal in most Mediterranean countries. Nevertheless, illegal drift nets are still widely used in some countries and the amount of bycatch is hypothetically expected to be very high. According to De Metrio and Megalofonou (1988), about 16,000 turtle are caught per year by Italian drift nets in the Ionian Sea. The use in recent years of legal drift net called “Ferrettara” similar in characteristics to the illegal “spadara” used to catch swordfish has been causing several problems for fisheries inspectors in Italian waters. For this reason illegal drift nets masked as legal “Ferrettara” are still being used. Moreover, several drift nets of EU countries were probably sold to non-EU countries, mainly Moroccan fleets, shifting the bycatch problems from the North to the South basins.

Bottom longlines are generally used at a depth of 200-700 m posing no problems (Bolten et al. 1994). However, there are some fishing grounds where this method is used at a much shallower depth causing numerous captures of marine turtles, mainly juveniles. Demersal longlines seem to be responsible for about 35,000 capture events. The most affected areas being the north African continental shelf (Tunisia, Libya, Egypt), the Alboran Sea (Morocco), the Levantine basin (Turkey) and the Aegean (Greece) (Casale 2008).

Studies on fixed nets in the Mediterranean are rare. Fixed nets represent a threat for sea turtles mainly in coastal areas (Lazar et al. 2004; Argano et al. 1992; Lazar et al. 1998), Figure (2) but the quantification of captures in these widely spread fisheries is very difficult to assess due to the very high number of small boats disseminated along all the Mediterranean coasts. Over 30,000 captures per year are estimated mainly in Tunisia, Libya, Greece, Turkey, Cyprus, Croatia, Italy, Morocco, Egypt, and France (Casale 2008), Figure 150). Turtle captures seem to be very high in certain areas like Croatia, Slovenia and Tunisia (Lazar et al. 2006; Bradai 1993). Juveniles are frequently caught nearby nesting areas in Greece, Turkey and Cyprus (Godley et al. 1998; Suggett and Houghton 1998). Lescure (1987) reported that in the Mediterranean French coast, loggerhead turtles are mainly captured with trammel nets and bottom trawl. These made some authors (Casale et al. 2005a) state that the interaction of sea turtles with the static net fishery in the Mediterranean could be very significant and comparable to other fisheries.

Very little information is available yet for pelagic pair trawl. In the North Adriatic, it’s estimated that about 1,550 turtles are caught accidentally per year (GFCM-SAC 2008) with all turtles being released alive. The purse seines seem to represent a minor problem since the annual captures are probably very low and turtles are released alive.

9.2. Main Fishing Gear Parameters Affecting Marine Turtle Bycatch and Mortality

The capture of loggerhead turtles strongly depends on different parameters. The most important factor is the fishing effort. Number of vessels, engine power (KW, HP), gross tonnage (GT, TSL), and time at sea (hours per day, days per year, etc.) are all essential parameters to be kept under control. Nevertheless, the mortality rate is variable and it depends largely on gear type, practices on board, and capability of surviving forced apnoea.

Three main processes must be considered in longline turtle bycatch; namely, attraction of turtle by bait, hooking, and escape during hauling of the gear. The hooks can be identified by several parameters like general shape ('J' or 'G'), dimensions (total length, length diameter, gap between point and shank, length shank, width, throat, barb size, etc.), material (steel, inox), point (with barb or not), shape of eye flat or twisted (generally the shank and the point are not on the same plain but on plains with a difference of 10-25° degree offset to enhance catching efficiency). The most important parameters in the hooking process are hook's overall width which connects with turtle mouth dimensions as well as its gap which ensures deeper penetration of the point and better holding power of the fish and its shape which influences the hooking position.

Direct mortality induced by longlines appears to be low (Table 5) but delayed mortality is a matter of concern because it is widely unknown and expected to be very high (Camiñas and Valeiras 2001; National Marine Fisheries Service 2001; Lewison et al. 2004; Deflorio et al. 2005; Kapantagakis and Lioudakis 2006). The low direct mortality can be justified by the fact that hooked loggerhead turtles have enough power to rise with lines to the surface and breathe. About 80% of hooked turtles is released alive but post-release mortality strongly depends on hook position in different parts of the digestive tract (mouth, oesophagus, stomach, intestine, etc. (Camiñas and Valeiras, 2001). If the hook is swallowed in the lower oesophagus or the stomach, the turtle has a very low chance of survival. On the contrary, the mortality of turtles with a hook in the mouth or higher oesophagus seems to be less likely though a hook in the mouth can affect the feeding performance,particularly if it impedes closure of the mouth (Casale et al. 2007b). Studies made by rescue centres showed high post-release mortality both in the short- and long-terms. A capture in a line with hook timer showed that a turtle was able to survive for 18 hours and 9 minutes. The estimates of potential mortality caused by drifting longline obtained in various areas by different authors (Aguilar et al. 1993; Casale et al. 2007b; Lewison et al. 2004) are similar and ranged from 17-20% to 30-40%. Barb size and hook length can affect the capacity of the turtle to disengage. Some studies showed that the use of barbless hooks implied a small reduction in unhooking time leading to an improvement of survival rate (Alòs et al., 2008).

In addition to the hook, the secondary lines of a longline; i.e., branchlines which is the piece of line attached to the hook, can easily cause turtle death especially if it is long enough to be affected by intestinal peristalsis (Bjorndal et al. 1994; Casale et al. 2007b; Oros et al. 2004). In these cases, death typically occurs after days. Unfortunately, a common practice of fishermen is to cut the branchline from the deck while the captured turtle is still in the water to gain time and save effort as turtles are often very heavy so leaving most turtles released with potentially lethal branchlines longer than 1 m (Guglielmi et al. 2000). Thus, it appears that hooks cause death in the short-term and branchlines in the long one (Casale et al. 2007b).

Other parameters affect sea turtle catch and mortality. Albacore and Bluefin tuna longlines generally produce higher direct mortality in hooked turtles than swordfish longlines. This is probably because of gear structure and hook size but it could be also due to the fishing depth or to distance from the coast or to higher catch amount that could sink the gears causing turtle mortality by reducing breathing possibilities. Turtle mortality in longline could be strongly affected by the depth of the main line setting. Loggerhead turtles spend most of their time at less than 40 m and they do not dive deeper than 100 m (Polovina et al. 2003). Juveniles of loggerhead turtles are capable of diving to depths below 200 m but they generally stay in the upper 30 m (Dellinger 2000; Dellinger and Ferreira 2005). Thus, the main interaction depth with longline is the upper 20 m of the water column (Dellinger and Ferreira 2005). Several studies confirmed the very low direct mortality rate due to shallow-set longline gear activities (Deflorio et al. 2005; Gilman et al. 2007; Jribi et al. 2008; Pinedo and Polacheck 2004; Piovano et al., 2009). The mortality rate in bottom longlines seems to be concentrated on juveniles in particular since bigger individuals are able to drag the main line with its weights up to the surface to breathe. Therefore, demersal longlines seem to be responsible for a potential delayed mortality of about 40% since hooked turtles, particularly juveniles, cannot reach the surface to breathe (Casale 2008). Finally, turtle mortality seems to be correlated with setting time; the longer the setting time, the more turtles captured die (Camiñas et al. 2006).

In bottom trawl, incidental catch of loggerhead turtle probably occurs during towing operations when turtles are foraging on the bottom. When captured by a trawl, loggerhead turtles may drown becoming first comatose and eventually dying. Mortality by trawling is due to forced apnoea, thus longer or faster tows are responsible of higher mortality rates. Therefore, towing time is one of the main factors affecting the mortality rate (Henwood and Stuntz 1987) but additional factors might be involved especially in bottom trawl (Stabenau et al., 1991). Sasso and Epperly (2006) found that turtle mortality differs depending on season being higher in winter than in summer. They also found that tows of short duration (<10 min) have negligible effect on mortality. Thus, it might not be reasonable to allow longer fishing times of 60 minutes in winter and 40 minutes in summer as recommended by the US National Research Council (1990). The observed direct mortality in Turkish waters was really low (0.3%), while some specimens (4.6%) were figured comatose or weak when released from the nets (Oruç 2001).

Passive nets are potentially more harmful than pelagic longlines and bottom trawl to marine turtles since the mortality rate associated with this gear is estimated to be about 50% and even higher (Table 1). Studies on gillnets and trammel nets showed that these gears are responsible of high direct turtle mortalities (50-100%). Drowning is the main reason for these high mortalities since turtles get entangled in nets when trying to feed on fish caught in them and, hence, are forced to stay underwater for unsustainable periods of time (Laurent 1991; 1996; Table 1). Environmental parameters could also contribute to marine turtle mortality. High water temperatures like such those off North Africa countries associated with high metabolic rates can strongly reduce the resistance to forced apnoea.

Table 5. Assessment of turtle bycatch and direct mortality in Mediterranean by gear type (adapted from Lucchetti and Sala, 2010)

GEAR TYPE	AREA	CATCH / YEAR	DIRECT MORTALITY	REFERENCE
Drifting longline	Entire Mediterranean	50,000	0-4% (40% potential)	Casale 2008
	Entire Mediterranean	60,000-80,000	Potential mortality 17-42%	Lewison et al. 2004; NMFS 2001
	Western-central Mediterranean	35,000		Panou et al. 1992
	Spain	22,000-35,000	0.36-7.7% (20-30% potential)	Aguilar et al. 1995; Tudela 2000, Carreras et al. 2004
	(Balearic Islands: 15,000-18,000; Aguilar et al. 1995; Camiñas 1988; Camiñas et al. 2001; Mayol et al. 1988)			
	Spain		0.54 - 4.24%	Camiñas et al. 2006
	South Sicily (Italy)	2,148	>30% potential	Casale et al. 2007a,b
	Lampedusa Island (Italy)	245		Casale et al. 2007a
	Ionian sea (Italy)	1,084-4,447	0% (potential mortality high)	De Florio et al. 2005
	Ionian sea (Greece)	280-3,181		Panou et al. 1999; SGRST-SGFEN 2005
	Aegean and South Ionian seas (Greece)	1,145-5,474		Kapantagakis and Lioudakis 2006
	Cephalonia (Greece)	50		Panou et al. 1992
	Malta	1,500-2,500		Gramentz 1989
	Cyprus	2,000		Godley et al. 1998
	Morocco	3,000		Laurent 1990
	Algeria	250-300		Laurent 1990, Camiñas 2004
	Tunisia	486-4,000	0% (9.1% potential)	Echwikhi et al. 2006; Salter, 1995; Demetropoulos, 1998; Jribi et al. 2008
Bottom trawl	Entire Mediterranean	30,000	5% (20-25% potential)	Casale et al. 2004; Casale 2008; Laurent et al. 1996; Lazar and Tvrtkovic 1995; Oruc 2001
	Italy	8,500	14% (57% potential)	Casale et al. 2004; 2007a; De Metrio and Casale 2001
	Lampedusa Island (Italy)	4,056		Casale et al. 2007a
	North Adriatic	4,273	9.4% (43.8% potential)	Casale et al. 2004
	France		3.3-3.7%	Laurent 1991; Delauguerre 1987
	Croatia	2,500	low	Lazar and Tvrtkovic 1995
	Thracian sea	0-418		Margaritoulis et al. 2001
	Ionian sea (Greece)	0-448		Margaritoulis et al. 2001
	Tunisia (whole continetal shelf)	14,000		Jribi and Bradai 2008
	Gulf of Gabés (Tunisia)	2,500-5,500	3.3%	Bradai 1992; Jribi et al., 2004
	Egypt	2,269-high	1-10%	Nada and Casale 2008; Laurent et al. 1996
	Turkey	High	1.6% (13% potential)	Oruç A (2001); Oruç et al. 1996
Drift Nets	Italy	16,000	20-30%	De Metrio and Megalofonou 1988
	Ligurian and Tyrrhenian sea (Italy)	low		Di Natale 1995
	Spain	117-354	3.3%	Aguilar et al. 1995
	Spain	236		Silvani et al. 1999
Bottom longline	Entire Mediterranean	35,000	Potential mortality 10%	Casale 2008
	Lampedusa Island	257		Casale et al. 2007a
	Tunisia	733-2,000	0.53%-12.5% (33% potential)	Echwikhi et al. 2006; Jribi et al. 2008; Bradai 1993
	Egypt	2,218		Nada and Casale 2008
Fixed nets	Entire Mediterranean	30,000	>50% (60% potential)	Casale 2008
	Balearic islands (Spain)	209	50-100%	Carreras et al. 2004
	Corsica (France)	low	93.3-75%	Laurent 1996; Delaguerre 1987
	France	10-100-low	50-100%	Laurent 1991
	Italy		50%	Argano et al. 1992
	Slovenia-Croatia	657-4,038	50% - 73%	Lazar et al. 2006
	Cyprus	500	10%	Godley et al. 1998
	Tunisia	920-2,000	5%	Bradai 1993
	Egypt	754		Nada and Casale 2008
	Turkey	1,328	10%	Godley et al. 1998
Pelagic pair trawl	North Adriatic	1,550		GFCM-SAC 2008
	Turkey	high (5 trawlers catch around 100 loggerheads)		Oruç A (2001)
Purse seine	Egypt	37		Nada and Casale 2008
Small scale fishery fixed nets, purse seines, bottom and surface longlines etc.)	Tunisia	5,000		Bradai 1995

9.3. Mitigation Measures

The identification of methods to reduce or prevent marine turtles bycatch is a high priority for fishery managers and a necessary component of conservation efforts. The most obvious and simplest approach to solve the bycatch problems is the reduction of fishing effort even if this is often not a practical nor viable option for economic, social and sustenance reasons. A summary of the main mitigation measures tested in the Mediterranean is shown in Table 6).

Table 6. Principal bycatch mitigation measures tested in the Mediterranean sea. SWO-LL: Swordfish longline; ALB-LL: Albacore longline; BLL: Bottom longline (adapted from Lucchetti and Sala, 2010)

GEAR TYPE	MITIGATION MEASURE	AREA	ACTION	REFERENCE
Drifting longline	Hook size	Mediterranean	Smaller turtles are caught by BLL (smaller hooks)	GFCM-SAC 2008
		Ionian Sea (Italy)	Larger turtles are caught by SWO-LL (larger hooks)	De Florio et al. 2005
		Balearic Islands (Spain)	Smaller hooks caused deep-hooking	Alòs et al. 2008
		Tunisia	Smaller hooks (BLL) caused deep-hooking	Jribi et al. 2008
	Hook shape: J vs circle hook	Atlantic - Mediterranean	Results do not support move from J to circle hook	project UE-FISH/2005/28-A
		Italy	Circle hooks reduced turtle bycatch; circle hooks were not easily swallowed	Piovano et al. 2009
		Spain	Hook type is not the main factor for turtle bycatch reduction	De La Serna et al. 2008
		Spain	Inconclusive results	Parga 2008
		Spain	Circle hook can shift the problem from turtle to cetaceans or sharks	Camiñas and Valerias 2001
		Western Mediterranean	No evident differences	Rueda and Sagarminaga 2008
			Circle hook only reduce leatherback turtle bycatch	Casale 2005 (review of Watson et al. 2003; 2004)
			Circle hook effective in reducing turtle bycatch in several countries, circle hooks reduce the rate of hook ingestion	Several authors (i.e. Watson et al., 2004; NGO web-sites etc)
	Bait Type	Spain	Mackerel bait instead of squid bait reduce turtle bycatch	Rueda and Sagarminaga 2008
		Atlantic - Mediterranean	Mackerel bait instead of squid bait reduce turtle bycatch	project UE-FISH/2005/28-A
			Mackerel bait instead of squid bait reduce turtle bycatch	GFCM-SAC 2008
	Depth setting	Atlantic - Mediterranean	Deep longline: reduction in turtle bycatch, reduction in target species	Rueda and Sagarminaga 2008
		Ionian Sea (Italy)	Most of turtle bycatch between 10-15 m depth	Project Life Nature 2003 - NAT/IT/000163
		Mediterranean	SWO-LL catch turtle at a depth < 60m; ALB-LL catch turtle at a depth < 20m	Laurent et al. 2001
	Sensory stimuli	Italy (tank tests)	Floats presence or absence does not influence turtle behaviour	Piovano et al. 2002
		Italy (tank tests)	Odour play a key role in the bite-no bite decision	Piovano et al. 2004
		Spain	Lightsticks attract sea turtle	Rueda and Sagarminaga 2008
		Italy (tank tests)	Acoustic tests: inconclusive results	Piovano et al. 2002
Bottom trawl	TED (Turtle Excluder Device)	Italy	Effective in reducing turtle bycatch, discard and debris; possible problems with loss of large fish	Lucchetti et al. 2008
		Turkey	Effective in reducing turtle bycatch	Atabey and Taskavak 2001

SWO-LL: Swordfish longline
ALB-LL: Albacore longline
BLL: Bottom longline

9.3.1. Drifting Longline

Several technical parameters of a longline can affect the incidental catch of marine turtles (Figure 151). Some studies have been carried out on mitigation measures to reduce turtle bycatch in longline fisheries within a number of different longline fisheries around the world but few of them have been conducted in the Mediterranean. Drifting longline in the Mediterranean targets large pelagic fish such as albacore (*T. alalunga*), Bluefin tuna (*T. thynnus*) and swordfish (*X. Gladius*). Several measures have recently been proposed to reduce marine turtle capture and mortality in pelagic longline (see the review of Gilman et al., 2006). These include:

- Using small circle hooks (≤ 4.6 cm narrowest width) in place of smaller J and Japan tuna hooks;
- Setting gear below turtle-abundant depths;
- Using single hooking fish bait vs. multiple hook threading;
- Reducing gear soak time and retrieval during daytime; and
- Avoiding bycatch hotspots through fleet communication programmes and area and seasonal closures.

Among these solutions, reducing hook size, setting hooks deeper in water column, and changing hook shape are particularly promising.

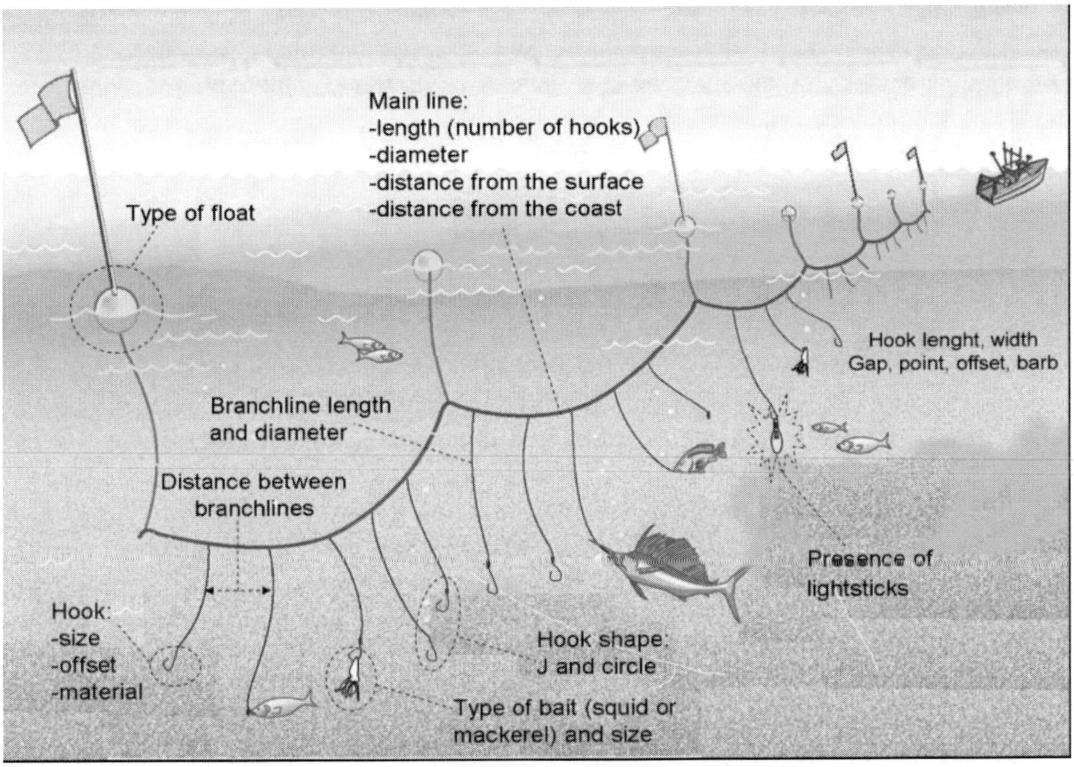

Figure 151. Technical parameters of a longline affecting the turtle bycatch.

Hook Size

Hook size affects the probability that the hook is swallowed and the turtle becomes hooked internally increasing delayed mortality potential after release. Some studies on captivity showed that loggerhead turtles had a much lower tendency to ingest hooks larger than 51 mm in width than smaller hooks (Løkkeborg 2004). The reduction of hook size from "18/0 J" to "9/0 J" is the easiest solution because larger hooks are less likely to be swallowed by turtles due to physical constraints of the mouth (Watson et al. 2003). Larger loggerheads are probably more likely to ingest hooks than smaller loggerheads. In the Balearic Islands,

Alòs et al. (2008) found that hook size is the most important cause for deep-hooking which is reduced by using large hooks. Moreover, large turtles tend to be caught in deep-hooking locations. Finally, they found that large hooks reduce the incidence of hooking injuries with a small reduction in catch rate. Similar results were obtained by Jribi et al. (2008) in Tunisia where the smaller hooks of bottom longlines seemed to be more easily swallowed in the digestive tract in comparison with hooks of surface longlines that remained in the mouth. In the Ionian Sea, the hook used for albacore fishing (4 cm long) is notably smaller than that used for swordfish (10 cm long) and it is baited using small bait that makes biting easier for small turtles (Deflorio et al. 2005).

During the GFCM-SAC Transversal Working Group on bycatch/incidental catches in 2008, participants stressed the point that bottom longlines catch a higher number of juvenile loggerheads due to the smaller hook size.

Hook Shape

Hook shape greatly affects its position in mouth and, hence, the capacity of turtle to disengage from the hook. Hook location also plays a fundamental role in post-release survival of turtles particularly when hooks remain lodged in the body with trailing fishing line. Regarding the Mediterranean, one of the most important mitigation measures tested to reduce turtles bycatch is the change in hook shape from traditional "J" shape to a "circular" one (Figure 152). Different manufacturers produce hooks in many different sizes and designs. In the case of J hooks, the point is parallel to the shank, while it is perpendicular to the shank in circle hooks. There is growing evidence that circle hooks tend to be located mostly externally in the jaw or mouth as opposed to deeper hooking (Watson et al. 2005; Gilman et al. 2006; Gilman et al. 2007; Read 2007). Nevertheless, the efficiency of circular hooks in reducing turtle bycatch and throat hooking is clearly demonstrated in certain fishing grounds only.

Piovano et al. (2009) tested hooks with different shapes but the same gap (about 2.6 cm width circle hook size 16/0 vs. J hook size 2) to determine the potential effectiveness of hook design on reducing marine turtles capture and maintaining acceptable levels of target species capture rates in a shallow-set longline swordfish fishery. They found that circle hooks can effectively reduce the number of immature loggerhead turtles accidentally captured by up to 70% without affecting the capture rate of target species (swordfish). They also found that all hooked turtles were brought on board and released alive irrespective of the type of hook. This seems to confirm the very low direct mortality rate due to shallow-set longline gear activities. They also showed that 81% of hooking occurred in the mouth, while 19% were swallowed all of which were on J hooks (zero swallowed with circle hooks).

On the contrary, the results obtained in a project carried out in the Atlantic and Mediterranean areas (UE-FISH/2005/28-A) do not support the promotion of shifting from J hooks to either of the two circle hooks (16/0 0° offset and 18/0 10° offset) as these hooks did not consistently reduce turtle catch rates although there was an indication that 18/0 circle hooks were less likely to be swallowed than J hooks or 16/0 circle hooks and had negative impacts on swordfish catches. De La Serna et al. (2008) found that the type of hook was not viable reducing marine turtle bycatch and suggested that probably other factors were more important and involved in this. Moreover, circular hooks particularly in combination with squid bait were found to reduce the catch rates of most important fish target species in general.

In the western Mediterranean, turtle catch rate of circle hooks was slightly lower than on J hooks. Rueda and Sagarminaga (2008) compared 12/0 circle hooks and J shape ones having similar dimensions in Spanish longline vessels but they did not find any significant difference in turtle bycatch and in tuna catches. Moreover, Camiñas and Valerias (2001) noticed that circular hooks seemed to shift the bycatch problem from turtles to cetaceans and sharks. Casale (2005) in reviewing Watson et al. (2003; 2004) emphasized that the overall effect of circle hooks in reducing turtle bycatch is limited to the soft-shelled leatherback turtle. Furthermore, NOAA studies on loggerhead turtle in the Atlantic suggest that catch rate of these species is affected by hook size and bait rather than by hook shape. Parga (2008) tested small circle hooks instead of traditional J hook in Spanish surface longline targeting *T. alalunga*. Results were inconclusive except for a significant reduction of immature *T. thynnus*.

Some differences in fishermen reactions were also observed. Circular hooks are not well accepted by Spanish fishermen (Baez et al. 2006) as their use is considered to diminish yields of target species (Gilman et al. 2006). On the contrary, Italian fishermen did not find any appreciable difference (project life nature 2003 - NAT/IT/000163). For this, Piovano et al. (2009) support the enforcement of circle hooks in the Mediterranean swordfish longline fishery because it can represent a simple and inexpensive technical solution to decrease the capture of turtles.

Bait Types

Bait type is one of the main factors affecting longlines efficiency. In non-Mediterranean experiments (U.S. and Pacific data), mackerel bait reduced turtle bycatch compared to squid bait (82% of all loggerheads caught were caught with squid bait) and increased the catch of swordfish compared to squid bait (Watson et al. 2005). This was probably due to the fact that mackerel bait tends to come free of the hook as the turtle takes small bites of it, while squid remains more firmly attached requiring the turtle to take larger bites to swallow the bait increasing the chances of becoming hooked (Gilman et al. 2006). Moreover, the CPUE of swordfish was reduced by using circle hooks with squid bait in Atlantic waters. In captivity studies carried out in Italy, Piovano et al. (2005) found that mackerel (Scomber spp.) bait smell was an important component for the detection of bait by turtles while smell-less artificial baits were generally unattractive for turtles. Experiments conducted on board Spanish Mediterranean swordfish longliners showed that turtle bycatch was lower with mackerel baits with no significant difference in target species (swordfish) catch (Rueda et al., 2006; Rueda and Sagarminaga, 2008). Spanish experiments allowed concluding that the shift in bait type to mackerel during summer periods when the incidence of turtles in the fishing grounds of EU vessels is the highest, may be greatly useful for the conservation of loggerhead turtle conservation though Bluefin tuna catches were lower on mackerel compared to squid bait. Results obtained in the EU project UE-FISH/2005/28-A showed that the combination of hook and bait type that resulted in the lowest bycatch of turtles and the highest catches of swordfish was J hooks with mackerel bait.

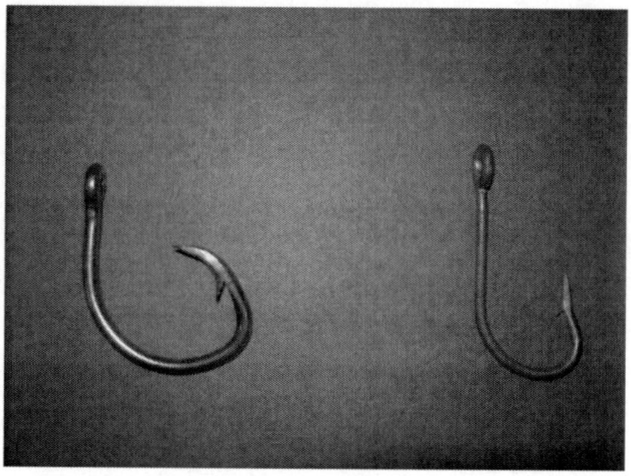

Figure 152. Traditional J hook (right); Experimental circle hook (left).

The participants of the GFCM-SAC Transversal Working Group on bycatch/incidental catches in 2008 concluded that the most successful measure to reduce turtle bycatch in swordfish longline is the change of bait from squid to mackerel which decreased the incidental catch of turtles by 80% without affecting the catch of swordfish.

Meanwhile the positioning of the bait on the hook can affect the turtle bycatch (Figure 153), the use of horizontal baited mackerel can reduce the catch of unwanted turtles similar to a mackerel bait while maintaining the catch of the target species swordfish similar to a squid bait.

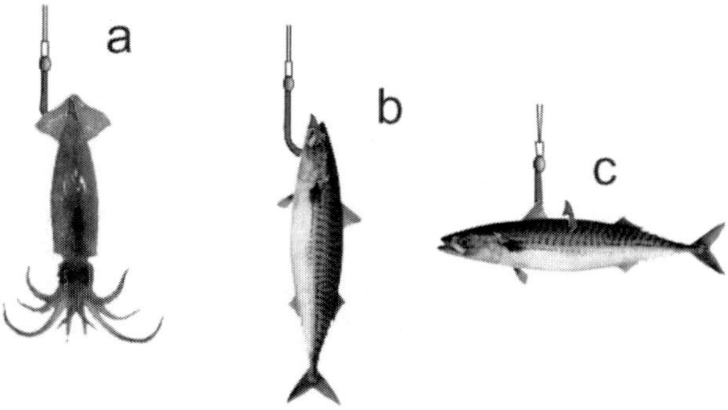

Figure 153. Different types of bait: a) Squid; b) mackerel (vertical set); c) mackerel (horizontal set).

Depth Setting

The depth setting of a longline strongly affects turtle bycatch. Studies of turtle behaviour indicate that they swim in shallow waters and usually get tangled in or hooked by longlines at depths of 100 meters or less. Tunas and swordfish usually swim at much lower depths. One of the possible solutions to reduce turtle bycatch is to set the main line deeper. The mainline is weighted with lead weights and released or "set" in such a way that its section holding 20-40 baited hooks goes down to and remains below 100 meters which is safe and out of turtles range yet within target species range allowing longline fishermen to minimize encounters with turtles while maximizing their tuna catch (WWF Smartgear competition, 2005; Figure 154). However, this solution might be difficult to apply in the Mediterranean.

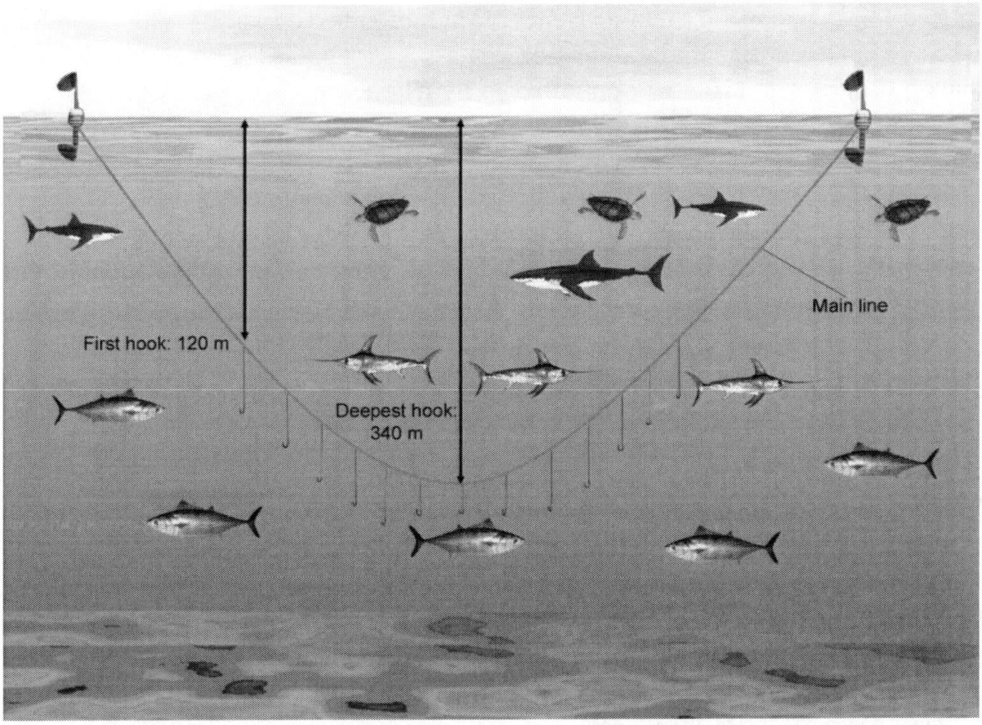

Figure 154. Bycatch species above the line at 100 m include sea turtles, sharks and some billfish; target species below the line at 100 m include bigeye tuna and day swimming broadbill swordfish. All baited hooks are below the 100 m line.

Rueda and Sagarminaga (2008) carried out tests with longlines positioned at different depths in the Balearic Islands. They found a reduction in turtle bycatch in the deeper longline but found also a reduction in target species catches. Preliminary results of project Life Nature 2003 – NAT/IT/000163 in the Ionian Sea seemingly indicate that most turtle bycatch occur when hooks are set between 10 and 15 m deep but more data are needed to confirm this. Other studies (Laurent et al. 2001) showed that the maximum depth at which marine turtles were caught was 60 m for swordfish longline and 20 m for albacore one.

Baez et al. (2007) in a research project carried out in the Balearic Islands (Western Mediterranean), observed that the probability of catching at least one loggerhead was related not only to the setting depth but also to distance of the fishing ground from the coast.

Jribi and Bradai (2008) stated that bottom longline has the potential to be more harmful than pelagic longlines given the fact that the caught animals do not have any chance to reach the surface to breath.

Sensory Stimuli

One potential approach to reduce marine turtle interactions with longline fisheries can be through the behaviour of turtles and the factors that lead them to interact with fishing gear. Southwood et al. (2008) overviewed sound, chemical, and light detection in turtles.

Fishermen have often suggested that hooks set closer to the floats have a higher probability of catching turtles accidentally. Experiments made at the "Delphynursery" of "Fondazione Cetacea" in Cattolica and at the Centro Recupero Tartarughe Marine in Linosa,

Italy aimed at evaluating floats attractiveness to turtles (Piovano et al., 2002). They tested turtle behaviour to lit and unlit white floats and the results seemed to indicate that the distribution of turtles in tanks was not affected by the presence of floats either lit or unlit. Therefore, the presence or absence of floats does not affect turtle behaviour. Moreover, the effects of bait colours (yellow, red and blue) and bait odour were tested in experiments with 27 loggerheads (22 immature and 5 adults) in open tanks. Colour attractiveness showed to be partially age dependent. Juveniles react differently to bait colours than sub-adults which even showed sharp individual differences. Young individuals never attacked a red-coloured sheath, while sub-adults seemed to apparently prefer this colour. Some other studies (Swimmer et al., 2002; Swimmer et al., 2005; Watson et al., 2002) showed that blue bait was not effective in reducing marine turtle capture rate than untreated bait.

Swimmer and Brill (2006) highlighted the importance of olfactory and acoustic stimulus in turtle reaction. Chemical cues can play a key role in marine turtle bite/no bite decision once a food item has been visually located. Piovano et al. (2004) showed that loggerheads had the ability to distinguish between fishing lures based on odour. In their experiment, two identical squid-shaped plastic lures were used with a small piece of fish hidden in one of them. The lure containing the fish showed greater attack frequency suggesting that olfactory cues are involved in prey recognition mechanisms.

Recent behavioural experiments indicate that lightsticks used in many longline fisheries attract turtles, and Rueda and Sagarminaga (2008) observed that most turtles were caught on hooks after sunrise. In non-Mediterranean experiments, Wang et al. (2006) found that lightsticks that flash intermittently, have wavelengths between 540 nm and 600 nm, or project only downward may be potential strategies that limit the attractiveness of lightsticks to juvenile turtles. The participants to the Working Group on bycatch/incidental catches (Rome, Italy, 15-16 September 2008) recommended to seriously considering the opportunity of banning lightsticks and any light source in pelagic longline fishery.

Regarding acoustic stimulus, deterrent experiments were carried out on 4 juveniles and 7 sub-adults of loggerhead turtles in a 10-m diameter open round tank at Cattolica (Italy) "Delphynursery". At frequencies between 50 and 400 Hz, some avoidance behaviour was observed with the maximum level of avoidance (20%) at 50 Hz. A "neutral" behaviour, where turtles react to the sound but not moving towards or away from the source, was observed in the range 50-700 Hz with highest levels (40%) between 50 and 100 Hz. No response was always observed at frequencies above 700 Hz. In any case during the experiments, the most frequent behaviour at all frequencies was "no response". These results though based on a small sample together with the high level of acoustic pollution in the Mediterranean did not encourage the continuation of this type of experiments.

9.3.2. Bottom Trawl

Mortality by trawling is mainly due to the forced apnoea during towing activity. The recovery of turtles after submergence takes several hours (Stabenau and Vietti, 2003) sometimes as long as 20 hours (Lutz and Dunbar-Cooper, 1987). The long recovery time is one of the main problems as turtles that experience multiple captures are likely to be even more susceptible to lethal acidosis (Lutcavage and Lutz, 1997). In order to reduce the time of submergence and then turtle mortality, the Turtle Excluder Device (TED) was proposed in the early 1980s. TED is a sort of grid that diverts large objects or animals like turtles towards a special exit positioned before the codend (Epperly, 2003; see chapter 9.1.1.6 for more details). TEDs have proved to be very effective mainly in prawn trawl fisheries and for this reason, several countries adopted TEDs as a mandatory management measure and bottom trawlers of Southeast Asia, South-America and Africa are obliged to use TEDs in their nets in order to export prawns to USA.

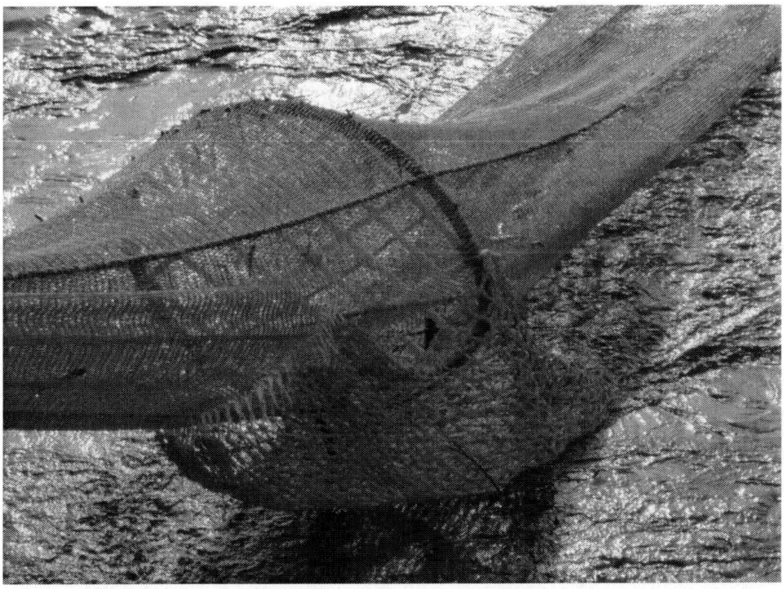

Figure 155. TED experiments in the Adriatic sea, a lower escape window was fitted to the grid.

Some authors (see Casale et al., 2004 for review) believe that TEDs currently available are probably not a realistic solution for reducing turtle bycatch in the Mediterranean mixed bottom trawl fisheries because they are designed for shrimp trawl fishery and they would also exclude the larger commercial specimens. In two recent experiments conducted in the Mediterranean trawl fisheries by Atabey and Taskavak (2001) and Lucchetti et al. (2008), however, Sala et al. (2011) found that TEDs can be properly proposed as a management tool for the conservation of marine turtle population in the Mediterranean, at least in certain periods and areas. Atabey and Taskavak (2001) tested the Supershooter TED in the shrimp fishery of Turkey and obtained very good results as not only loggerhead- and green turtles were excluded by the modified Supershooter but also unwanted incidental catches like jellyfish, sharks, and rays were also excluded. They found that most catches occur at depths between 11 and 30 meters and that the proportion of dead and comatose turtles resulting from trawls increases with towing time. Lucchetti et al. (2008) investigated the effect of different TED designs and materials on loggerhead turtle bycatch frequently captured in bottom trawl in the Adriatic fisheries. The best results were obtained with the TED called "Supershooter" which reduced the debris and improved fish quality in the catch. This might imply a reduction of additional sorting operations on board increasing time and costs. Authors in agreement with the doubts raised by Casale et al. (2004) stressed the fact that additional tests should be conducted in order to evaluate the loss of large fish.

Recently a flexible TED was tested in the Adriatic Sea by Lucchetti et al. (2014, unpublished data); the results are promising since this TED seems to avoid the catch of turtles while commercial loss are negligible.

9.4. REMARKS

Surface longlines, bottom trawls and illegal driftnets are the major threats to the survival of this species but the impact of fixed nets (gillnets and trammel nets) due to their high direct mortality should be carefully considered. Bottom trawls mainly impact turtle population in the wintering habitats, while longlines and fixed nets have their main impacts from spring to autumn with most captures occurring in summer.

Regarding surface longline, the efficiency of circle hooks in reducing turtle bycatch has been widely investigated in the Mediterranean fisheries. However further investigations of post-hooking mortality in loggerhead turtles would be very useful to determine the real effects of J and circle hook types on the overall impact of these fisheries on turtle populations. Fishermen highlight some difficulties in releasing turtles from circle compared to J ones; this issue might affect turtles survival and needs further investigation. Thus, it is possible to conclude that circle hooks have the potential of reducing turtle mortality only in certain fisheries and areas, and rigorous field tests are needed before requiring circle hooks as a mandatory measure in longline fisheries. However the use of circle is undoubtedly the most promising solution to reduce turtle bycatch in longline fisheries.

In addition to hooks, branchlines appear to be a major cause of marine turtle mortality. A relatively short piece of branchline left in the mouth of a released turtle can easily cause death if ingested and further research on branchline characteristics such as material, length, thickness etc. should be made to reduce turtle mortality.

The type of bait seems to be one of the main factors affecting turtle bycatch in longline. Changing the bait from squid to mackerel has been found to be effective in reducing incidental capture of loggerhead turtles with no significant effects on target species catch. Results obtained in different areas seem to suggest that the greatest reduction in turtle bycatch rates with the least effect on target species catch would be achieved by using only mackerel bait and J hooks instead of squid bait and circle hooks. A very simple, relatively cheap and effective method for reducing incidental turtle catches would be through using mackerel bait at least during periods when turtles are most abundant. It has also been suggested that using larger bait might deter or prevent turtles from swallowing it and, hence, the attached hook but this has to be tested (Gilman et al., 2006).

Moreover, some authors strongly suggested considering the banning of lightsticks and any light source that can strongly attract loggerhead turtles in pelagic longline fisheries but this extreme solution should be carefully tested in the field.

Tests on dyeing baits carried out in non-Mediterranean countries were effective with turtles in captivity under laboratory conditions but they do not appear to be effective as a mitigation measure in reducing turtle bycatch in longline fisheries since the reaction to different colours strongly depends on individual age and other factors like smell and it does not seem worthy to continue this kind of tests. Furthermore, the effects of physical factors; i.e., light penetration and colour absorbance with the depth, currents, oceanographic factors, temperature etc., make it very difficult to adopt different bait colours as a mitigation measure. Finally, these experiments were conducted within the confines of shallow pools where there was very little light attenuation and their results on colour and odour can't be extended to open sea conditions anyway. The isolation of a single turtle in a captive environment is another important factor to be taken in consideration here also. In conclusion, tank tests should be considered with very high caution.

Studies on bait odour effect on turtle bycatch highlight the importance of an integrated approach towards sensory deterrents as both visual and chemical cues are likely to attract marine turtles to longline gear and contribute to potentially harmful interactions (Piovano et al. 2004).

Experiments with acoustic deterrent show that habituation to acoustic signals is an important issue with respect to feasibility and long-term effectiveness of an acoustic deterrent (Moein et al., 1994). Results obtained in the Mediterranean but more generally worldwide taking into account the problems and difficulties associated with the use of acoustic deterrents showed that there is a low possibility that an acoustic signal could selectively deter marine turtles from interacting with longline gear without affecting commercial species. This does not encourage continuing these types of experiments.

Regarding bottom and mid-water trawl, TEDs could probably represent a suitable solution in these fisheries but only when properly matched to fishing conditions; i.e., TED angle, construction materials, floatation, position and size of the exit hole, webbing flap etc. The introduction of TEDs may reduce turtle mortality by avoiding the multiple submergences of a turtle through allowing an adequate rest interval at the surface where the animal might be able to recover from the acidosis by hyperventilating (Stabenau et al., 1991; Stabenau and Vietti, 2003). Additionally, TEDs could reduce the amount of discard and marine litter in codend catch leading to an improvement of fish quality and to a reduction of the sorting time. However, there might be a risk in increasing the losses of large commercial flat species such as turbot (*Psetta maxima*), flounder (*Scophthalmus r*hombus) and angler fish (*Lophius* spp). Lucchetti et al., (2014, unpublished data) recently tested a flexible TED that resulted highly efficient in reducing the catch of turtles while commercial loss are negligible.

Towing time is one of the main factors affecting mortality rate. Towing time would probably need to be to 10 min or less in order to achieve the negligible mortality of <1%. Obviously this is not a practical solution in our opinion. Regarding bycatch estimates, it's recommended to assess the impact of pelagic trawl on turtle population more in depth. Moreover, it's suggested to test TEDs in special fishing areas and fisheries like "sea cucumber grounds" of the middle Adriatic Sea and the prawn fisheries in Sicilian, Turkish and Tunisian waters.

Regarding passive nets, Casale et al. (2005a) stated that fixed nets may have the same impact as the bottom trawl nets. On the other hand, the direct mortality associated with these types of gears seems to be very high in comparison with other fishing gears but no practical solutions other than changing mesh size or twine thickness are available for set nets in the Mediterranean at this time. Nevertheless, the use of high hanging ratio that makes the nets tighter and the use of gillnets instead of trammel nets could reduce turtle entanglements. Further studies to develop excluder devices in set nets would be necessary.

One promising bycatch reduction technology relies on the differences in vision between sea turtles and many species of fish. Researchers have found that nets equipped with green light-emitting diodes (LEDs) or chemical lightsticks can reduce catch rates of the green sea turtle (*Chelonia mydas*). Research has also shown that green loggerhead (*Caretta caretta*) and leatherback (*Dermochelys coriacea*) sea turtles are sensitive to UV light, and although some fish species have UV vision, several economically-important fish species do not.

However, measures other than technical solutions such as spatial and temporal measures should be considered to reduce interactions with this gear. For example, some regions like the coastal areas of Turkey and Greece are characterised by the abundance of young turtles in certain seasons; i.e., after nesting, and avoiding such areas in these seasons would be a good practice. Finally, driftnets should be banned in all countries of the Mediterranean due to the very high hypothesized bycatch rate of these nets.

In general, the direct mortality on gear retrieval is often very low for most fishing practices except fixed nets but post-release mortality is expected to be very high. It is believed that further studies focused on the delayed mortality with video cameras, satellite tags etc. would be very useful understanding the efficiency of mitigation measures. For this reason particularly the question of turtle damage and behaviour after escape or release, international cooperation is important for achieving efficient management that ensures the survival of marine turtles while sustaining the economic benefits provided by fisheries.

Furthermore, some simple modifications to common fishing practices would be useful in reducing *C. caretta* bycatch. The increase of setting depth for longlines has been found to decrease the overall catch rates of turtles but it has led to an increase in mortality of turtles that are still hooked because they cannot reach the surface to breathe and die through drowning. Moreover, a reduction in turtle bycatch but coupled with a reduction in the catches of target species were observed in deep longlines. Distance from the coast is a parameter that fisheries managers should also take into account for the conservation of marine turtle populations. Limiting drifting longline effort to within 35 international nautical miles from the coast, at least in certain regions, could reduce loggerhead bycatch substantially without affecting swordfish catches and with little resistance from fishermen. For bottom trawl, the reduction of haul duration would be an effective operational measure for reducing direct mortality and occurrence of injured as well as weak and comatose individuals.

Fishermen agree on the importance of economic loss due to turtle interactions with fishing gear. Loss of hooks, bait, branch lines and other components of the gear and loss of time are economic concerns that need to be solved. The capture of turtles also leads to a decrease in fishing effort and yields on drift longlines as a result of reduction in the number of hooks and time necessary to repair or replace the gear. In any case, fishermen cooperation is essential for the survival of turtles after catch. Comatose individuals can survive or die depending on circumstances. Turtles caught are generally hauled on board in a very weak or comatose condition and should not be released immediately since they cannot swim to the surface to breathe and in these cases; the probability of drowning is very high. This problem can be substantially reduced by simply keeping the turtles on board and allowing them to recover. Therefore, it is fundamental to inform fishermen on the procedure to identify comatose turtles and release them only when they are healthy. Fishermen cooperation is also essential in removing the hooks from turtle's mouth. Hooks that are lodged externally like in jaw can be easily detected by fishermen and should be removed with the correct de-hooking device and actions. Fishermen intervention at this time can greatly affect the probability of survival (Piovano et al., 2009). Mediterranean guidelines for fishermen strongly recommend that fishermen cut the fishing line as close to the hook eye as possible (Gerosa and Aureggi, 2001). However, this is more likely for superficial hooking. Awareness campaigns, handling practices and dehooking devices given to fishermen so that they can cut branchlines and remove hooks whenever possible are valuable mitigation measures (Casale, 2008;).

In summary, reducing loggerhead bycatch in the Mediterranean will only be successful through a multidisciplinary approach including changes to fishing gear and practices, management policies (closed areas and seasons), turtle reaction behaviour to different stimuli and continued education and updating for fishermen. Therefore, a binding cooperation between the fishing industry, management bodies and research institutions is paramount in protecting this species especially for its survival after the catch.

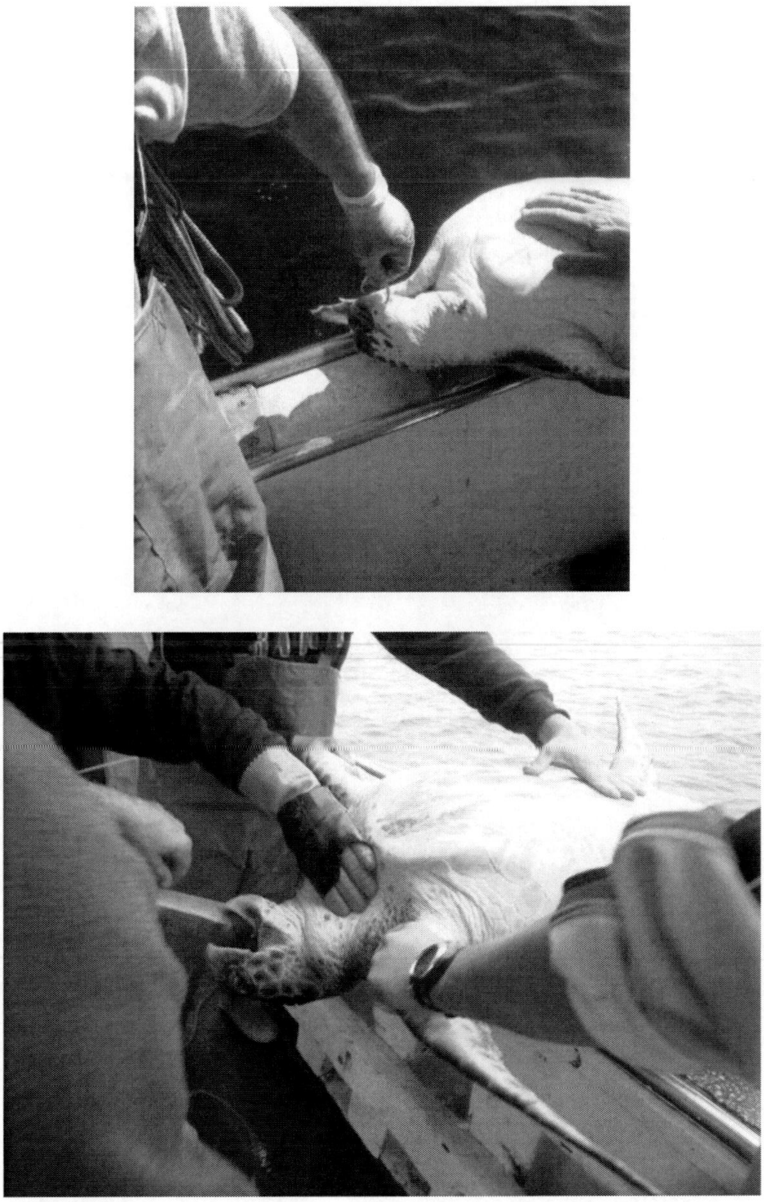

Figure 156. Sea turtle caught with a longline: de-hooking procedures.

Chapter 10

INTRODUCTION OF NEW TECHNOLOGY IN A FISHERIES

Two different types of advanced technology can be introduced in fisheries. Firstly, some technology can be adopted by fishermen on a voluntarily basis, e.g., fisherman wants to change its gears in order to catch more fish or to improve the energy performance (fuel reduction of the fishing vessel), that is a clear short-term benefit. Secondly, it may be necessary to introduce new technology to ensure a fishery is sustainable e.g., there could be a need to reduce the catch of juveniles by modifying the codend meshes or by introducing Bycatch Reducer Devices (BRD). In this case the benefits are usually assessed in the long terms, thus fishermen may need to be convinced since a longer-term benefit is not straight obvious.

The first type of change may need to be controlled or at least monitored (to avoid the possibility of over exploitation). The second type of change may usually need to be initiated by management; in this case the change is considered as an *up to bottom* imposition and often this is not the most effective solution, since fishermen need to be involved in the decision making process. The principle is that it is better to have rules that everybody agrees with, than to have rules that the majority ignores except the managers. Therefore, the best way to regulate a fishery is on a voluntary basis where everyone agrees to comply, even if enforcement is necessary.

Changes in gear and selectivity may have significant effects on catches, populations and hence landings. Therefore, three main issues should be taken into account before introducing a new technology:

- the practical problem of introducing a new gear;
- the immediate short-term effect on the economics of fishing (often a loss),
- the longer term effect of a change when the stock has responded (should be a benefit).

Furthermore when a new fishing method or practice needs to be introduced in a fishery, it should be considered that each fishery is different:

- Vessel design, size, power, e.g., there is a possibility that a new gear cannot be handled easily on small vessels, therefore the new gear dimensions should be calibrated in relation with the boat size, power etc.;
- Onboard technology, e.g., a new gear or practice might be introduced in a fishing vessel only if certain deck machinery is installed (winch, echosounder etc.);
- Gear design and size: major changes to existing net designs would cause difficulties in making new nets and repairing them;
- Fishing practices: the introduction of a new gear might imply a major changes to the common practice (fishermen must be trained up);
- Target species: the introduction of a new gear might involve a change in the species targeted; this affects both the resources as well as the economic revenue;
- Market conditions: changing the quantities of each species caught (increasing some and decreasing others) can affect the economics of the operation.

The introduction of a new technology can be affected by all these factors.

The following issues should be considered when introducing an innovation in fishing gears in order to address the technical change, to assess the efficacy of the new solution, to evaluate its reliability:

- what are the problems to be solved: eg high discards of young fish (if any which species), high discards of non-target species;
- What are the targets of this modification (eg reducing the catch of juveniles, reducing bycatch, reducing bottom impact etc.);
- Which of the possible technical solutions is best: new gears (e.g., less impacting gears, more selective gears), modified gears (e.g., change to the codend, bycatch reducer devices etc.)
- what are likely the consequences:
- Environmental impact (physical impact, type of bottom impacted)
- Selectivity (juveniles conservation, bycatch reduction, fishing mortality etc.)
- What are the consequences for each fish stock affected
- What are the consequences for the total effort applied
- What are the economic benefit (working time, fish quality etc.)
- What are the social consequences (e.g., number of fishermen employed)
- What is the time scale for the introduction of the new technology

Therefore, the introduction of new technologies or the modification of the traditional gears implies the knowledge of the actual situation, this can be done by reviewing the available information or by collecting new data on a fishery (commercial gears, fishing practices, market procedures etc.). The basic concept is that if there is a need to change to some better technology it is essential to know current technology e.g., if we want to introduce more selective gears we need to know which kind of codend and meshes are currently used (real situation, sometimes managers can have a biased view).

The new technology should fit some simple requirements to be accepted by fishermen and to be effective:

- Practical at sea (do not involve major changes to the common practices, easy to use and cheap to maintain)
- Acceptable for fishermen (economically viable)
- Acceptable for management (achieves the management and biological targets)
- Enforceable (easy to be controlled by fisheries inspection bodies)

Therefore, before introducing a new gear or practice there is a need to open a discussion among stakeholders (fishermen, net makers, suppliers, scientists, enforcement officers as well as managers and the environmental bodies), in order to know what they think about the new solution and what they suggest. Stakeholders need to be clearly informed on the capital costs (maintenance and running costs) on the effects on fishing operations. Meetings and publicity are considered as an essential step in order to make the new technology more acceptable.

Finally a reasonable timetable should be set in order to smoothly introduce a new technology; eg it is not feasible to introduce a new gear modification (such as a modification of the codend meshes) with only few months' notice. Fishermen need to "digest" even mentally the new solutions and the new management measures.

Chapter 11

FISHING VESSEL TECHNOLOGY

Awareness of the need to increase the efficiency of boats is reflected in the fact that more naval architects are being employed in design and construction of fishing vessels, and that professors of naval architecture and shipbuilding research organizations are starting to work on improving designs of fishing boats.

The actual panorama of fishing industry is complicated due to the combination of many factors. Overfishing and the economic crisis affect revenues, while the high fuel consumption of existing fishing vessels coupled with the continuous rise up of fuel price increase management costs so that it's non-profitable for many fishermen at the moment. Fishermen can only try to reduce management costs mainly affected by fuel consumption and cost. Fuel consumption can only be reduced by reducing fishing vessels energy usage. The energy used by a fishing vessel varies with its size and engineering, weather conditions, type and size of fishing gears, location, skill and knowledge. Similar fishing vessels could have different energy usage, due to different engineering such as different hull design and propulsion systems, different propeller type and size, different techniques and tactics.

From the point of view of fishing vessel dynamics, it is possible to define two major fishing activities on the basis of the fishing gear used and fishing technique. It is possible to discriminate between towed gears and static gears. All vessels need to sail to fishing grounds and then back to the harbour but fishing vessels that use towed gears (trawlers) must also make the trawling phase which is a very different dynamically compared with sailing phase.

Fishing vessel behaviour and performance in terms of power requirements for sailing and trawling phases are influenced by many parameters. After a brief description of most important dynamic parameters, some aspects of ship dynamics will be discussed. Power requirements on the basis of hull hydrodynamic resistance and power requirements for trawlers during trawling are defined. Furthermore, the main characteristics of propulsion system are discussed and the most important factors for optimum design are highlighted. Finally, some contributions and suggestions are proposed to optimize the design of fishing vessels.

11.1. Geometric and Mechanical Parameters of the Hull

A brief description of the most significant geometric and mechanical parameters of a fishing vessel hull is given below.

The *length overall (LOA)* is the distance between the extreme points of the bow and the stern and is the maximum length of the hull. The *length on the waterline (LWL)* is the distance between the extreme points of the bow and stern wet from the water. It is possible to define different LWL depending on the depth of the hull, related to the floating plan, which is influenced by the weights on board. The maximum LWL is measured when the vessel is fully loaded. The *length between perpendiculars (LBP)* is the distance between the perpendicular AV (bow) and the perpendicular AD (aft). The perpendicular bow is the one that passes through the point where the plane of the load waterline project intersects the stem. The aft perpendicular passes through the axis of rotation of the rudder. The length measurement is represented in Figure 157. For Mediterranean fishing vessels, LOA is 10% higher than LWL and LWL is 15% higher than LBP. An increase in length will increase frictional resistance but reduces wave-making resistance usually though this is complicated by the interaction of the bow and stern wave systems. Thus, fast ships will benefit overall from being longer than slow ships but there will be bands of length in which the benefits will be greater or less.

The *beam (B)* is the maximum transverse dimension of the hull, in m (Figure 158). Considering that higher beam allows for more space in the main deck, for high performances of the hull the beam should be 1/3 of the the waterplane length. The *depth (D)* is the vertical distance measured at half length between the building line and the straight line of the beam of the bridge deck (Figure 158). It is possible to define a bow depth and a aft depth, which can be different on the basis of the trim of the vessel. The difference between bow and aft depth is between 1.5 m and 2.5 m, depending on the lengh of the hull.

The cross-section at the maximum width is defined as *midship section* and *midship section area (A_M)* is the immersed midship section surface, m^2. The *waterplane area ($A_{W)}$* is the area of waterplane and it depends on the. The *volume of the underwater hull (∇)*, in m^3.

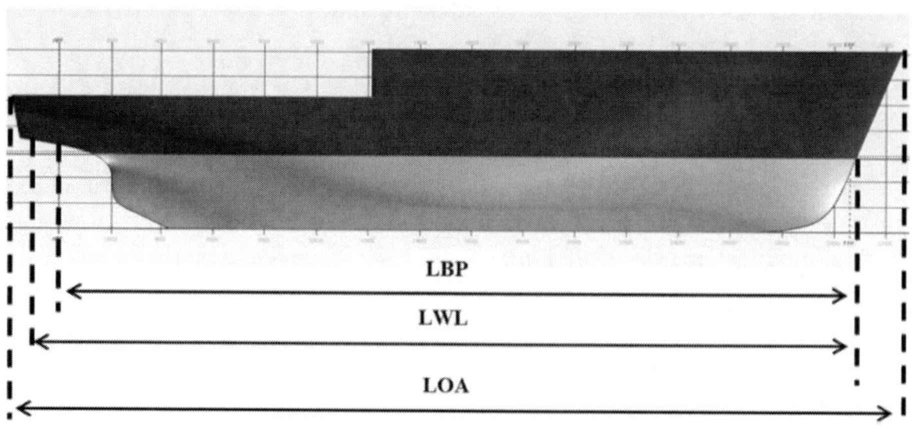

Figure 157. Length overall (LOA), length of the waterplan (LWL) and length betweeb perpendiculars (LBP) of the hull.

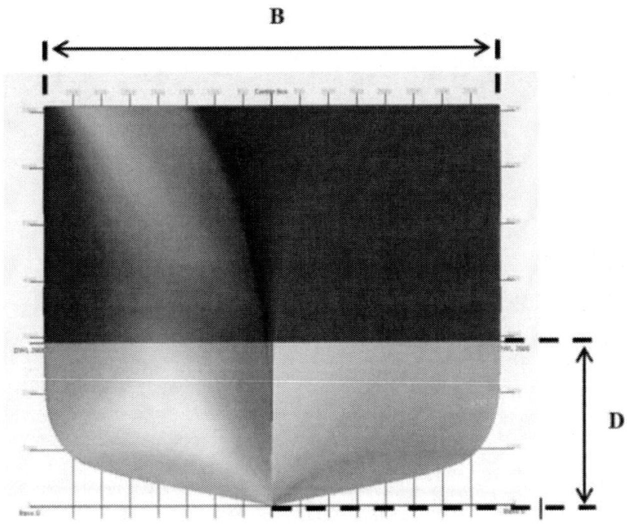

Figure 158. Beam and depth of the hull.

Mechanical Characteristics

The *displacement (Δ)* is the weight of the corresponding volume of water moved by the submerged part of the hull. On the basis of Archimedes principle, the weight of the ship is equal to the weight of the volume of water displaced by its hull. The relationship between volume and displacement of the ship (*Δ*):

$$\Delta = \gamma \nabla$$

Where γ is seawater specific weight ($\gamma = 1.026$ t/m^3).

Ship's displacement significantly affects its behaviour at sea. Displacement is a force and is expressed in newtons but the term mass displacement can also be used. Even when the main hull parameters have been fixed, it is possible to vary the distribution of displacement along the ship length. This distribution can be characterized by the longitudinal position of the centre of buoyancy (LCB). For a given block coefficient, the LCB position governs the fullness of the ends of the ship. As the LCB moves towards one end that end will become fuller and the other finer. There will be a position where the overall resistance will be minimized. This generally varies from just forward of amidships for slow ships to about 10% cent of the length aft of amidships for fast ships.

The *centre of buoyancy* is the mass centre of water displaced by the hull and is the point of application of hydrostatic vertical force (Archimedes thrust). It is different from the center of gravity. For fishing vessel it is in general 1 m upstair and in the same longitudinal section

The *tonnage* represents the capacity of the vessel in terms of volumes on board. Ton is derived from tun, which was a wine cask. The number of tuns a ship could carry was a measure of its capacity. The tonnage is a volume measure and for many years the standard ton was taken as 100 cubic feet. Two 'tonnages' are of interest to the international community, one to represent the overall size of a vessel and the other to represent its carrying capacity.

Gross Tonnage $$GT = K_1 \cdot V$$

Net Tonnage $$NT = K_2 \cdot V_C \left(\frac{4T}{3D}\right)^2 + K_3 \left(N_1 + \frac{N_2}{10}\right)$$

Where:

$V =$ total volume of all enclosed spaces of the ship in cubic metres

$K_1 = 0.2 + 0.02 \log_{10} V$

$V_C =$ total volume of cargo spaces in cubic metres

$K_2 = 0.2 + 0.02 \log_{10} V_C$

$K_3 = 1.25 \cdot \frac{GT + 10000}{10000}$

$D =$ moulded *depth* amidships in metres

$T =$ moulded draught amidships in metres

$N_1 =$ number of passengers in cabins with not more than eight berths

$N_2 =$ number of other passengers

$N_1 + N_2 =$ total number of passengers the ship is permitted to carry.

In using these formulae:

1. When N_1+N_2 less than 13, N_1 and N_2 are to be taken as zero.
2. The factor $(4T/3D)^2$ is not to be taken as greater than unity and the term K_2VC $(4T/3D)^2$ is not to be taken as less than 0.25 GT.
3. NT is not to be less than 0.30 GT.
4. All volumes included in the calculation are measured to the inner side of the shell or structural boundary plating whether insulation is fitted or not in ships constructed of metal. Volumes of appendages are included but spaces open to the sea are excluded.
5. GT and NT are stated as dimensionless numbers. The word ton is no longer used.

Due to differences between systems adopted by different countries in making allowances for machinery spaces, etc., there are many anomalies. Sister ships could have different tonnages merely because of their different flagships. To remove these anomalies and establish an internationally approved system, the International Convention on Tonnage Measurement of Ships was adopted in 1969. It came into force in 1982 and became fully operative in 1994. The Convention was held under the auspices of the International Maritime Organisation (IMO) to establish a universally recognized system for tonnage measurement. The two parameters of gross and net tonnage are used. Gross tonnage is based on the volume of all enclosed spaces. Net tonnage is the volume of the cargo space plus the volume of passenger spaces multiplied by a coefficient to bring it generally into line with previous calculations of tonnage. Each is determined by a formula.

Special tonnages are calculated for ships operating through the Suez and Panama Canals. They are shown on separate certificates and charges for the use of the canals are based on them.

Form Coefficients

The values of these coefficients can provide useful information about the ship form.

The *block coefficient (C_B)* is the ratio between hull volume and overall hull dimensions. It represents the ratio between the effective volume of the hull and the volume coming from the hull maximum dimensions (Figure 159). The block coefficient indicates whether the form is full or fine, and whether the waterlines will have large angles of inclination to the middle line plane at the ends. The angle at the bow is termed the angle of entry and influences resistance. For Mediterranean fishing vessels the block coefficient is about 0.45-0.55 (Messina and Notti, 2007).

The *waterplan surface coefficient (C_{WL})* is the ratio between the the effective horizontal surface of the hull at the maximum depth and the horizontal maximum dimensions. This coefficient is mainly related to the volumes and spaces on board.

The *prismatic coefficient (C_P)* is the ration considers the effective hull volume at the maximum depth and the volume generated considering the submerged part of the midship cross section and the waterplane length. This parameter is effective in the hull behaviour and for Mediterranean fishing vessels it is usually from 0.50 to 0.65 (Messina and Notti, 2007). The main effect is on wave-making resistance and, therefore, the choice of prismatic coefficient is not so important for slow ships where it is likely to be chosen to give better cargo carrying capacity. For fast ships, the desirable prismatic coefficient will increase with the speed to length ratio.

$$C_B = \frac{\nabla}{L_{WL} \cdot B_{WL} \cdot T}$$

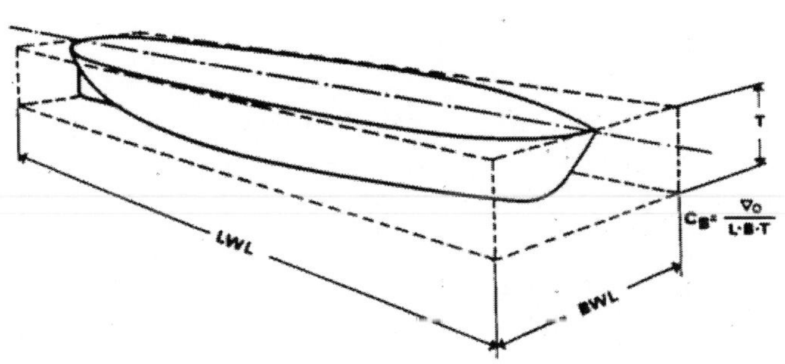

Figure 159. Block coefficient (from Tupper et al., 1996).

$$C_{WL} = \frac{A_{WP}}{L_{WL} \cdot B_{WL}}$$

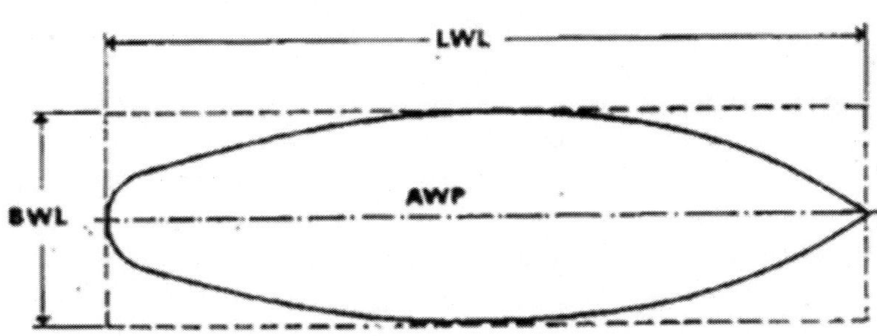

Figure 160. Waterplane coefficient (from Tupper et al., 1996).

Prismatic coefficient

$$C_P = \frac{\nabla}{A_M \cdot L}$$

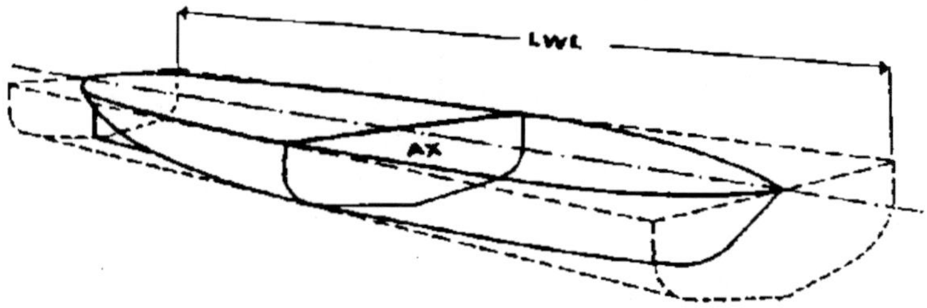

Figure 161. Longitudinal coefficient (from Tupper et al., 1996).

11.1.1. Interaction between Main Hull Metrics and Resistance

There can be no absolutes in terms of optimum form. The designer must make many compromises. Even in terms of resistance, one form may be better than another at one speed but inferior at another one. Another complication is the interdependence of many form factors including those chosen for discussion below. In this discussion, only general comments are possible. Frictional resistance is directly related to the wetted surface area and any reduction in this will reduce skin friction resistance. This is not, however, a parameter that can be changed in isolation from others. Other form changes are likely to have most effect on wave-making resistance but may also affect frictional resistance because of consequential changes in surface area and flow velocities around the hull.

Fullness of Form

Fullness may be represented by the block or prismatic coefficient. For most ships, resistance will increase as either coefficient increases. This is reasonable as the full ship can be expected to create a greater disturbance as it moves through the water. There is evidence of optimum values of coefficients on either side of which resistance might be expected to rise. This optimum might be in the working range of high-speed ships but is usually well below practical values for slow ships. Generally, the block coefficient should be reduced as the desired ship speed increases. In moderate-speed ships, power can always be reduced by reducing block coefficient so that machinery and fuel weights can be reduced. For given overall dimensions, however, a lower block coefficient means lower payload. A balance must be struck between payload and resistance based on study of the economics of running the ship.

Slimness

Slimness can be defined as the ratio of length to the cube root of volume of displacement; i.e., this is Froude's circular M, or in terms of a volumetric coefficient which is the volume of displacement divided by the cube of the length. For a given length, greater volume of displacement requires steeper angles of entrance and run for the waterplane endings. In high-speed forms with low block coefficient in general, the displacement length ratio must be kept low to avoid excessive resistance but this is not so important for slow ships,. Fast ships require larger length to beam ratios than slow ships.

Breadth to Draught Ratio

Generally, resistance increases with increase in breadth to draught ratio within the normal working range of this variable. This can again be explained by the angles at the ends of the waterlines increasing and causing a greater disturbance in the water. With very high values of beam to draught ratio, the flow around the hull would tend to be in the vertical plane rather than the horizontal one. This could lead to a reduction in resistance.

Length of Parallel Middle Body

In high-speed ships with low block coefficient, there is usually no parallel middle body. In ships of moderate and high block coefficient, parallel middle body is needed to avoid the ends becoming too full. For a given block coefficient, the ends become finer as the length of parallel middle body increases and vice versa. Thus, there will be an optimum value of parallel middle body for a given block coefficient.

Section Shape

It is not possible to generalize on the shape of section to adopt, but slow to moderate speed ships tend to have U-shaped sections in the fore body and V-shaped sections aft. It can be argued that the U-sections forward keep more of the ship's volume away from the waterline and so reduce wave-making.

Bulbous Bow

The principle of the bulbous bow is that it is sized, shaped and positioned so as to create a wave system at the bow which partially cancels out the ship's own bow wave system, so reducing wave-making resistance. This can only be done over a limited speed range and at the expense of resistance at other speeds. It was originally applied to moderate to high speed ships but has also been found to be beneficial in relatively slow ships such as tankers and bulk carriers and these ships now often have bulbous bows. The effectiveness of the bulb in the slower ships where wave-making resistance is only a small percentage of the total suggests that the bulb reduces frictional resistance as well. This is thought to be due to the change in flow velocities which it creates over the hull. Sometimes the bulb is sited well forward and it can extend beyond the fore perpendicular.

11.2. Energetic Implications in Fisheries

As ship moving through water and air it experiences both water and air forces. Water and air masses may themselves be moving; water due to currents and air due to wind. These will in general be of different magnitudes and directions. Unless winds are strong, water resistance will be the dominant factor in determining the speed achieved. Resistance determines the thrust required of the propulsion system. Then, propulsion deals with providing this thrust and the interaction between the propeller and flow around the hull.

11.2.1. Hull Resistance

When a moving body is near or on the free surface of a fluid, pressure variations around it are manifested as waves on the surface. Energy is needed to maintain these waves and this leads to resistance. Also, all practical fluids are viscous and movement through them causes tangential forces opposing the motion. Because of the ways they arise, the two types of resistance are known as "*wave-making resistance*" and "*viscous or frictional resistance*". Because flow variations and eddies are created by the particular ship form, this resistance is sometimes linked to the *form resistance*. The ship has a number of appendages each of which has its own characteristic length and it is best to treat their resistances separately from those of the main hull since they can generate each type of resistance associated with the hull. Collectively, they form the *appendage resistance*.

Wave-making Resistance

A body moving on an otherwise undisturbed water surface creates a varying pressure field that manifests itself as waves because pressure at the surface must be constant and equal to atmospheric pressure. When the body moves at a steady speed, the wave pattern seems to remain the same and move with the body. The energy for creating and maintaining this wave system must be provided by the propulsion system. At the same time waves, cause a drag force that must be opposed by the propulsion system if the ship is not to slow down. This drag force is the *wave-making resistance*.

The Wave Pattern

The pressure field around the ship can be approximated by a moving pressure field close to the bow and a moving suction field near the stern. Both forward and after pressure fields create its own wave system as shown in Figure 162.

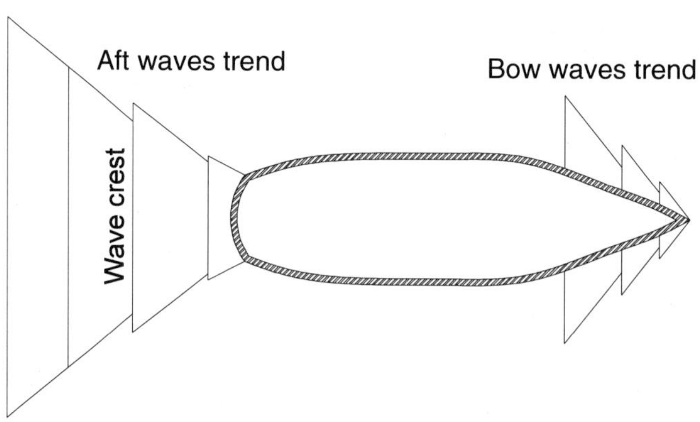

Figure 162. Bow and stern wave systems.

Interference Effects

In addition to waves created by the bow and stern, others may be created by local discontinuities along the ship's length. However, the qualitative nature of interference effects in wave-making resistance is illustrated by considering just the bow and stern systems. The transverse waves from the bow travel aft relative to the ship reducing in height. When they reach the stern-generated waves they interact with them. If crests of the two systems coincide, the resulting wave is of greater magnitude than either because their energies combine. If the crest of one coincides with a trough in the other, the resultant energy will be less.

Frictional Resistance and Hull Roughness

Due to viscosity, the particles immediately adjacent to the hull adhere to it and move at the speed of the ship. At a distance from the hull the water is at rest. There is a velocity gradient that is greatest close to the hull. The volume of water that moves with the body is known as the boundary layer. Its thickness is usually defined as the distance from the hull at which the water velocity is 1% of the ship speed. Frictional resistance is associated with Reynolds who studied flow through pipes. He showed that there are two distinct types of flow. In the first known as laminar flow, each fluid particle follows its own streamlined path with no mass transfer between adjacent layers. This flow only occurs at relatively low Reynolds numbers. At higher numbers, the steady flow pattern breaks down and is replaced by a more confused flow pattern called turbulent flow.

Calculations have been made for laminar flow past a flat surface at speed V, length L and wetted surface area S, and these have led to the formula developed by Blassius, as:

$$Specific\ resistance\ coefficient\ (C_F) = \frac{R_f}{\frac{1}{2}\rho S V^2}$$

Depending on the particular surface, it is possible to find different C_F.

It is clear that apart from the wetted surface area and speed, the major factor in determining the frictional resistance is the roughness of the hull. This is why so many researchers have devoted so much time to this factor. For slow ships, frictional resistance is the major proportion of total resistance and it is important to keep the hull as smooth as possible. Due to increase in boundary layer thickness, the ratio of a given roughness amplitude to boundary layer thickness decreases along the length of the hull. Protrusions have less effect at the aft end than forward. Roughness affects C_F coefficient and it can be considered under three headings:

1. *Structural roughness.* This depends upon design and method of construction. In a riveted ship, plate overlaps and edges and rivet heads constituted roughness. These are avoided in modern welded construction but in welded hulls, plating exhibits waviness between frames, particularly in case of thin plating, which is also a form of roughness.
2. *Corrosion.* Steel corrodes in seawater creating a roughened surface. Modern painting systems are reasonably effective in reducing corrosion all the way while the coating remains intact. If it is abraded in one area then corrosion is concentrated at that spot and pitting can be severe. This is bad from the structural point of view as well as for frictional resistance. Building ships on covered slipways and early plate treatments to reduce corrosion both reduce initial hull roughness on completion of build. To reduce corrosion during build and in operation, many ships are now fitted with cathodic protection systems either active or passive.

3. *Fouling.* Marine organisms such as weed and barnacles can attach themselves to the hull. This would represent a very severe roughening if steps were not taken to prevent it. Traditionally, the underwater hull has been coated with anti-fouling compositions. Early treatments contained toxic materials such as compounds of mercury or copper which leached out into the water and prevented the marine growth taking a hold on the hull. Unfortunately these compounds also pollute the general ocean and other treatments are now used. Fouling is very dependent on time a ship spends in port relative to its time at sea and the ocean areas in which it operates. Fouling increases more rapidly in port and in warmer waters. During towing trials, it was found that frictional resistance increased by about 5% over 40 days, that is by about of 1% per day. This is a common allowance applied to a ship's frictional resistance to allow for time out of dock.

For an operator, the deterioration of the hull surface with time results in a slower speed for a given power or more power being needed for a given speed. This increases running costs which must be set against the costs of docking, cleaning off the underwater hull and applying new coatings.

Eddy Making Resistance or Viscous Pressure Resistance

In a non-viscous fluid, the lines of flow past a body close behind it creating pressures that balance out those acting on the forward part of the body. With viscosity, this does not happen completely with pressure forces after the body being lesser than those on its fore. The flow also breaks away from the hull where there are rapid changes of section and eddies are created. These effects can be minimized by streamlining the body shape so that changes of section are more gradual. However, many features are likely to generate eddies. Transom sterns and stern frames are examples. Other eddy creators can be appendages such as bilge keels, rudders and so on. Bilge keels are aligned with smooth water flow lines as determined in a circulating water channel to minimize the effect. At other loadings and when the ship is in waves, bilge keels are likely to create eddies. Similarly, rudders are made as streamlined as possible and the breakdown of flow around them is delayed by this means till they are put over to fairly large angles. Flow break away can occur on an apparently well rounded form. This is due to die velocity and pressure distribution in the boundary layer. Velocity increases where pressure decreases and vice versa. Bearing in mind that water is already moving slowly close into the hull, pressure increase towards the stern can bring the water to a standstill or even cause a reverse flow to occur. That is the water begins to move ahead relative to the ship. Under these conditions, separation occurs.

Appendage Resistance

Appendages include rudders, bilge keels, shaft brackets and bossing, and stabilizers. Each appendage has its own characteristic length and, therefore if attached to the model, would be running at an effective Reynolds' number different from that of the main model. Thus, although obeying the same scaling laws, its resistance would scale differently to the full scale. That is why resistance models are run naked. This means that some allowance must be made for the resistance of appendages to give the total ship resistance. Allowances can be obtained by testing appendages separately and scaling to the ship. Fortunately the overall additions are generally relatively small, say 10 to 15 % of the hull resistance and errors in their assessment are not likely to be critical.

Wind Resistance

In conditions of no natural wind, air resistance is likely to be small relative to water resistance. When wind is blowing, fore and aft resistance forces will depend upon its direction and speed. If coming directly from ahead, the relative velocity will be the sum of wind and ship speed. The resistance force will be proportional to the square of this relative velocity.

Trawling Resistance

For fishing vessels that use towed gears, fishing activity is characterized by two basically different operating conditions corresponding to sailing and trawling phases. Hauling and shooting phases are not relevant from the point of view of the vessel design in terms of power and main dimensions. One of the most important characteristics of fishing vessel is the trawling phase. Trawling is generally made at low speed (2.5–5 kn) with vessel towing the fishing gears at the same speed. The overall drag is mainly depending on gear drag as hull resistance is not predominant at low speed and it could be assumed as up to 10% of the gear drag.

In Figure 163, the resistance against speed for sailing and trawling phases of a 30 m Loa trawler are represented in terms of specific resistance against specific vessel speed. For both trawling and sailing phases, the value 1 in the horizontal axis represents the maximum speed for the vessel while for the vertical axis it represents the resistance. It is possible to note that the propulsion system should face the same resistance ad very different vessel speed among the trawling and sailing conditions.

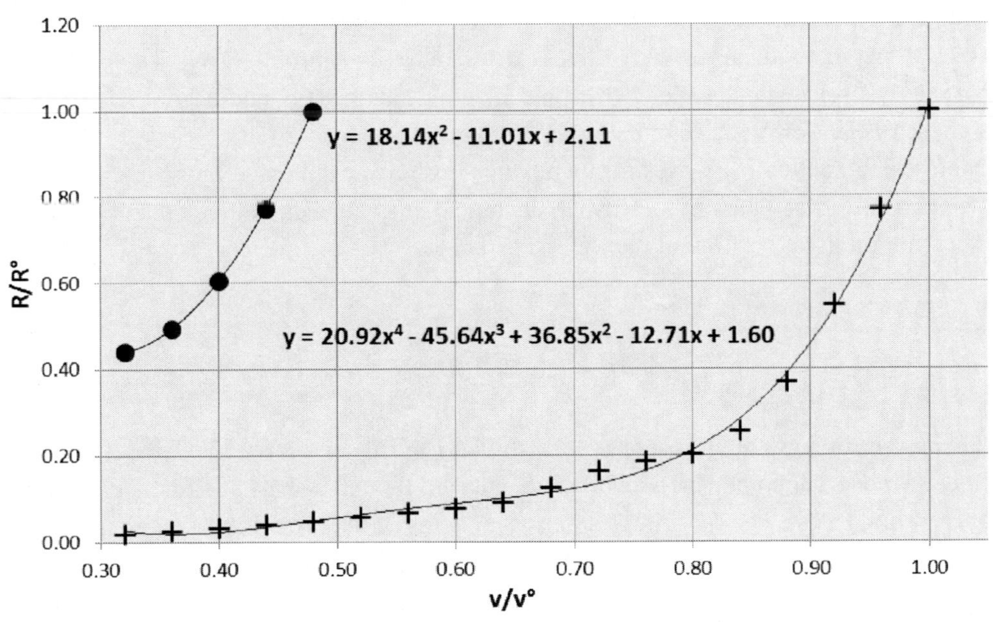

Figure 163. Resistance curve during trawling and sailing of a bottom otter trawler. The resistance R over the maximum resistance R* is variable with the speed v over the design speed v*(Notti and Sala, 2012).

11.2.2. Resistance Estimation

The model or data obtained from model experiments is still the principal method used. The principle followed is that stated by Froude. Ship resistance can be obtained from that of the model by:

1. Measuring total model resistance by running it at the corresponding Froude number;
2. Calculating frictional resistance of the model and subtracting it from the total leaving the residuary resistance;
3. Scaling the model residuary resistance to the full scale by multiplying by the ratio of the ship to model displacements;
4. Adding a frictional resistance for the ship calculated on the basis of the resistance of a flat plate of equivalent surface area and roughness;
5. Calculating or measuring separately the resistance of appendages;
6. Making an allowance if necessary for air resistance.

Apart from tests of individual models, a great deal of work has gone into ascertaining the effect of hull form on resistance. The tests start with a parent form and then varying systematically a number of form parameters that are considered to be likely significant. Such a series of tests is called a methodical series or a standard series. The results can show how resistance varies with the form parameters used and are useful in estimating power for new designs before the stage has been reached at which a model can be run. To cover (n) values of (m) variables (m^n) tests would be required; so work and time involved can be enormous. In planning a methodical series, great care is needed on deciding the parameters and range of variables. One methodical series is that carried out by Admiral D.W. Taylor. He took variables as the prismatic coefficient, displacement to length ratio, and beam to draught ratio. With eight, five and two values of the variables respectively, he tested 80 models.

More recent methodical series for merchant ships have been made by BSRA and DTMB. The former considered block coefficient, length to displacement ratio, breadth to draught ratio, and longitudinal position of the LCB.

11.3. THE PROPULSION SYSTEM

The propulsion system generates the thrust needed to propel the vessel during all activities. An internal combustion engine is connected though a shaft line to the propeller that runs generating thrust. The most important components of the propulsion system are the main engine and the propeller. Depending on the type of engine adopted, a reduction gear could be necessary to move the propeller. An auxiliary engine is connected for emergency uses but cannot run the propulsion in case of problems with the main engine.

11.3.1. Propulsion System Layout

In Figure 164, a high-speed marine diesel engine (1) drives the intermediate shaft (5) through a reduction gear (4). Generally, a fixed pitch propeller (7) running at different speed produces the required thrust for the different fishing phases. The power take-off forward is used for other energy users, such as hydraulic or electric users. The requested thrust is obtainable controlling the engine revolution speed, thus the propeller rotational speed.

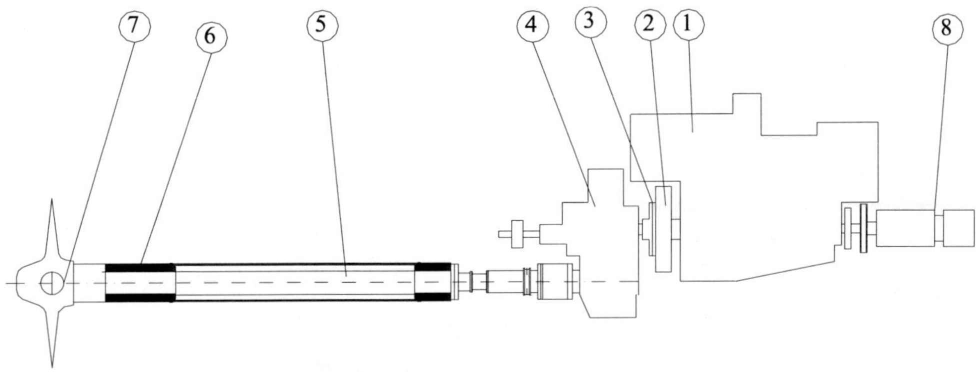

Figure 164. Standard propulsion system of a fishing vessel: 1) diesel engine, 2) flywheel, 3) coupling flange, 4) reduction gear, 5) intermediate shaft, 6) stern tube, 7) fixed pitch propeller; 8) Power take-off forward (Notti and Sala, 2012).

The engine room for older fishing vessel is shown in Figure 165 where the main engine is the primary engine providing power for all the energy users. An auxiliary engine is ready for emergency services such as water pumps, electricity for navigation instrumentation and light on board.

In new vessels (Figure 166) power generation for propulsion and hydraulic users, such as rope and net winches and bow thrusters, comes from the main engine, while an auxiliary engine provides for electricity and other users on board. A second auxiliary engine is available to replace the first one and for emergency services.

11.3.2. Propulsion System Efficiency Chain

In Figure 167 is described how the power is transferred from the main engine to the propeller:

P_A is the brake power, the overall mechanical power generated by the main engine for propulsion;

P_B is the power measured after the reduction box; it is affected by the reduction box efficiency;

P_D is the net power delivered to the propeller;

P_T is the net power developed by the propeller, affected by the propeller efficiency;

P_E is the effective power generated, considering the R_T and the V_A, abovementioned.

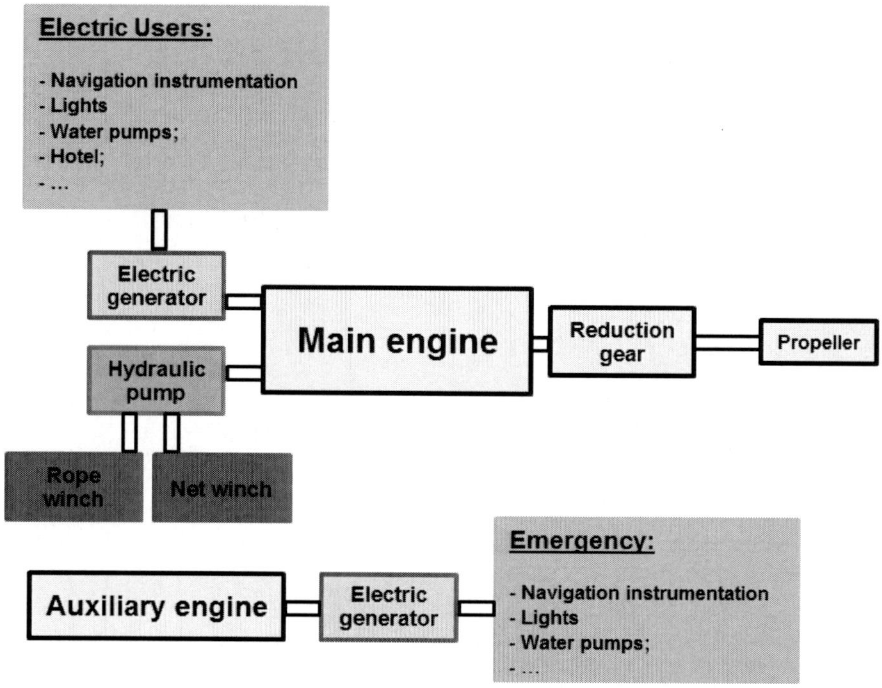

Figure 165. Standard engine room layout. In older vessels the main engine provides power for propulsion and for hydraulics such as rope winch or net winch, and for electric users like hotel, water pumps, light etc. an auxiliary engine is ready for emergency services.

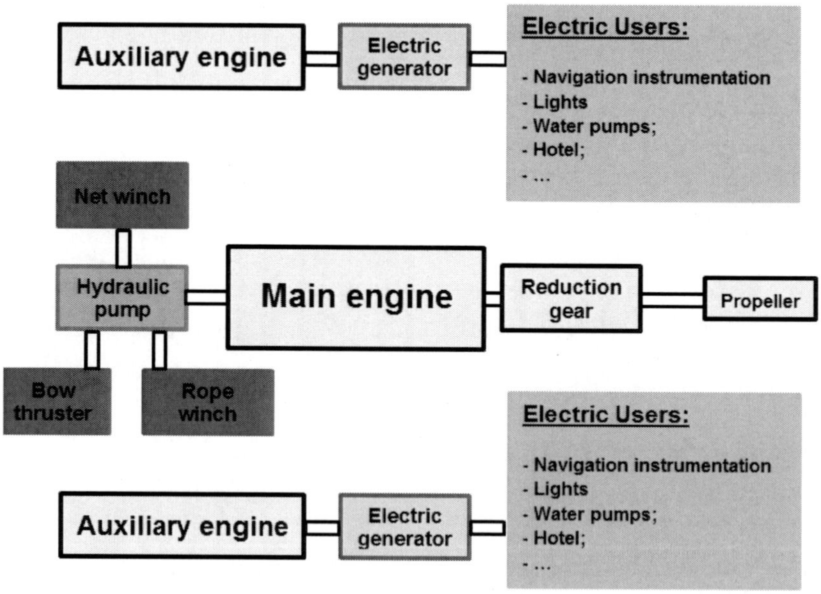

Figure 166. New engine room layout. New vessels usually are designed with the main engine providing power for propulsion and for hydraulics such as rope winch or net winch, while an auxiliary engine is devoted to electric users like hotel, water pumps, light etc. Another auxiliary is ready for replacing the first one and for emergency services.

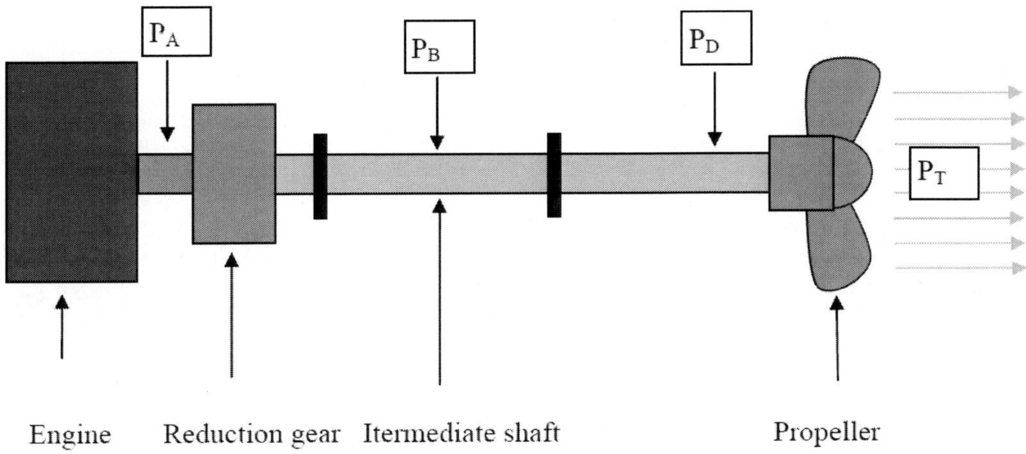

Figure 167. Power transmission through the elements of the propulsion system.

The effective power P_E is defined by the relation:

$$P_E(hp) = \frac{R_T \cdot V_A \cdot 0.514}{75}$$

where:

R_T is the overall resistance in kgf;
V_A is the advance speed in kn.

The efficiency chain helps to define effective power expressed by the propulsion system P_E depending on the power installed on board P_B and the propulsion system layout.

$$\mathbf{P_E} = \eta_H \cdot P_T = \eta_H \cdot \eta_P \cdot P_D = \eta_H \cdot \eta_P \cdot \eta_m \cdot P_A = \eta_H \cdot \eta_P \cdot \eta_m \cdot \eta_r \cdot P_B$$

$$\mathbf{P_E} = \eta_T \cdot P_B$$

where:

η_H is the hull efficiency, as defined above;

η_P *is* the propeller efficiency: normally it *is* 0.55 ÷ 0.6;

η_m is the shaft efficiency and It takes into account bearing and contact connections; normally the range is 0.98 ÷ 0.99;

η_r is the reduction gear box (if any) and it is about 0.97 ÷ 0.98;

η_T is the overall propulsion system efficiency, generally 0.53 ÷ 0.58.

The propulsion system efficiency is mainly affected by the propeller efficiency.

11.3.3. Main Engine

Today diesels provide the overwhelming majority of propulsion and electrical generating power for the world's fishing fleets. With such complete dependence on diesels, it is little wonder that the record high diesel fuel prices of 2008 caused a near panic in the commercial fishing industry (Alaska Sea Grant Marine Advisory Program asg-52 2010).

In general, fishing vessels use 4–strokes high speed diesel engines instead of low or medium speed diesel engines used in the past. Modern engines are overcharged with a range of power from few dozens to thousand kW according to size of vessel. Technological advances have resulted in an overall increase in performance up to 50%. One of the main advantages of high speed diesel engines is their lower weights and sizes compared to low speed ones.

Supercharging is arranged upstream of the intake valve through an air compressor that can be controlled mechanically (mechanical supercharging) or moved by a turbine (turbocharger) fed with flue gas coming from cylinders to flow into a manifold where pressure tends to assume a constant value, then escapes to atmosphere. The result is a higher volume of air generating higher compression in cylinders, thus more power is generated at each cycle.

Increase of air pressure, however, leads to an increase in temperature of the fluid associated with a decrease of its density. To overcome this, the charge is cooled through a suitable heat exchanger (intercooler) placed downstream of the compressor offering the following advantages:

- Reducing temperature levels reached, thus lowering thermal loads on the engine; and
- Improving mechanical efficiency since power derivative is increased without significantly changing pressure levels.

As a rule of thumb, power increases 3.5% for a 10°C decrease of air temperature in a medium-size diesel engine.

On the basis of official data sheets provided by major engine manufacturers (A, B, C, D), Figure 168 shows that the maximum fuel consumption (FC) is directly proportional with the maximum brake power (BP) which represents the maximum power deliverable to the propeller. It is obvious that the higher the power requested, the higher the fuel consumption.

As shown in Figure 168, the relation between power installed on board and fuel consumption at the maximum power suggests that the power to be installed should be minimum as possible, considering that powerful engines running at partial loads has lower efficiency. It is crucial to reduce as much as possible the amount of power needed and to choose the engine that is able to deliver the power request at an high efficiency condition. Despite this obvious conclusion, many owners prefer to install more power in order to make possible to increase propulsive performances, without considering the lowering of engine efficiency at partial loads.

For each of the engine manufacturers considered, Table 7 shows the range of brake power (BP) available for heavy duty engines, together with the maximum fuel consumption declared (FC) and the specific fuel consumption (SFC). On the basis of the minimum specific fuel consumption, higher specific fuel consumption are reported as percentage, (Notti et al., 2013). The specific fuel consumption represents the efficiency of the engine, as the higher the

SFC, the higher the fuel consumption required by the engine in order to generate the same power.

From Table 7 it could be observed that the range of specific fuel consumption is 152 to 207 g/kWh. The opinion of engine manufacturers interviewed is that, after a huge development which allowed for a reduction in specific fuel consumption in a range of 15-25%, actual efforts in the development of new engines are focused on issues like emissions reduction engine durability and reliability. In their opinion further sensible improvements in terms of fuel consumption (more than 3–5% at each new engine concept) is not comprised in the short term. This approach could represent an opportunity for the development of new engine generations, such as LNG fuelled engines and for the introduction of new fuels like diesel from algae and biodiesel.

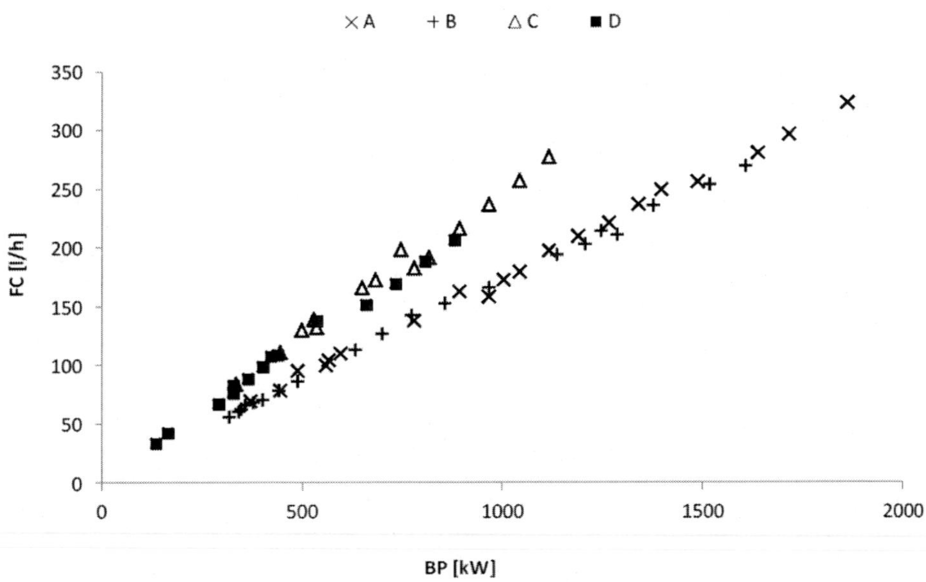

Figure 168. Relationship between brake power and fuel consumption for the major marine diesel engines installed on board fishing vessels (BP= brake power in kW; FC= fuel consumption in l/h; A, B, C, D represents four major engine manufacturers, identified with generic symbols).

11.3.4. Reduction Gearbox

A medium-sized diesel engine runs at a speed that, in most cases, exceeds that required by the propeller. For the reduction of rotational speed, a reduction gear is inserted between the main engine and the intermediate shaft. Reduction gear boxes are necessary to couple high speed engines with propeller. As a mechanical transformer, the reduction gear box is necessary for modify torque and rotational speed of the main engine in order to allow the propeller to operate properly. The parameter that describes such transformation is an a-dimensional parameter named reduction ratio (R). From the main engine to the propeller, the torque is increased, while the rotational speed is decrease, by a factor corresponding to the reduction ratio. Reduction gearboxes usually have a unique reduction ratio, obtained as a compromise between trawling and sailing conditions. Such reduction devices are commonly named "one-speed" reduction gear boxes. The efficiency is generally high (0.98 – 0.99) so the they not affect sensibly the propulsion system efficiency from the point of view of the energy loss. Nevertheless a proper reduction gear is important for a good matching of the engine and the propeller.

Table 7. Fuel consumption values for the range of power considered

Engine Manufacturer	PB	FC	SFC	%
	[kW]	[l/h]	[g/kWh]	
A	373 - 1864	69 - 323	152	2%
B	320 - 1610	56 - 269	149	0%
C	339 - 1118	84 - 277	214	44%
D	136 - 883	33 - 206	207	39%

For each engine manufacturer the range of power of marine diesel engines mostly used on board fishing vessels is compared with the fuel consumption (FC) at the maximum power and the specific fuel consumption (SFC). Referred to the minimum specific fuel consumption, the higher specific fuel consumption of other engines is reported as percentage (A, B, C, D represents four major engine manufacturers, identified with generic symbols).

11.3.5. Propeller

The propeller is the terminal element of the propulsion system. Through the propeller, the mechanical energy provided by the main engine is transformed to thrust needed to overcome resistance. The blades are generally made of bronze alloy, copper and zinc, with some other elements like nickel, aluminium, manganese etc. Robustness and resistance to corrosion and wear are essential given their continuous operation in seawater. Shape, profiles and the number of the blades of a propeller can be highly variable depending on their intended use. The choice of shape to be assigned to the blade in its radial extension and the profile of the section of the blade itself are a remarkably complex problem. Generally, a fixed blade propeller is used in many fishing vessels. The term "fixed blade" (sometimes also defined “fixed pitch”) is related to the fact that by assimilating the propeller to a body that moves screwed in the water, its pitch remains constant, and is a characteristic imprinted propeller same being created and not more modifiable during the exercise, but only with interventions to modify the shape of the blades.

Propeller Main Parameters

Depending on applications, marine propellers (Figure 170) have three, four or even more *blades* (1) arranged at equal angular intervals around a *hub* (4) to which they are connected by fusion. An observer located aft propeller faces only the *active face* of each blade and the opposite side is *called back*. The two faces converge in the *rim* of the blade that is distinguished in *input flap* (2) and *outlet flap* (3). The intersection of the active face of the blade with the diametric plane passing through the point of the blade farthest from the axis is called *generator* (6). This generator can be straight or curved. In general, it is inclined with respect to the vertical axis by a certain angle (5), said *rake*.

Figure 169. Fixed blade propeller.

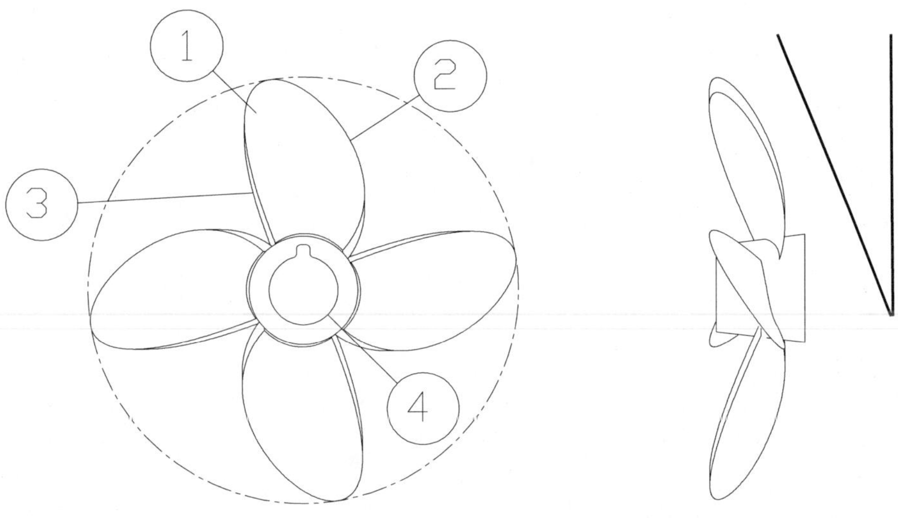

Figure 170. Geometric characteristics of the propeller.

The main geometric dimensions are:

- Diameter [D] is the diameter of the circle propeller;
- Area of the disk [AD]: the area of a circle of diameter;
- Projected area of the blades [AP]: is the area of the projection of the blades on a plane perpendicular to the axis propeller;
- Developed area of the blades [AS]: Dissecting the propeller blades with a cylinder coaxial with the hub, loops are obtained. For each radius r it is defined the segment equivalent to loop obtained. Thus, the developed area of the blade is obtained.

Summing the developed areas of each blade, the developed area of the propeller is calculated;

- Direction: The direction of rotation can be clockwise or counter-clockwise; and
- Number of blades [Z]: number of blades depending on thrust characteristics of the propeller.

11.3.6. Interaction Hull–propeller

Once the effective power (EP) of the hull at various speeds is determined, the main engine power to be transferred to the propeller to generate thrust able to propel the ship at those speeds should be also determined. The power generated by the main engine to the propeller is used in an amount proportional to its performance and that of the hull itself.

Once the propeller is in interaction with the hull, the performance of the fluid flow that involves the propeller is disturbed by the presence of the hull and its appendages. The hull, in its movement, entrains a strip of water due to the viscous friction of the fluid; to it, in the stern area, overlap shear waves and noticed as a turbulent wake and swirling. The wake of water, which affects the area of operation of the propeller, has a speed V_W, referring to the mainland, generally less than that of the vessel V, and is opposed to it. The speed of the wake V_W is usually expressed in function of the speed of the ship in accordance with the following formulation:

$$V_W = w \cdot V$$

Thus the propeller advance speed V_A is lower than the speed V of the ship. In particular we have:

$$V_A = V - V_W = V - w \cdot V = V \cdot (1 - w)$$

A second factor that must be taken into account is the influence of the propeller on the hull. Experiments on models have shown that the resistance of the hull when the propeller is running behind (*autopropulsion tests*) is larger than that when the model is dragged without propeller. There is therefore an increase in resistance due to the presence of the propeller, which could be attributed to the suction or reduced pressure to the rear of the hull originated from the propeller and also to the increase of friction resulting from the increase of the speed of the water that goes to the propeller. This means that the propeller thrust is decreased up to a quantity T-t on the basis of the relation:

$$R_T = T - t \cdot T = (1 - t) \cdot T$$

It is possible to combine the parameters described as follows:

$$\eta_H = \frac{1-t}{1-w}$$

11.4. Some Opportunities of Improving Fishing Vessels Efficiency

11.4.1. Sailing Speed

Fuel consumption is closely linked to the delivered engine power which, on turn, depends on ship's resistance and speed.

A typical feature of the vessel resistance curve is of moderate increase at low speed with increasing steepness in the higher speed regions. At the top of the speed range, the resistance increases with speed in the 6th to 8th power. In order to reduce the resistance, thus the fuel consumption, it would be enough to make the ship to operate at a lower speed/length ratio. Two main factors determine the shape of the resistance curve for a vessel:

- vessel displacement;
- vessel length.

Ship resistance is roughly proportional to its displacement. Some investigations shown a 35 ÷ 45% resistance increase for displacement increases by 50%, (Messina and Notti, 2007). The vessel length determines the steepness of the resistance curve at different speeds and, in practice, the maximum attainable speed of the vessel.

For a displacement type hull, there will be a practical upper speed limit which cannot be exceed, despite the increase in power installed. Therefore, a reduction in speed when the ship is sailing from one fishing area to another and from there to the harbor and vice versa, could allow a large fuel saving. This could be accomplished by two different ways, i.e.:

- by lengthening the ship to realize as much length as possible, according to its requirements in terms of stability, seaworthiness and working efficiency;
- by reducing the speed.

Generally speaking, lowering by 10% the free running speed reduces by 30-40% the fuel consumed per mile steamed. Most of the fuel is consumed by applying the last rpm of the engine. When the rpm are increased from 80% to 100% fuel consumption is doubled. A flow meter should be installed on board the trawlers so to make the fisherman to carefully monitor the fuel consumption and to practice more economic trawling trips.

11.4.2. Fuel Quality

Fishing vessels can be inefficient for a number of reasons. One of these reasons can be the diesel fuel itself. Diesel fuel quality is declining throughout the world, and we expect this trend to continue. The designs of diesel engines striving to increase performance have made a lot of advancements in engine fuel delivery to the combustion chamber. The diesel engines of today are much quieter, smoother, and also more powerful.

The quality of diesel fuel on the other hand has not advanced at the same rate as the improvements of engines. As soon as it is produced, diesel fuel begins to deteriorate. Less than 30 days from refining, all diesel fuel, regardless of the brand, goes through a natural process called oxidation. In fact, a recent survey suggests that 70% of diesel fuel sold at does not meet the Society of Automotive Engineers' (SAE) standard for diesel fuel lubricity. The three leading factors affecting fuel are:

1. *Crude oil sources:* In the past, crude oils were selected with the end product in mind because certain crudes produce high quality fuels and others do not. As crude oils become scarce, the ability to carefully select specific crudes is less economically feasible and compromises are made.
2. *Refining Techniques:* In order to increase the amount of a certain end product that can be refined from a barrel of crude, refineries now use a process called catalytic cracking. The process is economical and produces a high quality product for gasoline, but not for diesel fuel. Due to the cost advantages of this process and the increased demand for gasoline, "cracking" is likely to be the preferred refining method in the future, resulting in lower quality diesel. Another refining issue is the process used to reduce the sulfur content in diesel fuel. (Manufacturers are required to meet lower sulfur content limits due to environmental regulations.) The process has the side effect of reducing the lubricity of the fuel.
3. *Blending:* The third factor affecting fuel quality is blending. In cold weather conditions, for example, many refiners blend kerosene (or number 1 diesel) with number 2 diesel to lower the temperature at which the fuel starts to gel. This has the undesirable effect of reducing the energy content of the fuel, resulting in a decrease in fuel economy. It also may adversely affect the fuel's lubricating and performance properties resulting in potential wear on vital engine components. In many diesel fuel injection systems, the fuel itself is the only lubricant for the precision engineered injection pump.

Proper filtration of diesel fuel is critical to maintaining the performance and long life of a diesel engine. In order to produce and control the extremely high injection pressures common in a diesel, the injection pump components and nozzles are machined to extremely close tolerances - often measured in microns (one micron is 40 millionths of an inch). To prevent the premature failure of these vital components, it is critical that diesel fuel be filtered to remove extremely small particles of foreign matter. The particles that a secondary or final fuel filter captures, for example, are in the range of 5-10 microns (.0002 - .0004"). To illustrate how small these tolerances are, consider:

- The naked eye cannot see particles smaller than 40 microns.
- A grain of sand is approximately 100 microns.
- A human hair is approximately 70 microns.
- A single grain of talcum powder is 10 microns.

The greatest enemy of diesel fuel injection components is water. Once water enters the fuel system, it will rapidly wear and oxidize steel components and lead to:

Rusting and corrosion of components;
Governor/metering component failure;
Sticky metering components (both pump and nozzle);
Injection component wear and seizure.

Water contamination can exist in diesel fuel in three forms:

1. Emulsified water: Water is suspended in the fuel like oil and vinegar in salad dressing.
2. Free water: Water is separated from the fuel and usually is found on the bottom of fuel/storage tanks.
3. Dissolved water: Water has been chemically dissolved in the fuel, like sugar in coffee. The warmer the fuel, the more water will be dissolved, but as temperatures drop, the water will come out of the solution in the form of free water.

Diesel engines vary widely in their cetane requirements, and there is no commonly recognized way to measure this value. In general, the lower an engine operating speed, the lower the CN of the fuel it can use. Large marine engines can tolerate fuels with Cn's as low as 20, while some manufacturers of high-rpm diesel engines specify 55 CN fuel. Using cetane improvers can increase Cetane Number (but not Cetane Index). These additives, usually organic nitrates, boost CN by 2 to 7 numbers, depending on the dosage and the type of base stock used.

Other influences that affect fuel efficiency:

1. *Viscosity:* Viscosity influences the spray pattern when the fuel is injected into the cylinder. Low-speed marine engines can use higher viscosity fuels than high-speed road-transport engines, and still run without excessive smoking. Minimum viscosity limits are imposed to prevent the fuel from causing wear in the fuel injection pump.
2. *Low Temperature Flow:* Unlike gasoline's, which have freezing points well below even the most severe winter temperatures, diesel fuels have pour points and cloud points well within the range of temperatures at which they might be used. This presents no problem on vessels where heated storage can be arranged. Seasonal blending to control cloud point (the temperature at which wax separates from the fuel) is the refiner's assurance against field problems. In the winter, there is an increasing tendency to use flow improvers, polymeric additives that modify the wax structure as it builds up during cooling. These additives keep wax crystals small, so they can pass through the fine pores of fuel filters en route to the injector pump.
3. *Storage Stability*: In storage, diesel fuels are attacked by atmospheric oxygen, which can cause deposits of varnish, and for marine fuels containing residual components, asphaltic material. Antioxidants and dispersants are added to prevent such problems, while copper metal deactivators reduce the catalytic effects of screens and other parts. In the presence of water, bacterial action can cause a build-up of slime in the storage system, leading to filter plugging. Biocides are added to inhibit bacterial growth. In cold-weather areas, there is the risk of static electric charges building up

during high-rate dispensing of distillate fuels. Refiners include antistatic additives in diesel blends to prevent explosions.

4. *Component Compatibility:* Diesel fuels are injected into the engine through precision pumps and fine injector nozzles. Dirt and water contamination must be avoided to protect these critical components. Specifications include tight limits on water and sediment, but some fuel marketers also install final filters at the nozzles of service station pumps to protect against dirt picked up in the distribution system. Depending on the crude source, diesel fuels contain various amounts of sulfur oxides on combustion. These can cause high rates of engine wear and rapid depletion of engine oil additives. Engine manufacturers often relate oil change intervals to the fuel sulfur content. Deposit build-up in engines is influenced by fuel quality. Fuels that leave a heavy carbon residue and contain excessive amounts of high boiling point materials are prone to cause engine deposits.
5. *Sulfur Content:* Sulfur content is the first diesel fuel property to be widely controlled by legislation aimed at limiting exhaust emissions. Sulfur is present in all crude oils and refined products. Sulfates are part of a diesel engine's particulate emissions; therefore, controlling fuel sulfur level reduces the level of sulfate pollutants.
6. *Water Content:* All diesel fuels contain small amounts of water. Hydro-carbon type and distribution, and bulk temperature control the amount that a fuel can hold. As temperature decreases, the amount of water dissolved in the fuel will also decrease and may lead to a water layer forming on the bottom of the storage vessel. Draining the tank regularly will minimize this layer and prevent bacterial contamination and the inadvertent pumping of water into the fuel system. Excessive water in a fuel system can cause corrosion, filter plugging and icing (in the winter).

11.4.3. By-Pass Oil Filtration

Marine engines and equipment operate in some of the most severe environments. It is paramount that engines operate efficiently and require as little down-time for maintenance as possible. Often, because the operating environment is so harsh, maintenance managers will have to change oil and filters earlier and more often to reduce engine wear. This, of course, increases operating expenses and increases the potential for environmental liability if a spill should occur. Technology will keep engine oil constantly clean while in operation dramatically extending the time between oil changes by hundreds of hours. Simply put, if the oil is clean and additives are maintained, the oil does not need to be changed. Benefits of By-Pass Oil Filtration include the following:

- Safely extended oil drain intervals
- Reduced frictional engine wear
- Reduced new oil purchases up to 90%
- Reduced oil-related maintenance costs up to 90%
- Reduced downtime/improved productivity
- Additive levels that remain well within acceptable limits
- Better-maintained oil viscosity

- Improved oil circulation
- Decreased sludge and varnish deposits
- Cooler running engines and/or equipment
- Increased engine efficiency
- Payback typically in less than one year

In SAE Technical Paper 2001-01-1898 a test with a 1 micron bypass filter showed that "The recycler achieved an 8.9% reduction in fuel consumption. The effect of the recycler on the fuel consumption was considered to be mainly due to the action of the recycler in reducing formation of in-cylinder deposits and solid contaminants thus reducing engine friction and fuel consumption." The authors explain: "Lubricating oil is... a significant contributor to combustion chamber deposits (CCD). Engine deposits exist on the piston surface, cylinder head, and intake-exhaust valves. The primary source of piston deposits is the lubricant, and oxidation of the lubricant is the primary cause of deposit formation. In cylinder, deposits consist of ash from the lube oil additives, carbon and absorbed unburned fuel and lubricating oil. The CCD can be a source of wear in engines, increased friction and hence increased fuel consumption. Piston crown land heavy carbon has been shown to increase oil consumption and deposits have been shown to increase as piston temperature increases above 250 degrees Celsius."

Figure 171. By-Pass Oil Filtration System.

In this SAE paper there is a comparison with a test of the same type of engine where a 6 micron filter was used and a 5% reduction in fuel consumption was measured (SAE Technical Paper 1999-01-1139). The authors conclude that the lower fuel consumption could be explained by the finer filtration in the 1 micron filter.

In SAE Technical Paper 2001-01-0699 a test of trucks equipped with 6 or 1micron bypass filters can be read: "It clearly shows that the lube oil recycler has reduced the fuel consumption rate. This is attributed to the filtration of the recycler, which reduced the accumulation of the solid contaminants and formation of the deposits in cylinder. Thus the friction was reduced and energy loss due to friction was decreased."

In Figure 171 a typical installation of a By-Pass Oil Filtration System on a Caterpillar C-15 engine. The gray canister contains a 1 micron filter cartridge and is mounted high on the engine next to the heat exchanger. Mounting the canister at this level aids in changing the filter cartridge as the oil in the canister drains back to the engine. Changing our oil on a time basis is safer than just guessing when to change it, but it is not very cost effective. To date, it has been common practice to change out your oil on a prescribed basis, based on manufacturer's recommendations - like at 250 hours of run time. With the costs of oil going up all of the time, this can be an expensive practice, especially when there are and have been ways to take the guess work out of when you should actually be changing out your lubricant. Now, instead of a complete oil change including primary engine filters, simple change the canister cartridge at the 250 hour interval, take an oil sample and send it to be analyzed. The oil analysis report will give the recommendation when it's time to perform an oil drain.

Engine oil is comprised of two parts – 85% of it is simply base stock and the other 15% of it is the chemistry that the oil company's add, that actually performs all of the critical functions of the lubricant. The reason we change out our engine oil is either because the oil has become too contaminated with particles and by-products of the combustion process of the engine, or because the chemistry in the oil has depleted to such a level that it's no longer in good enough shape to lubricate and help protect the internal components of the engine.

The chemistry in your oil is sacrificial in nature – it is designed to perform a function and then be depleted. It is normal that at the typical oil change interval on most engines, even with standard filtration, that the chemistry is still very much intact and most people actually change it when it still has much of its useful life remaining. What you would see though, if you tried to extend your oil change intervals out with no added filtration, is that the chemistry would hold up for a short period of time before depleting extremely rapidly and then your engine would be at risk of premature wear and tear.

When using a much finer secondary filtration system, the chemistry in the oil will actually last for a much longer time period. This happens because the bypass filter is actually keeping all of the products of combustion (soot, ash, dirt particles, acid, etc…) under control much better than the OE filters and as a result, the chemistry doesn't have to work as hard and actually does a much better job protecting the internal components for a much longer time period. The only change that you would hope to see on an oil analysis when using a 1 micron filter as opposed to the OE 40 micron filters is the ISO 4406 code drop down to levels that are actually cleaner than even brand new, bulk tank oil, even after 500-1,000 hours of run time.

The superior bypass filters will hold more contaminants than any other oil filter on the planet and will also remove moisture from your oil (which is a severe enemy of oil life), but it won't enhance or add to the oil's current additive package. The filter's whole job is to keep contaminants under control so that the chemistry can last safely for as long as possible. Expect to see oil change intervals extended out anywhere from 5-10 times. Bypass filters will keep the used lubricant anywhere from 6 – 10 times cleaner than brand new oil. This would be impossible to achieve without better filtration for all of the reasons mentioned above.

Extending oil changes minimizes waste oil and that is a great thing for the environment. Bypass systems eliminate a whole lot of man hours per oil change as it is much quicker to just change a cartridge than to drain an entire sump; this also means more uptime for a vessel and having superior filtration on your engines and hydraulics gives operators peace of mind just knowing that they've got that extra bit of insurance on their equipment at all times.

Systems work just as well on hydraulic systems, which often times cost more to maintain than engines. Best of all, this technology is not expensive. Return-On-Investment (ROI) is the key to increasing profits. Bypass Oil Filtration systems can pay for themselves within the first year. Most companies already have the price of the paid for in their maintenance budget, requiring no additional capital investment.

11.4.4. Bottom Coatings

It is well known that a fouled hull reduces a ship's speed considerably. Nowadays this equates to increased fuel consumption and cost of operation. Owners, operators and managers are well aware of this and are taking steps to reduce fouling as much as possible. However, fouling is too often overlooked at the beginning in the weed or grass stage. Usually, by the time it is dealt with barnacles, clams, sponges, and kelp is growing in full force. What is all too often missed or disregarded is the effect of the very early stages of biofouling: microscopic fouling or biofilm, commonly referred to as SLIME.

Within hours of a clean hull being submerged in the sea, bacteria begin to accumulate on that hull, whether or not is it coated with biocidal antifoulants (AF) or silicone or other foul-release coatings (FRC) or anything else. In its early stages, this slime is hardly visible. Light slime has been shown to increase fuel consumption by 8% or more and a heavy slime can result in fuel consumption increases of 18% or more. The increase in power required to maintain a cruising speed can cause excessive wear to engines. Modern ship propulsion plants are not typically designed with the large power margins that earlier propulsion plants had.

As an example, a vessel that requires 200 gallons of fuel per day to maintain a cruising speed of 10 knots with a completely smooth and unfouled hull, the way it was at its first speed trials. If that vessel were to build up a thin layer of slime in a month and a thick layer of slime in two months, by the end of those two months of sailing, it would be requiring 220 gallons of fuel per day to maintain the same cruising speed. Taking fuel at $6 per gallon, the slime build-up would cost an additional $120 per day just to keep operating at the same service speed. Even if the fouling remained at that level, in a month it would have used $3,600 more fuel than it would have if the hull were clean. Based on 250 days per year, the costs could add up to over $30,000 in additional fuel costs.

Increased fuel costs cannot be ignored if the ships are to be kept running economically. These are the kind of numbers that are being ignored when an owner or operator repaints a ship with traditional antifouling paint or a foul-release coating, after launching or relaunching then pays no further attention to the underwater hull until it's time to drydock again in 2 ½ years. Ignoring slime is a costly mistake for any vessel owner or operator. It is a matter of concern to those responsible for budgets, costs, profit and loss, long term investment value and total ownership cost as well as the environmental impact.

11.4.5. Bulbous Bows

Speaking about bulbous bows it is mandatory to assume that, especially for existing vessels, this appendage can be added only for steel hulls. The bulbows bow should be considered as an appendage to increase hydrodynamic performances of the hull (Figure 172). At high speed, i.e., at sailing speed, wave making resistance accounts for the greater portion of the drag, due to the wave making resistance. The bulb is able to reduce such resistance due to the production of a own wave trend which minimize the hull wave trend. At low speed bulb will even cause an increase in drag because of its greater wetted surface area, even if at a low speed the added power consumption is negligible and generally little time is spent in this speed range. The reduced power demand provided by the adoption of the bulb in comparison with the power requested by the same hull without bulbows bow has been tested and results are shown in Table 8. At high speed, the reduction in power request by the hull vary between 8 and 15%. In addition, sea keeping increases due to dampening of the pitching motion. When charging into head seas there is the chance of slamming the bulb on the troughs, but this is limited to a very narrow range of wave train and heading. A slight change in direction and/or speed will cure this ill-effect. Their proportions are derived from the features and dimensions of the vessel itself. The diameter (volume) is a direct result of the hull midship area. The length is determined by the stem profile, as the farther forward the bulb extends the more leverage it has but is generally kept shorter than the bow overhang. The section shape may be a modified ellipse to reduce pounding in head seas. The vertical placement is calculated so the bulb is just below the surface where it will create a wave in front of the ship interfering with the natural wave train of the vessel, creating a wave hollow where a crest should be. In this way the vessel will run flatter and the overall wave height will be reduced. The vessel will pitch less which will cause fewer disturbances in the water. Any time a vessel can be moved through the water with less waves and overall disturbance to the surface less power has been transmitted to the water to create those waves.

Vessels over sixty feet should consider the design and implementation of a bulb. It can be a relatively inexpensive addition considering the potential gains in fuel efficiency. Steel vessels have the advantage as the cutting and welding can be completed quickly. Other materials such as fiberglass may require cutting and bonding to the hull with additional fairing work that may be required such as in the picture below.

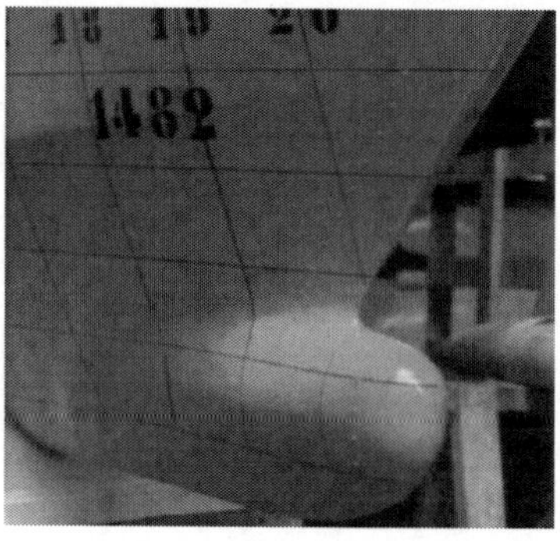

Figure 172. Bulbous bow for steel hulls.

11.4.6. Improvements in the Propulsion System

As the propulsion system efficiency is mainly affected by the propeller efficiency, it is important to pay attention at his right design depending on the request of thrust and the main engine characteristics. Most of the fishing vessel usually use propellers from systematic series; one of the most important is the Troost systematic series (Kuiper, 1992) studied in the flume tank in Wageningen (Netherlands). One of the most important and effective parameters is the diameter.

To show how the propeller diameter modifies the efficiency of the propeller, a typical B 4-55 series propeller is studied at different diameters. In Table 9 the major parameters are obtained for a range of diameter from 1.5 to 2.0 meters to generate the same thrust. In Figure 173 is shown the trend of such parameters. It is possible to note that the power requested decreases increasing the diameter.

This means a fuel saving proportional to the power reduction. At the same time the increasing in torque delivered by the reduction gear has no negative effects for the propeller efficiency. Finally, as the revolution speed decreases too, it results in a benefit for the durability of the propeller which is less stressed by the rotational speed.

Ducted propellers (Figure 174) consist of a duct structure put ahead of the propeller. It will modify the ship's wake. Adopting a nozzle can be suitable to reduce power demand and fuel consumption, for the same thrust, especially during trawling. Nevertheless, model tests with and without such device revealed energy savings (5-10%) due to lesser hull resistance during sailing. So that it is important to establish the ratio of trawling in the overall fishing activity. Comparison between two B 4-70 series propellers with same diameter (D = 2 m) and pitch over diameter (P/D = 0.8) (Table 9). Differences in dynamic parameters, depending on the propeller diameter, for the same thrust.

Table 10 in terms of power requested PD by for the same thrust shows a reduction in the power requested PD of about 25% by the ducted propeller.

Table 8. Comparison in hull resistance between twin hulls, one with and the other without bulbous bow

V	R_T	P_E	R_T	P_E	Variation in power request
	Without bulb		With bulb		
[kn]	[kg]	[hp]	[kg]	[hp]	[%]
8	914	50	845	46	**-8.70**
9	1241	77	1083	67	**-14.90**
10	1636	112	1409	97	**-15.50**
11	2369	179	2235	169	**-5.91**
12	4171	343	3895	321	**-6.85**
13	7562	674	6523	582	**-15.80**

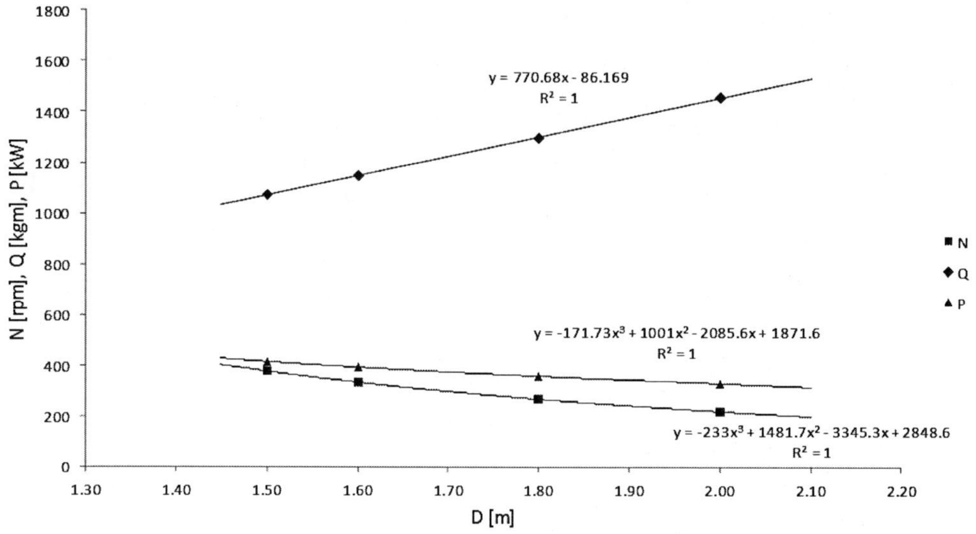

Figure 173. Trend of major dynamic parameter at various diameters of a B 4.55 series propeller.

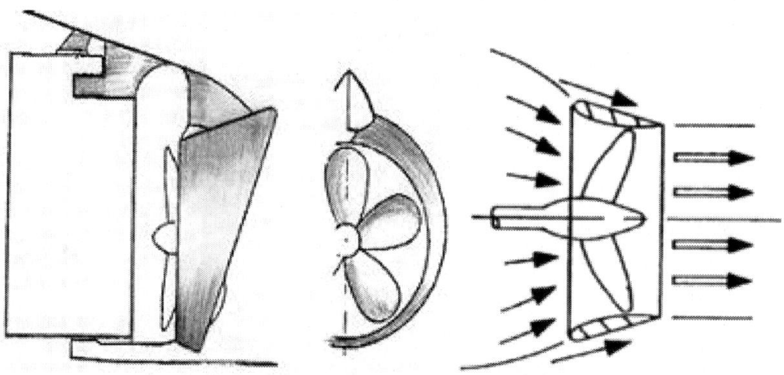

Figure 174. Ducted propellers (Messina et al., 2007).

Table 9. Difference in dynamic parameters, depending on the propeller diameter, for the same thrust

D	N	Q	P
[m]	[rev/min]	[kgm]	[kW]
1.50	378	1071	416
1.60	335	1146	394
1.80	269	1300	359
2.00	221	1456	330

Table 10. Comparison of the power delivered (P_D) between a B 4-70 open propeller and a B 4-70 ducted in nozzle 19a NACA profile propeller; both propellers with a 2 m diameter (d) and 0.8 pitch over diameter ratio (P/D). For the same thrust (T) the ducted propeller requires less power (P_D) than the open propeller

T (kg)	PD (kW)		Ratio
	B 4-70 (D = 2.0 m)	B 4-70/19a (D = 2.0 m)	
4000	204	172	84%
5000	276	227	82%
6000	355	285	80%
7000	440	346	79%
8000	530	411	78%
9000	625	480	77%
10000	726	551	76%

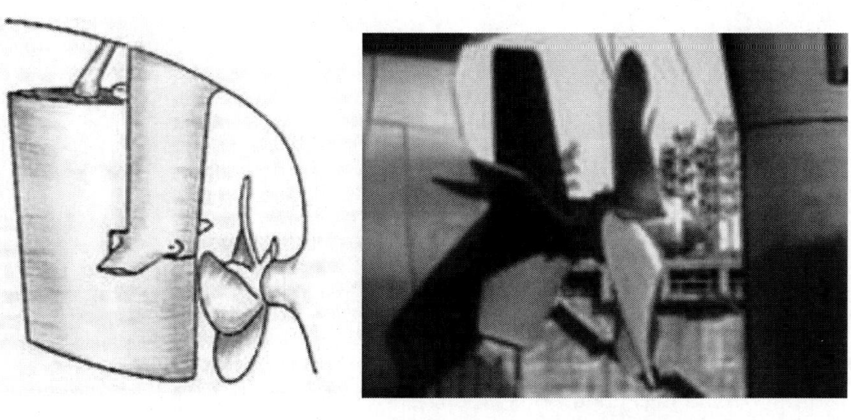

Figure 175. Stator for increasing propeller thrust. After the propeller the water flow waste energy due to turbulence. The stator reduces the wasting in energy and increase propulsion system efficiency.

Figure 176. Controllable pitch propeller. A gear into the hub moves the blades around radial axis respect to the propeller rotational axis.

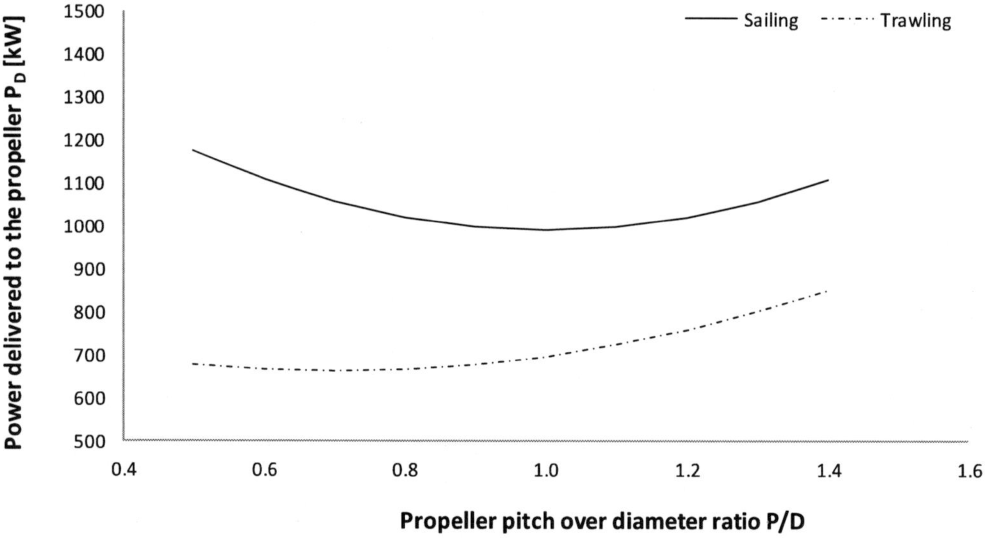

Figure 177. Power requested during trawling and sailing phases by a controllable pitch propeller at different P/D values.

This structure, consists of putting some lifting flaps on the stern strut in order to reduce the loss of kinetic energy due to the rotation of the propeller wake. It could be applied together with a ducted stern, the efficiency could be improved by 2 - 5%.

So, different operating conditions such as trawling and sailing in fishing activity, especially for trawlers, lies to a propulsion system able to optimize both fishing phases (Figure 176). Using a controllable pitch propeller the main engine is used at constant speed and the propulsion is managed by changing the propeller pitch, thus it is possible to settle main engine operating condition near the minimum specific fuel consumption, saving fuel. The controllable pitch propeller optimizes the propeller performance on the basis of thrust and speed required. Figure 177 represents the power requested by a controllable pitch propeller of a fishing vessel during trawling and sailing conditions. For the same diameter, the P/D changes with the pitch. The power requested for the two phases is minimized at different P/D values, thus at different propeller pitch.

11.5. Deck Machinery

Typically the deck machinery equipment in a fishing vessel consists on devices for manoeuvring the fishing gear, for manipulate the fish and for auxiliary and electrical power generation. The hydraulic winch for wire rope (Figure 178) is one of the common deck machinery consisting on a multi-drum which moved by hydraulic system, for the hauling of the fishing gear after a shoot. It is usually equipped with a rope guide (Figure 179) which is a device that allows for a proper recovery of the cables, especially when a huge amount of cable is shot. It is moved by the hydraulic system on the left and on the right with a speed dependant on the speed impressed to the rope winch during hauling.

Other common winches are used for gilson ropes (Figure 180) when it is necessary to move weights on board, or when the codend, during hauling, has to be elevated in order to release the fish inside the tanks.

Purse seiners adopt a powerblocks (Figure 181) for the recovery of the surrounding net. It is similar to a rope winch but it is installed to a mechanical arm that is able to move till the water surface and recover the net elevating the initial part and supporting the power block.

The use of an auxiliary engine can be a possible solution when space in machine room is available and when it is necessary a huge amount of stable electrical power.

Figure 178. Hydraulic rope winch. Each of the two parts of the winch is made by two drums that can be moved independently, by means of by-pass valvles to address the hydraulic power onto the desired drum.

Figure 179. Rope guide. It is moved by the hydraulic system on the left and on the right with a speed dependant on the speed impressed to the rope winch during hauling.

Figure 180. Winch for Gilson rope.

Figure 181. Power block on a purse seiner.

A more feasible solution is the adoption of an hydraulic power system to generate electric power. The main engine provides mechanical power to an hydraulic pump. An electronic system controls pressure and flow rate of the hydraulic oil to maintain the electric generator at an optimal operational point. Depending on the electric power requested electronic system manages hydraulic oil flow and pressure to assess the right hydraulic power. So that the main engine is never overloaded, providing every time the right mechanical power, saving fuel. Eliminating the need for an auxiliary engine to run the vessel´s electrical system produces also a fuel saving. Furthermore the electric power on board is stable and of good quality (fixed frequency and voltage) many electric users could be included, for example, electric winches.

In Figure 182 a typical hydraulic power system layout is showed. The main engine delivers the requested power through a power take off (1) to a multiplicator (2) necessary to adapt the revolution speed of the main engine to the speed of the three hydraulic pumps connected (generally the multiplication ratio is 3-1.5 to make the pumps running at 1500-200 rpm). Pump number 3 is devoted to rope and net winch and for bow thruster, pump number 4 is to move an electric generator and pump number 5 is redundant and it is able to substitute the pump number 3, for safety reasons.

Pump number 4 is connected to a hydraulic driven generator system (Figure 183). The main engine provides mechanical power to a hydraulic pump. An electronic system controls pressure and flow rate of the hydraulic oil to maintain the electric generator at an optimal operational point. Depending on the electric power requested electronic system manages hydraulic oil flow and pressure to assess the right hydraulic power. So that the main engine is never overloaded, providing every time the right mechanical power.

Figure 182. Hydraulic power system layout. The forward power take off of the main engine (1) is connected trough a multiplicator (2) to three hydraulic pumps (3, 4, 5). Pump 3 is devoted to rope and net winch and for bow thruster, pump number 4 is to move an electric generator and pump number 5 for safety reasons.

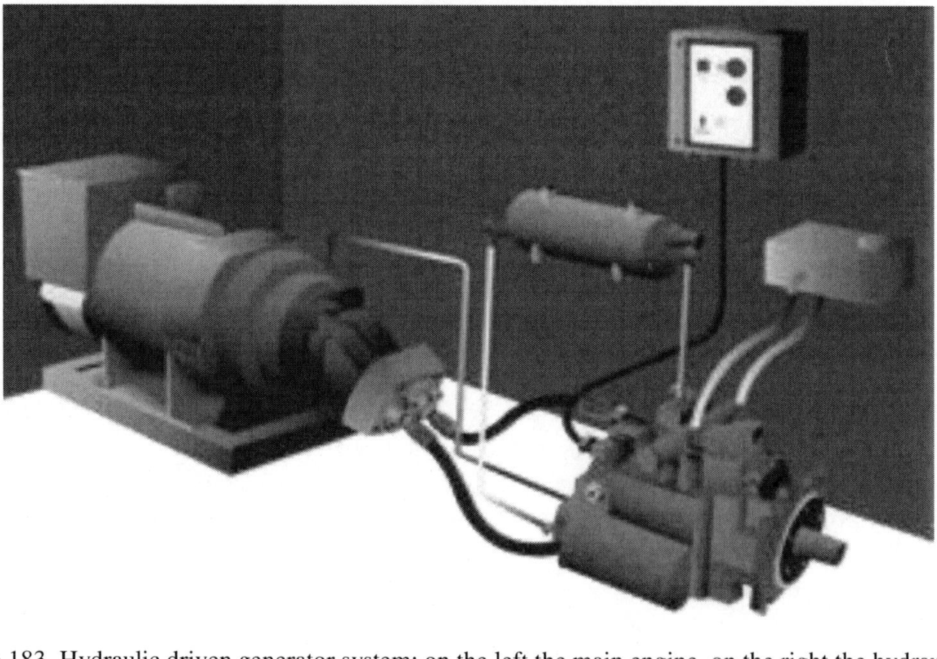

Figure 183. Hydraulic driven generator system; on the left the main engine, on the right the hydraulic pump connected to a distribution header. On the top the electronic system control the operating point of the pump on the basis of the hydraulic power request.

Especially for deck machinery such as winches, even if actually it is easier to generate mechanical power trough hydraulic systems due to a minor "cost-weigth-space" relationship, there are many reasons for favouring an electrically driven system. We should also note the following:

- Lower installation costs as the machine is free of tubes;
- Lower energy consumption, maintenance and noise levels, thereby improving security and operator well-being;
- Winch ignition through the alternating current includes a frequency converter, an alternating current motor (AC), a frequency converter on the winch, a refrigeration system for two motors and an energy regeneration system;
- Higher efficiency on the machine (input/output power) and energy regeneration during the free pay out improve vessel's energy consumption quota;
- The efficiency of electric ignition is around 85%-90% as opposed to hydraulic system efficiency that may vary from 50% to 70%;
- Provides better control than hydraulic starting systems with a variable speed control of 2% to 100% of maximum rpm and the whole torque on any rpm level;
- Environmentally friendly;
- Extremely silent operation;
- No emissions or oil spills/leakages, no risk of fire, no need for oil filters and oil dirties disposing.

Electric drive winches (Figure 184) can provide better control than hydraulic drive winches. It is possible to set speed operation from 2% to 100% of maximum rated RPM and full torque at any RPM. Control system allows also automated management during shooting and hauling, controlling also tension and speed. As the trawling winches are moved by warps or by the net, one of the most important characteristics of electric winches is the possibility to recover energy during shooting. The energy recovered depend on the total length of the warps used for each haul and the speed of the vessels. Compared to hydraulic winches (Figure 185), electric winches can reduce power required for trawling winch because of more energy efficiency (80% for electrical against 55% for hydraulic winches). On other advantage is the opportunity to use management systems of the drag, the length of the cable and the torque applied to each winch drum (Figure 186).

Ice Machines for Chilling Fish

Not all fish in Egypt is produced to the appropriate standards of quality. A major cause of inferior fish is poor handling on fishing vessels, at aquaculture sites and during unloading, handling and transportation to markets. Constraints include insufficient use of ice and the use of containers (wooden boxes, chiefly) which provide inadequate protection during handling and transportation. In some cases, the initial handling is undertaken in facilities, which are not adequately sanitary or suitable.

There are three important ways of preventing fish going bad too quickly - care, cleanliness and cooling. Care in handling is essential because unnecessary damage can provide access through cuts and wounds for the spoilage bacteria, thus hastening their effect on the flesh. Cleanliness is important in two ways:

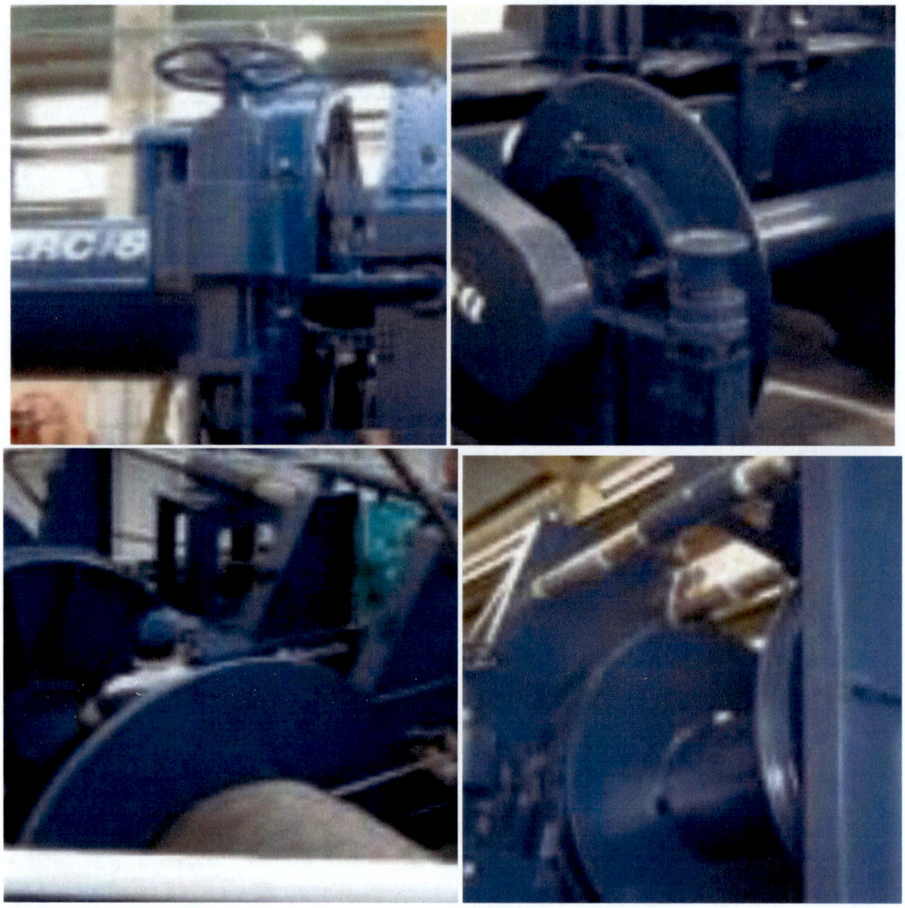

Figure 184. Electric winches.

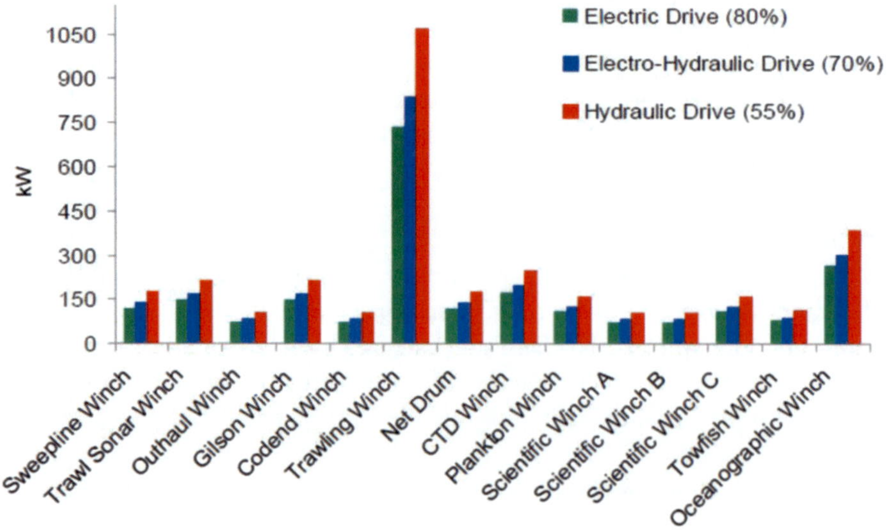

Figure 185. Comparison between hydraulic and electric winches, for different types of winches.

Figure 186. Electronic winch management.

- the natural sources of bacteria can largely be removed soon after the fish is captured by taking out the guts and washing off the slime from the surface of the fish;
- the chances of contamination can be kept to a minimum by ensuring the fish is always handled in a hygienic manner.

But most important of all, *the fish must be chilled quickly and kept chilled.* The speed with which bacteria grow depends on temperature. The higher the temperature, the faster the bacteria multiply, using the flesh of the dead fish as food. When the temperature is sufficiently low, bacterial action can be stopped altogether; frozen fish stored at a very low temperature, for example - 30° C, will remain wholesome for a very long time because bacteria are either killed or completely inactive at this temperature, and other forms of spoilage progress only very slowly.

But, even at -10°C, some kinds of bacteria can still grow, although only at a very slow rate. Therefore, for long-term storage, of many weeks or months, freezing and cold storage are necessary. It is not possible to keep unfrozen fish at a temperature low enough to stop bacterial action completely, because fish begin to freeze at about -1°C. However, it is desirable to keep the temperature of unfrozen fish as close to that level as possible in order to reduce spoilage; the easiest and best way of doing this is to use plenty of ice which, when made from clean fresh water, melts at 0°C. At temperatures not much above that of melting ice, bacteria become much more active and fish consequently goes bad more quickly. For example, fish with a storage life of 15 days at 0°C will keep for 6 days at 5°C and only about 2 days at 15°C before becoming unacceptable.

Figure 187. Tank with ice for chilling the fish just caught before the sorting and packaging.

Figure 188. Compressors for ice makers and chiller. The owner of this 28 m otter trawl fishing vessel requested two compressors for the ice maker and for refrigerate the ice room and other two for redundance in case of failure of main devices.

Ice as a cooling medium for fish has a great deal in its favour; it has a very large cooling capacity for a given weight or volume, it is harmless, portable and relatively cheap. It is especially valuable for chilling fish, since rapid cooling is possible. When fish are being cooled with ice, heat transfer is achieved by direct contact with the ice, by conduction through adjacent fish and by melt-water flowing over the fish. Cold melt-water takes up heat from the fish and when it flows over ice again it is recooled. Thus, intimate mixing of fish and ice not only reduces the thickness of the layer of fish to be cooled but also promotes this convective cooling interaction between ice melt-water and fish. Any further melting that occurs is due to heat from other sources, such as the warm surrounding air during the subsequent storage period. There are other ways of chilling fish besides using ice. For example, they can be immersed in chilled water or cold air can be blown over them. Sea water, cooled by mechanical means, Refrigerated Sea Water (RSW) or by the addition of ice, Chilled Sea Water (CSW), is a suitable alternative means of rapidly chilling large quantities of small whole fish, especially on board a fishing vessel. When cold air alone is used, as in a ice room, the heat taken from the fish will rapidly warm the air. The warm air rises and is cooled by contact with the coils of the cooler and then moves by natural convection or fan circulation back to the fish. It does not take much heat to warm the air; it takes 10,000 times as much heat to melt a given volume of crushed ice as it does to warm the same volume of air from 0 to 0.5 C. Thus, it is important to remember that for air cooling to be effective, a good circulation of cold air must be blown over the fish. However, even when a fan is fitted in a ice room it is difficult to achieve the rapid cooling rates possible with ice and chilled sea water. Another disadvantage of air chilling is that, without the use of ice, the fish becomes dry. Continuous air movement evaporates water from the fish surface and deposits it as frost or condensate on the coils of the evaporator. In addition, the air in some parts of the ice room will be colder than in others. Fish in the cold spots, for example close to the evaporator, may in time become partially frozen, even though the thermostat is set above freezing point at another location within the ice room. Slow freezing of the fish can be detrimental, since appearance, flavour and texture of the fish may be affected. In Figure 188 is shown an array of compressors for ice makers and chiller. The owner of this 28 m otter trawl fishing vessel requested two compressors for the ice maker and for refrigerate the ice room and other two for redundance in case of failure of main devices.

11.5.1. Some Devices to Increase Safety on Board

In trawling, the steel wires freely go to the sea, while the vessel is going at 4-6 knots. The drums of the winch are not broken and the paying out of the wires is checked by means of a regulation of the braking intensity. Due to friction and consequent generated heat, the band breaks, become longer than their initial length. When the trawling begins, the winch operator breaks the drums, the bands being still hot. Over the trawling, the break bands become cold and go slowly back to their initial shorter length. Between the screwed bar and its nut screw very high stresses will arise, to prevent the operator from turning the bar in order to release the breaks of the drums. In case the trawling equipment will grasp at a submerged obstacle, the impossibility of suddenly releasing the breaks will endanger the safety of both the vessel and the crew. The device of Figure 189, gives a solution to this problem. A small electro-hydraulic pump sends oil to the cylinder D increasing the braking torque, the pressure inside the cylinder will become higher but a relief valve will modify the pressure to the preset value.

When the fishing gear is hauled in and the otterboards come out from the sea, the fishermen must not stay close to the gallows to avoid accidents in case of a rope break-up. The otterboards should be handled only when on board.

Among the incidents which could be verified during the manoeuvre of the otterboards the following could be cited:

- Bumps caused by moving masses;
- crushing for pressure of the otterboards against the bulwark;
- dragging by chains or hooks;
- falls on and out board.

The latter in particular could take place when the fishermen must hook the otterboards to the gallows to avoid banging. In fact, after completing the fishing operations and wires have been wound on the winch drums, the otterboards are dangling from the pulleys (4) of the stern gantry (Figure 190) connected to two arms (7) sticking out from (6). If both the hydraulic arms are so made that they could go in and out their guide (6), the otterboards will be brought inside the shape of the deck and then by the winch, they will be settled between the bulwark and the gantry with a simpler but surer manoeuvre.

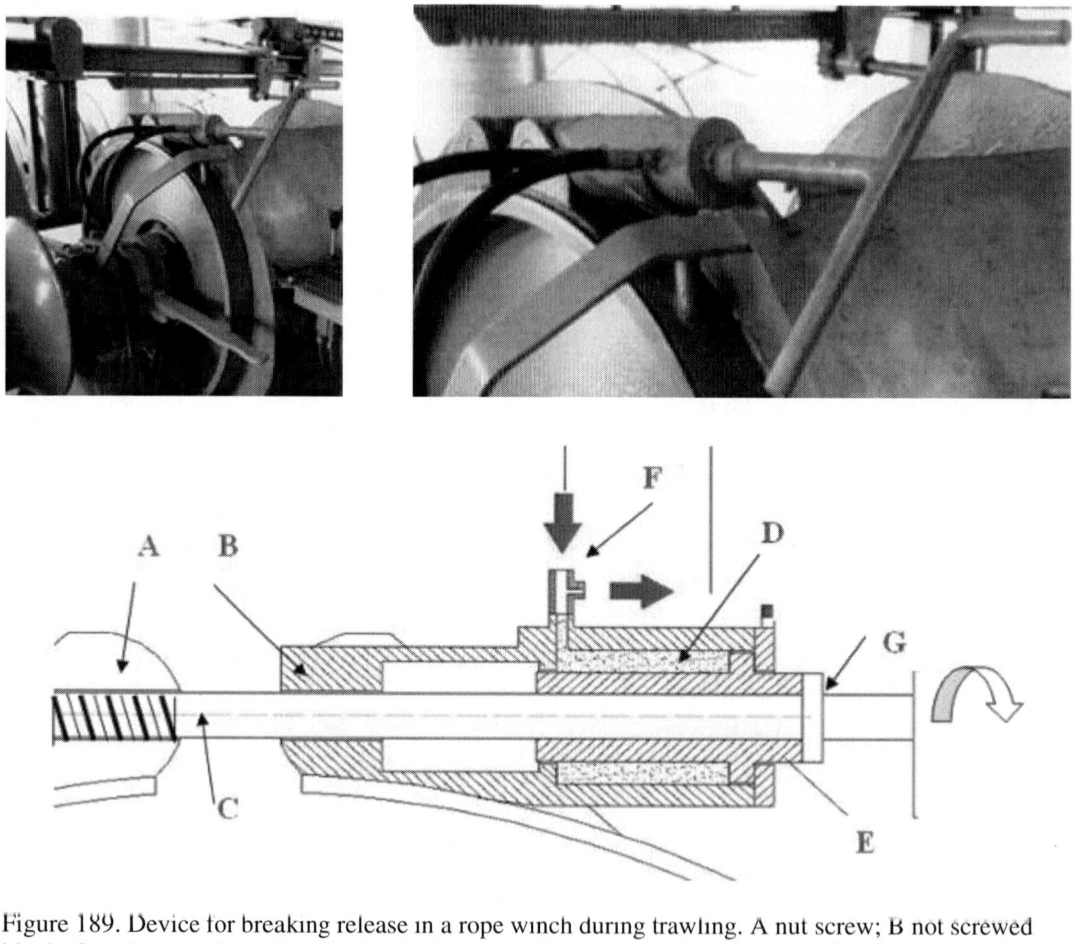

Figure 189. Device for breaking release in a rope winch during trawling. A nut screw; B not screwed block; C end screwed bar; D cylinder; E piston; F relief valve; G disc of the bar C.

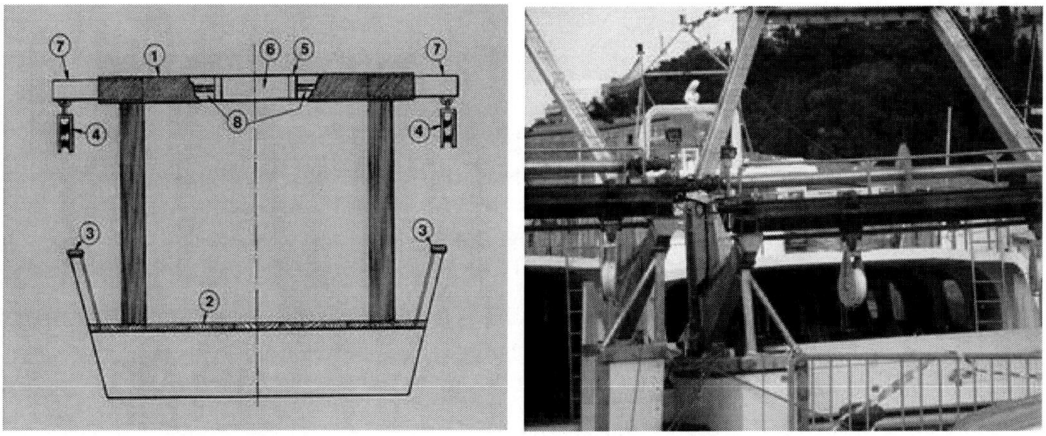

Figure 190. Electric or hydraulic movable pulleys for shooting and hauling the doors in otter trawl fishery (Messina and Notti, 2008).

11.6. Conclusion

Currently, we have the ability to reduce diesel fuel consumption dramatically, perhaps by as much as 45% in some cases. This is achieved by the combinations of fuel emulsion, hydrogen injection, and on-board fuel management analytic systems. These technologies also have capability to reduce NOX emissions by as much as 90%. By-pass oil filtration systems can extend oil drains by as much as 5 or more times and help protect the engine guarding against costly maintenance or break downs. Vegetable based lube and hydraulic oils reduce risks to the environment when spilled and are actually a higher grade of lubrication than petroleum based oils and grease. Bilge water oils are collected by a material that becomes a solid that becomes non-hazardous for disposal. And, natural materials and oil eating microbe's for cleaning and degreasing. There are problematic barriers to market. Fuel supply companies do not want the vessels using less fuel. They will promote all kinds of propaganda such as fuel emulsion and hydrogen will destroy the engine and void engine warranties. The service companies that change oil do not want By-Pass Oil Filtration as they may lose much of their revenue performing regular oil changes. Engine companies lobby the governments for emission standards in new model engines so they can sell new replacements. Governmental approval of devices that may bring an older engine to compliance may not happen. These new technologies are not even considered by the US EPA as an emerging technology at this point, mainly due to they are too simple and very cost effective.

With all this said there are those that are moving ahead and approaching the barriers one project at a time. There are tremendous opportunities for the vessels and the companies that own them to increase their bottom line by going "Green". It may only take a few vessels and one or two companies (fleets) to implement the fuel saving technologies and methods and a whole new industry would emerge with the creation of new "Green" jobs.

It is possible to summarize main issue for increase efficiency in fisheries with:

1. Slow down. In a displacement vessel even a small decrease in boat speed will save fuel. Reducing power as little as 12% from full throttle will save 30%.
2. Look at your exhaust. Exhaust from a well-maintained diesel engine is virtually invisible. If yours is black, white or blue the engine needs attention.
3. Check the propeller. When out of the water check the prop for bent blades, dings, or eroded edges. While underway, check the prop wash for excess turbulence and bubbles that suggest a prop that‘s too small or has too little pitch.
4. Maintain the bottom. Get the weeds and barnacles off and keep them off with proper anti-fouling paint. Sponsors, struts, sea chests, keel coolers, transducers and stabilizers all increase hull drag so if there‘s something below the waterline you don‘t need, get rid of it.
5. Re-think your electrical system. Do you need to run a diesel gen-set around the clock or can you use batteries and an inverter for your ―hotel‖ power? It‘s more efficient to use shore power at the dock than run a generator.
6. Check your steering. If there‘s play in your steering, adjust it to eliminate as much as possible. A good autopilot can steer straighter than any helmsman. The pilot‘s control head probably has adjustments that change steering parameters and allow you to minimize delayed or over-steering.
7. Plan your route and timing. Remember when vessels used to depart ―on the tide‖? Tide & current tables and oceanographic current charts can indicate ways to get a boost from nature. Good weather forecasts help you avoid headwinds or delaying sea conditions, as well as suggest chances to get a boost from tail winds.
8. Reduce vessel weight. More important on a planning or semi-displacement vessel, weight control reduces the amount of power needed to achieve a given speed.
9. Keep good records. You only know if you‘re making an improvement (or making things worse) if you have good numbers on vessel performance both before and after changes.
10. Do the math. Fuel is only one of the costs of your operation. Capital expenditure (the price of new equipment) is another, as is the value of your time.

REFERENCES

Aldebert, Y. (1997). Demersal resources of the Gulf of Lions (NW Mediterranean). Impact of exploitation of fish diversity. *Vie Milieu* 47: 275–284.

Aldebert, Y., Recasens L. (1996). Comparison of methods for stock assessment of European hake *Merluccius merluccius* in the gulf of Lion (Northwestern Mediterranean). *Aquat. Liv. Res.,* 9: 13-22.

Alós, J., Cerdà M., Deudero S., Grau A.M. (2008). Influence of hook size and type on short-term mortality, hooking location and size selectivity in a Spanish recreational fishery. *J. Appl. Ichthyol.* 24, 658–663.

AMCC (Alaska Marine Conservation Council) (2001). *Bycatch: Wasting Alaska's Future.* Special Update – June 2001. Anchorage, AK: Alaska Marine Conservation Council. 11 pp.

Ardizzone, G.D., Tucci P., Somaschini A., Belluscio A. (2000). Is bottom trawling partly responsible for the regression of Posidonia oceanica meadows in the Mediterranean Sea? *In*: Kaiser, M. J. & de Groot, S. J. (eds) Effects of Fishing on Non-target Species and Habitats. Blackwell Science, London. pp 37-46.

Atabey, S., Taskavak E. 2001. A preliminary study on the prawn trawls excluding sea turtles. *Urun Derg J Fish Aquat Sci* Vol. 18, no. 1-2, pp. 71-79. 2001.

Bahamon, N., Sardà F., Suuronen P. 2006. Improvement of trawl selectivity in the NW Mediterranean demersal fishery by using a 40mm square mesh codend. *Fish. Res.* 81: 15-25.

Ball, B., B. Munday, Fox G. 1999. The impact of a Nephrops otter trawl fishery on the benthos of the Irish Sea. *Journal of Shellfish Research*, 18(2):708.

Baranov, F.I. 1948. *Theory and assessment of fishing gear.* Chap. 7. Theory of fishing with gillnets. Pishchepromizdat, Moscow. (Translated from Russian by the Ontorio Department of Lands and Forests. Maplc. Dnt.).

Bas, C., Maynou F., Sardà F., Lleonart J. (2003). *Variacions demogràfiques a les poblacions d'espècies demersals explotades: els darrers quaranta anys a Blanes i Barcelona.* Institut d'Estudis Catalans, Sec. Ciènc. Biol., 135: 202 pp.

Bergman, M. J. N., Van Santbrink, J. W., (2000). Mortality in megafaunal benthic populations caused by trawl fisheries on the Dutch continental shelf in the North Sea in 1994. *ICES J. Mar. Sci.,* 57: 1321-1331.

Bertrand, J. (1988). Selectivity of hooks in the handline fishery of the Saya de Malha banks (Indian Ocean). *Fish. Res.,* 6: 249–255.

Bjordal, Å. (1988). Recent developments in longline fishing – catching performance and conservation aspects. In: *Proceedings of the World Symposium on Fishing Gear and Fishing Vessel Design,* St. Johns, Newfoundland, Canada, Nov. 21–24, 1988: 19–24.

Beverton, R. J. H., Holt, S. J. (1957). On the dynamics of exploited fish populations. *UK Min. Agric. Fish., Fish. Invest.* (Ser. 2) 19, p. 533.

Beverton, R. J. H., Holt, S. J. (1966). Manual of Methods of the Fish Stock Assessment Part II Tables of Yield Function. *FAO Fish. Tech. Pap.* No. 38 (4), Version 1, 67 pp.

Brambati, A., Fontolan, G. (1990). Sediment re-suspension induced by clam fishing with hydraulic dredges in the gulf of Venice (Adriatic Sea). A preliminary experimental approach. *Boll. Ocean. Teor. Appl.* 8: 113–121.

Broadhurst, M. K. (2000). Modifications to reduce bycatch in prawn trawls: a review and framework for development. *Rev. Fish. Biol. Fish.*, 10: 27-60.

Byrne, C. J., Fogarty, M. J. (1985). *Comparison of the fishing power of two fisheries research vessels.* Northwest Atlantic Fisheries Organization Scientific Council Meeting-Sept. 1985. NAFO SCR Dot. 85/90.14 p.

Caddy, J. F. (2000). Marine catchment basin effects versus impacts of fisheries on semi-enclosed areas. Effects of fishing on the structure and functioning of estuarine and nearshore ecosystems. *ICES J. Mar. Sci.,* 57: 628-640

Carr, H. A., Milliken, H. (1998). Conservation engineering: options to minimize fishing's impacts to the sea floor. In: *Effect of fishing gears on the sea floor of the New England* (eds E.M. Dorseyand J. Pederson). Conservation Law Foundation, Boston, Massachusetts, pp.100-103.

Churchill, J. (1989). The effects of commercial trawling on sediment re-suspension and transport over the Middle Atlantic Bight continental shelf. *Cont. Shelf Res.,* 9: 841-864.

Clark, G.H. (1931). *The California halibut (Paralichthys californicus) and an analysis of the boat catches.* Fish Bulletin No. 32. 54 pp.

Cloern, J. E. (2001). Our evolving conceptual model of the coastal eutrophication problem. *Mar. Ecol. Prog. Ser.* 210: 223-253.

Coll, M., Santojanni, A., Palomera, I., Tudela, S., Arneri, E. (2007). An ecological model of the Northern and Central Adriatic Sea: analysis of ecosystem structure and fishing impacts. *J. Mar. Syst.,* 67: 119-154.

Coll, M., Piroddi, C., Steenbeek, J., Kaschner, K., Ben Rais Lasram, F. (2010). The Biodiversity of the Mediterranean Sea: Estimates, Patterns, and Threats. *PLoS ONE* 5(8): e11842. doi:10.1371/journal.pone.0011842.

Collie, J. S., Hall, S.J., Kaiser, M.J., Poiner, I.R. (2000). A quantitative analysis of fishing impacts on shelf-sea benthos. *Journal of Animal Ecology* 69, 785-798.

Crespi, V., Prado, J. (2002). In: *FAO Fisheries and Aquaculture Department [online]. Rome.* Updated 9 August 2002. http://www.fao.org/fishery/equipment/fisheye/en.

Czerwinski, I. A., Erzini K., Gutiérrez-Estrada J.C. Hernando J.A. 2009. Deep-water longline selectivity for black spot seabream (Pagellus bogaraveo) in the Strait of Gibraltar. *Fisheries Science,* 75: 285-294.

Czerwinski, I. A., Gutierrez Estrada J. C., Soriguer M. C., Hernando Casal J.A. 2010. Hook selectivity models assessment for black spot seabream. Classic and heuristic approaches. *Fish. Res.* 102: 41-49.

Davism, F. M. (1934). Mesh experiments with trawls 1928-1933. *Fishery Investigations,* London, series 2, 14, 56 pp. London, HMSO.

Dayton, P. K., Thrush S.F., Agardy M.T., Hofman R.J. 1995. Environmental effects of marine fishing. *Aquatic. Conserv.: Mar. Freshw. Ecosyst.,* 5: 205-232.

Demestre, M., Sánchez, P., Kaiser, M. J. (2000). The behavioural response of benthic scavengers to otter-trawling disturbance in the Mediterranean. In: Kaiser, M.J., de Groot, S.J. (Eds.), *Effects of Fishing on Non-target Species and Habitats.* Blackwell Scientific Publications, Oxford, pp. 121–129.

Deval, M. C., Boek, T., Ates, C., Ulutuerk, T., Tosunoglu, Z. (2009). Comparison of the size selectivity of diamond (PA) and square (PE) mesh codends for deep-water crustacean species in the Antalya Bay, eastern Mediterranean. *J. Appl. Ichthyol.* Vol. 25 (4): 372-380.

Di Natale, A., Notarbartolo di Sciara, G. (1994). *A review of the passive fishing nets and trap fisheries in the Mediterranean Sea and of cetacean bycatch.* Report of the International Whaling Commission, Special Issue 15, pp. 189–202.

Duplisea, D. E., Jennings, S., Malcolm S. J., Parker, R., Sivyer, D. B. (2001). Modelling potential impacts of bottom trawl fisheries on soft sediment biogeochemistry in the North Sea. *Geochem Trans*., 2: 112.

Eayrs, S. (2007). A guide to bycatch reduction in tropical shrimp-trawl fisheries. *Food and Agricultural Organisation of the United Nations*. Rome. ISBN 987 92 5 105674 5. 105 pp.

Eigaard, O. R., Munch-Petersen, S. (2011). Influence of fleet renewal and trawl development on landings per unit effort of the Danish northern shrimp (Pandalus borealis) fishery. *ICES Journal of Marine Science*, 68: 26-31.

Engel, J., Kvitek, R. (1998). Effects of otter trawling on benthic communities in Monterey Bay National Marine Sanctuary. *Cons. Biol.*, 12: 1204-1214.

European Commission. 2002. *Communication from the Commission to the Council and the European Parliament on a Community Action Plan to reduce discards of fish.* COM(2002) 656 final. 21 pp.

FAO. 1984. *Papers presented at the Expert Consultation on the regulation of fishing effort (fishing mortality). Rome, 17–26 January 1983. A preparatory meeting for the FAO World Conference on fisheries management and development.* FAO Fish.Rep., (289) Suppl.2: 214 p.

FAO. 1995. *Code of Conduct for Responsible Fisheries*. Rome, FAO. 41 p.

FAO, Fisheries Department. 1996. Technical Consultation on Reduction of Wastage in Fisheries. Tokyo, 28 October–1 November 1996. *FAO Fish. Rep.,* 547. 28 pp.

FAO, Fisheries Department. 2003. The ecosystem approach to fisheries, F*AO Technical Guidelines for responsible fisheries,* 4, suppl.2. 112 pp.

FAO. 2005. *Increasing the contribution of small-scale fisheries to poverty alleviation and food security. FAO Technical Guidelines for Responsible Fisheries.* No. 10. Rome, FAO, 79 p.

FAO. 2011. CopeMed II – ArtFiMed, Guide des engins de pêche utilisés à Ghannouch et Akarit (golfe de Gabès, Tunisie). Malaga, Espagne, 29 pp

Farrugio H., Oliver P., Biagi F. (1993). An overview of the history, knowledge, recent and future research trends in the Mediterranean fisheries. *Sci. Mar.*, 57(2-3): 105-119.

Ferro, R. S. T., O'Neill, F. G. (1994). *An overview of the characteristics of twines and netting that may change codend selectivity.* ICES CM 1994/B:35.

Fiorentini, L., Dremière, P. Y., Leonori, I., Sala, A., Palumbo, V. (1999). Efficiency of the bottom trawl used for the Mediterranean international trawl survey (MEDITS). *Aquatic Living Resources*, 12: 187-205.

FORCE (Fisheries & aquaculture Oriented Research Capacity in Egypt). 2013. Final Report.

Franceschini, G., Raicevich, S., Giovanardi, O., Pranovi, F. (2002). T*he use of Side Scan Sonar as a tool in Coastal Zone Management. Littoral 2002,* The Changing Coast. (EUROCOAST / EUCC, Porto – Portugal Ed. EUROCOAST – Portugal, ISBN 972-8558-09-0). 11-14.

Fredj, G., Bellan-Santini, D., Meinardi, M. (1992). Etat des connaìssances sur la faune marine Méditerranéenne. *Bulletin of the Institute of Oceanography,* Special issue No. 9, pp. 133–145.

Freese L., Auster P. J., Heifetz J., Wing B. L. (1999). Effects of trawling on seafloor habitat and associated invertebrate taxa in the Gulf of Alaska. *Mar. Ecol. Prog. Ser.,* 182: 119-126.

Friedlander, A. M., Boehlert, G. W., Field, M. E., Mason, J. E., Gardner, J. V., Dartnell, P. (1999). Side scan-sonar mapping of benthic trawl marks on the shelf and slope off Eureka, *California. Fish. Bull.* 97: 786-801.

Fryer, R. J. (1991). A model of between-haul variation in selectivity. *ICES J. Mar. Sci.* 48:281-290.

Fryer, R. J. (1996). Estimating fl6et selectivity from vessel selectivity. In *Report of the Study Group on the use of selectivity measurements in stock assessment.* ICES CM 1996/B:4: 29-33.

Galbraith, R. D., Fryer, R. J., Maitland, K. M. S. (1994). Demersal pair trawl cod-end selectivity models. *Fish. Res.,* 20:13-27.

General Fisheries Commission for the Mediterranean. (1997). REC.CM-GFCM/22/1997/1: Limitation of the use of driftnets in the Mediterranean.

General Fisheries Commission for the Mediterranean. (2009). REC.CM-GFCM/33/2009/2 On the minimum mesh size in the codend of demersal trawl nets.

General Fisheries Commission for the Mediterranean. (2011). Recommendation GFCM/35/2011/4 on the incidental bycatch of sea turtles in fisheries in the GFCM Competence Area.

Giovanardi, O., Pranovi, F., Franceschini, G. (1998). "Rapido" trawl fishing in the Northern Adriatic: preliminary observations of the effects on the macrobenthic communities. *Acta Adriat.,* 39 (1): 37-52.

Glass, C. W., Wardle, C. S., Godsen, S. J., Racey, D. N. (1995). Studies on the use of visual stimuli to control fish escape from codends. I. Laboratory studies on the effect of black tunnel on mesh penetration. *Fish. Res.,* 23: 281 - 164.

Glass C. W., Wardle C. S., Gosden S. J., Racey D. N. (1995). Studies on the use of visual stimuli to control fish escape from codends. I. Laboratory studies on the effect of black tunnel on mesh penetration. *Fisheries Research,* 23:157–164.

Graham N., Ferro R. S. T. *Eds*. (2004). The *Nephrops* fisheries of the NE Atlantic and Mediterranean - A review and assessment of fishing gear design. *ICES Coop. Res. Rep.*, 270, 38 pp.

Gulland, J. A. (1956). *On the fishing effort in English demersal fisheries*. London, Fish. Invest.,41.

Guijarro, B., Massutí, E. (2006). Selectivity of diamond- and square-mesh codends in the deep-water crustacean trawl fishery off the Balearic Island (western Mediterranean). *ICES J. Mar. Sci.* 63: 52-67.

Hall, S. J. (1994). Physical hsturbance and marine benthic communities: life in unconsolidated sediments. *Oceanogr. Mar. Biol. Annu. Rev.* 32: 179-239.

Hall, S. J. (1999). *The effects of fishing on marine ecosystems and communities.* Blackwell Science Ltd., Oxford. 274 pp.

Hamley, J. M. (1975). Review of gillnet selectivity. *Journal of the Fisheries Research Board of Canada*, 32: 1943–1969.

Hansson, M., Lindegarth, M., Valentinsson, D., Ulmestrand, M. (2000). Effects of shrimp-trawling on abundance of benthic macrofauna in Gullmarsfjorden, Sweden. *Mar. Ecol. Prog. Ser.*, 198: 191-201.

Harmelin-Vivien, M. (1982). Ichthyofaune des herbiers de posidonies du Parc national de Port Cros: I. Composition et variations spatio-temporelles. *Travaux scientifiques Parc naturel Port Cros* 8: 69-92.

Henwood T.A., Stuntz, W.E. (1987). Analysis of sea turtle captures and mortalities during commercial shrimp trawling. *Fish. Bull.,* 85: 813–817.

Humborstad, O-B., Nottestad, L., Løkkeborg, S., and Rapp, H. T. (2004). RoxAnn bottom classification system, side scan sonar and video-sledge: spatial resolution and their use in assessing trawling impacts. *ICES J. Mar. Sci.* 61: 53-63.ICES, 2003a. Report of the ICES Advisory Committee on Fishery Management. 263. 229 pp.

Huse, I., Fernø, A. (1990). Fish behaviour studies as an aid toimproved longline hook design. *Fish. Res.,* 9: 287-297.

ICES. (2003). Environmental status of the European seas. A quality status report prepared by the international Council for the Exploration of the Sea for the German Federal Ministry for the Environment, *Nature Conservation and Nuclear Safety.* 75 pp.

ICES. (2004). Workshop on Discard Sampling Methodology and Raising Procedures. Charlottenlund (Denmark), 2 – 4 September, 2003. Final Report. 27 pp.

Ingólfsson, Ó. A., Jørgensen, T. (2006). Escapement of gadoid fish beneath a commercial bottom trawl: Relevance to the overall trawl selectivity, *Fisheries Research,* 79: 303-312.

Jennings S., Kaiser, M. J. (1998). The effects of Fishing on Marine Ecosystems. *Advance in Marine Biology.* vol..34.

Kaiser, M. J. (1998). Significance of bottom fishing disturbance. *Conservation Biology*, 12, 1230–1235.

Kaiser, M. J., Spencer, B. E. (1994). Fish scavenging behaviour in recently trawled areas. *Mar. Ecol. Prog. Ser.*, 112: 41-49.

Kawamura, G. (1972). Gillnet mesh selectivity curve developed from length-girth relationship. *Bull. Jap. Soc. Sci. Fish.*, 38: 1119–1127.

Kelleher, K. (2005). Discards in the world's marine fisheries. *FAO Fish. Tech. Pap.,* 470. 131 pp.

Kenchington, E. L. R., Prena, J., Gilkinson, K. D., Gordon, D.C. Jr., Macissac, K., Bourbonnais, C., Schwinghamer, P.J., Rowell, T.W., McKeown, D.L., Vass, W.P. (2001). Effects of experimental otter trawling on the macrofauna of a sandy bottom ecosystem on the Grand Banks of Newfoundland. *Can. J. Fish. Aquat. Sci.,* 58: 1043-1057.

Kennelly, S. J. (1995). The issue of bycatch in Australia's demersal trawl fisheries. *Rev. Fish Biol. Fish.*, 5, 13-21.

Kiessling, I. (2003). Finding solutions. *Derelict fishing gear and other marine debris in northern Australia.* 56 pp.

Kuiper, G. (1992). *The Wageningen Propeller Series.* ISBN 90-900 7247-0.

Kynoch, R. J., Ferro, R. S. T., Zuur, G. (1999). The effect on juvenile haddock by-catch of changing cod end twine thickness in EU trawl fisheries. *Mar. Technol. Soc. J.*, 33(2): 61-72.

Laurent, L., Lescure, J. (1994). L'hivernage des tortues caouannes *Caretta caretta* (L.) dans le sud Tunisien. *Rev. Ecol.Terr.Vie*, 49 : 63–86.

Lindeboom H. J., De Groot, S. J. eds. (1998). IMPACT-II: The effects of different types of fisheries on the North Sea and Irish Sea benthic ecosystems. Netherlands Institute for Sea Research. NIOZ-Rapport 1998-1. RIVODLO Report C003/98. Contract AIR2-CT94-1664 Financed by the European Community. 404 pp.

Lleonart, J., Maynou, F. (2003). Fish stock assessments in the Mediterranean: state of the art. *Sci. Mar.,* 67(Suppl. 2): 37-49.

Lindeboom, H. J., De Groot, S. J. eds. (1998). IMPACT-II: The effects of different types of fisheries on the North Sea and Irish Sea benthic ecosystems. Netherlands Institute for Sea Research. NIOZ-Rapport 1998-1. RIVODLO Report C003/98. Contract AIR2-CT94-1664 Financed by the European Community. 404 pp.

Løkkeborg, S., Bjordal A. (1992). Species and size selectivity in longline fishing: a review. *Fish. Res.,* 13: 311–322.

Lowry, N. (1995). The effect of twine size on bottom trawl codend selectivity. *Int. Counc. for the Explor. of the Sea.* CM 1995/B:6.

Lowry, N., Robertson, J. H. B. (1996). The effect of twine thickness on codend selectivity of trawls for haddock in the North Sea. *Fish. Res.* 26: 353-363.

Lucchetti, A. (2014). *Reduction of sea turtle mortality in the professional fishing.* Project LIFE12 NAT/IT/000937.

Lucchetti, A., Buglioni G., Conides A., Klaoudatos D., Sartor P, Sbrana M., Spedicato M.T., Stamatopoulos C. (2014). Technical measures without enforcement tools: is there any sense? A methodological approach for the estimation of passive net length in small scale fisheries. Mediterranean Marine Science DOI: 10.12681/mms.889.

Lucchetti, A., Piccinetti., Meconi U., Frittelloni C., Marchesan M., Palladino S., Virgili M. (2014). Transferable Fishing Concessions (TFC): A pilot study on the applicability in the Mediterranean Sea. *Marine Policy* 44: 438–447.

Lucchetti, A., Sala A. (2012). Impact and performance of Mediterranean fishing gear by side-scan sonar technology. *Canadian Journal of Fisheries and Aquatic Sciences.* 69(11): 1806-1816.

Lucchetti, A., Sala A. (2010). An overview of loggerhead sea turtle (*Caretta caretta*) bycatch and technical mitigation measures in the Mediterranean Sea. *Rev. Fish Biol.* Fish. 20(2): 141-161.

Lucchetti, A. (2008). Comparison of diamond- and square-mesh codends in the hake (*Merluccius merluccius* L. 1758) trawl fishery of the Adriatic Sea (central Mediterranean). *Scientia Marina* 72(3): 451-460.

Macdonald, D. S., Little, M., Eno, N. C., Hiscock, K. (1996). Disturbance of benthic species by fishing activities: a sensitivity index. *Aquatic Conserv: Mar. Freshw. Ecosyst.*, 6: 257–268.

Madsen, N., Moth-Poulsen, T., Holst, R., Wileman, D. (1999). Selectivity experiments with escape windows in the North Sea Nephrops (Nephrops norvegicus) trawl fishery. *Fisheries Research,* 42:167-181.

Madsen, N., Hansen, K.E., Moth-Poulsen, T. (2001). The kite cover: a new concept for covered codend selectivity studies. *Fisheries Research,* 49 (3): 219-226.

Malik, M. A., Mayer, L. A. (2007). Investigation of seabed fishing impacts on benthic structure using multi-beam sonar, sidescan sonar, and video. *ICES Journal of Marine Science,* 64: 1053-1065.

Marasco, R., Terry, J., (1982). Controlling incidental catch. An economic analysis of six management options. *Mar. Pol.,* 6:131-139.

Mattiello, S., Raicevich, S., Giomi, F., Botter, L., Di Muro, P., Pranovi, F., Beltramini, M., (2004). Resistance to stress and Hc functional modulation in *Liocarcinus* sp. *Micron*, 31 (1-2): 55-57.

McCaughran, D. A., (1992). Standardized nomenclature and methods of defining bycatch levelsand implications. *Proceedings of the National Industry Bycatch Workshop,* Newport, Oregon, February 4-6, 1992. Seattle WA: Natural Resources Consultants.

McConnaughey, R. A., Mier, K. L., Dew, D. B. (2000). An examination of chronic effects on soft-bottom benthos of the eastern Bering Sea. *ICES J. Mar. Sci.*, 57: 1377-1388.

Messina, G., Notti, E. (2007). Energy saving in trawlers: practical and theoretical approaches. *Proceedings ICMRT '07*, Ischia, Italy;

Messina, G., Notti, E. (2008). Some contributions to the integrated safety of fishing vessels. Proceedings Nor-Fishing Technology Conference, Trondheim, Norway.

Mitchell, J.F., Watson, J.W., Foster, D.G., Caylor, R.E. (1995). The Turtle Excluder Device (TED): *A Guide to Better Performance*. NOAA Technical Memorandum NMFS-SEFSC, 35 pp.

Moran, M. J., Stephenson, P. C. (2000). Effects of otter trawling on macrobenthos and management of demersal scalefish fisheries on the continental shelf of north-western Australia. *ICES J. Mar. Sci.*, 57: 510-516.

Morello, E. B., Froglia, C., Atkinson, R. J. A., Moore, P. G. (2005). Impacts of hdraulic dredgeing on a macrobenthic community of the Adriatic Sea, Italy. *Can. J. Fish. Aquat. Sci.,* 62: 2076-2087.

Murawski, S. A. (1992). A global assessment of fisheries bycatch and discards. *In* Alverson, D.L., Freeberg, M.H., Pope, J.G., Murawski, S.A. *FAO Fish. Tech. Pap.,* 339. Rome, FAO. 1994. 233 pp.

National Academy of Sciences. (1975). Marine Litter. In: *Assessing Potential Ocean Pollutants*. A report on the study of assessing potential ocean pollutants to the Ocean Affairs Board, Commission of the Natural Resources, National Research Council, National Academy of Sciences. Washington, DC. 465 pp.

Nedelec, C., Prado, J. (1990). *Definition and classification of fishing gear categories.* FAO Fisheries Technical Paper No. 222, revision 1. Rome: FAO, 92 pp.

Nike Bianchi, C., Morri, C. (2000). Marine Biodiversity of the Mediterranean Sea: situation, problems and prospects for future research. *Marine Pollution Bulletin,* Volume 40, Issue 5, 1: 367–376.

NMFS (National Marine Fisheries Service). 1998. Economics of bycatch: the case of shrimp and red snapper fisheries in the US Gulf of Mexico. In *Managing the nation's bycatch: priorities, programs and actions for the National Marine Fisheries Service.* 199 pp.

Nordic Council of Ministers. (2003). *Workshop on Discarding in Nordic Fisheries.* Sophienberg Castle, Rungsted, Denmark, 18–20 November 2002.

Notti E., Sala A. (2012). On the opportunity of improving propulsion system efficiency for Italian fishing vessels. *Paper proceedings of the Second International Symposium on Fishing Vessel Energy Efficiency E-Fishing,* Vigo, Spain, May 2012: 117-122. ISBN: 978-84-615-8943-2;

Notti E., Sala A. 2013. Information collection in energy efficiency for fisheries (ICEEF2012). Final Project Report to the European Commission - Joint Research Centre (JRC), Tender Contract IPSC/2011/04/01/NC Nr. 256660: 87 pp

Ordines, F., Massutí, E., Guijarro, B., Mas, R. (2006). Diamond vs. square mesh codend in a multi-species trawl fishery of the western Mediterranean: effects on catch composition, yield, size selectivity and discards. *Aquat. Living Resour.* 19: 329-338.

Özbilgin, H., Tosunoğlu, Z. (2003). Comparison of the selectivities of double and single codends. Fisheries Research, 63: 143-147.

Özbilgin, H. T., Tosunoğlu, Z., Metin, G., Tokaç, A. (2005a). Seasonal variation in trawl codend selectivity of annular sea bream (Diplodus annularis L., 1758). *Turkish Journal of Veterinary and Animal Sciences*, 29: 959–965.

Özbilgin, H., Tosunoğlu Z., Aydin C., Kaykaç H., Tokaç A. (2005b). Selectivity of Standard, Narrow and Square Mesh Panel Trawl Codends for Hake (Merluccius merluccius) and Poor Cod (Trisopterus minutus capelanus). *Turk. J. Vet. Anim. Sci.* 29: 967-973.

Palanques, A., Guille´n, J., Puig, P. (2001). Impact of bottom trawling on water turbidity and muddy sediment of an unfished continental shelf. *Limnol. Oceanogr.*, 46 (5), 1100–1110.

Palomera, I., Olivar M.P., Salat J., Sabatés A., Col, M., García, A. Morales-Nin, B. (2007). Small pelagic fish in the NW Mediterranean Sea: An ecological review. *Progr. Ocean.*, 74: 377-396.

Papaconstantinou, C., Farrugio, H., (2000). Fisheries in the Mediterranean. *Med. Mar. Sci.*, 1(1): 5-18.

Pelletier, D. (1998). Intercalibration of research survey vessels in fisheries: A review and an application. *Canadian Journal of Fisheries and Aquatic Sciences*. 55: 2672–2690.

Pilskan, C. H., Churchill, J. H., Mayer, L. M. (1999). Re-suspension of sediment by bottom trawling in the Gulf of Maine and potential geochemical consequences. *Cons. Biol.*, 12: 1223-1229.

Pitcher, T. J., Pauly, D. (1998). Rebuilding ecosystems, not sustainability, as the proper goal of fisheries management. In: Pitcher T.J., Hart P.J.B., and Pauly D. (Eds). *Reinventing fisheries management.* London: Chapman and Hall. 311–329.

POC (Pew Oceans Commission). 2003. America's living oceans: charting a course for sea change. A report to the nation – recommendations for a new ocean policy. Arlington, VA: POC. 144 pp.

Pope, J. A., Margetts, A. R., Hamley, J. M., Akyüz, E. F. (1975). Manual of methods for fish stock assessment. *FAO Fish.Tech. Pap.*, 41 (1). 46 pp.

Prado J. (1990). Fisherman's workbook. 1. Fishing equipment I. Title II. Dremière, P.Y. III. *Food and Agriculture Organization* 639.2. ISBN 0-85238-163-8.

Pranovi, F., Giovanardi, O. (1994). The impact of hydraulic dredging for short-necked clams, Tapes spp., on an infaunal community in the lagoon of Venice. *Scientia Marina,* 58(4) : 345-35.3

Pranovi, F., Raicevich, S., Franceschini, G., Farrace, M. G., Gionavardi O. (2000). Rapido trawling in the northern Adriatic Sea: effects on benthic communities in an experimental area. *ICES J. Mar. Sci.,* 57: 517-524.

Pranovi, F., Raicevich, S., Franceschini, G., Torricelli, P., Giovanardi, O. (2001). Discard composition and damage to non-target species in the "rapido" trawl fishery. *Mar. Biol.,* 139: 863-875.

Quignard, J. P., Farrugio, H. (1982). Quelques grands problèmes concernant l'eau, la faune et la pêche en Méditerranée. *La Pêche Maritime* 1248, 149–155.

Ramsay, K., Kasier, M. J., Hughes, R. N. (1998). Responses of benthic scavengers to fishing disturbance by towed gears in different habitats. *Journal of Experimental Marine Biology and Ecology* 224 73- 89.

Sacchi, J. (2008). *Impact des techniques de peche sur l'environnement en Méditerrane.* GFCM, Stud. & Rep., 84, 62 pp.

Sala, A., Lucchetti, A., Perdichizzi, A., Herrmann, B., Rinelli, P. (2015). Is square-mesh better selective than larger mesh? A perspective on the management for Mediterranean trawl fisheries. *Fisheries Research Volume 161,* January 2015, Pages 182–190.

Sala, A., D'Arc Prat Farran, J., Antonijuan, J., Lucchetti, A. (2009). Performance and impact on the seabed of an existing- and an experimental-otterboard: Comparison between model testing and full-scale sea trials. *Fish. Res.* 100: 156-166.

Sala, A., Buglioni, G., Lucchetti, A. (2010). *Fuel saving otterboards. Paper proceedings of the International Symposium on Energy use in Fisheries: Improving Efficiency and Technological Innovations from a Global Perspective,* Seattle, USA, November 2010: 4 pp.

Sala, A., Notti, E. (2012). *Information Collection in Energy Efficiency for Fisheries* (ICEEF). Final report project Service contract Nr. 256660.

Sala, A., Lucchetti, A., Affronte, M. (2011). Effects of Turtle Excluder Devices (TEDs) on bycatch and discards reduction in the demersal fisheries of Mediterranean Sea. *Aquatic Living Resources.* 24: 183-192.

Sala, A., Lucchetti, A. (2011). Effect of mesh size and codend circumference on selectivity in the Mediterranean demersal trawl fisheries. *Fisheries Research.* 110: 252– 258.

Sala, A., De Carlo, F., Buglioni, G., Lucchetti, A. (2011). Energy performance evaluation of fishing vessels by fuel mass flow measuring system. *Ocean Engineering,* 38 (5-6): 804-809.

Sala, A., Lucchetti, A. (2010). The effect of mesh configuration and codend circumference on selectivity in the Mediterranean trawl Nephrops fishery*". Fisheries Research.* 103: 63–72.

Sala, A., Lucchetti, A., Piccinetti, C., Ferretti, M. (2008). Size selection by diamond- and square-mesh codends in multi-species Mediterranean demersal trawl fisheries. *Fisheries Research* 93: 8–21.

Sala, A., Lucchetti, A., Buglioni, G. (2007). The influence of twine thickness on the size selectivity of polyamide codends in a Mediterranean bottom trawl. *Fisheries Research.,* 83: 192-203.

Sánchez Jerez, P., Ramos Esplá, A.A. (1996). Detection of environmental impacts by bottom trawling on *Posidonia oceanica* (L.) Delile meadows: sensitivity of fosh and macroinvertebrate communities. *J. Eco. Health,* 5: 239-253.

Santojanni, A., Cingolani, N., Arneri, E., Kirkwood, G., Belardinelli, A., Giannetti, G., Colella, S., Donato, F., Barry, C. (2005). Stock assessment of sardine (Sardina pilchardus, Walb.) in the Adriatic Sea with an estimate of discards. *Scientia Marina,* 69: 603–617.

Sardà, F., Lleonart, J., Cartes, J. E. (1998). An analysis of the population dynamics of *Nephrops norvegicus* (L) in the Mediterranean Sea. *Sci. Mar.* 62(Supl. 1): 135-143.

Sardà, F., Bahamon, N., Molí, B., Sardà-Palomera, F. (2006). The use of a square mesh codend and sorting grids to reduce catches of young fish and improve sustainability in a multispecies bottom trawl fishery in the Mediterranean. *Sci. Mar.* 70(3): 347-353.

Schwinghamer, P., Gordon, D. C., Rowell, T. W., Prena, J., McKeown, D. L., Sonnichsen, G., Guigné, J. Y. (1998). Effects of experimental otter trawling on surficial sediment properties of a sandy bottom ecosystem on the Grand Banks of Newfoundland. *Conserv. Biol.* 12: 1215-1222.

Sechin, Y. T. (1969). A mathematical model for the selection curve of a gillnet. *Rybn. Khoz.* 45: 56–58. (in Russian).

Smith, C. J., Papadopoulou, K. N., and Diliberto, S. (2000). Impact of otter trawling on an eastern Mediterranean commercial trawl fishing ground. *ICES J. Mar. Sci.* 57: 1340–1351.

Smith, C.J., Banks, A.C., and Papadopoulou, K. N. (2007). Improving the quantitative estimation of trawling impacts from side scan-sonar and underwater-video imagery. *ICES J. Mar. Sci.* 64: 1692-1701.

Stabenau, E. K., Heming, T. A., Mitchell, J. F. (1991). Respiratory, acid-base and ionic status of Kemp's ridley sea turtles (*Lepidochelys kempi*) subjected to trawling. *Comp. Bioch. Phys.,* 99: 107–111.

Sterling D., Eayrs S. (2006). Designs and assessment of two gear modifications to reduce the benthic impact and fuel intensity of prawn trawling in Australia. ICES Symposium. *Fishing Technology in the 21st Century: Integrating Fishing And Ecosystem Conservation,* Boston, 2006.

Stewart P. A. M. (2001). A review of studies of fishing gear selectivity in the Mediterranean. FAO COPEMED, 57 pp.

Stewart, P. A. M., Robertson J. H. B. (1985). Small mesh covers. *Scot. Fish. Res. Rep.,* 32.

Suuronen P. (2005). Mortality of fish escaping trawl gears. *FAO Fish. Tech. Pap.,* 478. Rome, FAO. 2005. 72 pp.

Thrush, S. F., Hewitt, J. E., Cummings, V. J., Dayton, P. K., Cryer, M., Turner, S. J., Funnel, G. A., Budd, R. G., Milburn, C.J., Wilkinson, M. R. (1998). Disturbance of the marine benthic habitat by commercial fishing: Impacts at the scale of the fishery. *Ecological Applications.*, 8: 866-879.

Tosunoğlu, Z., Aydın, C., Özaydın, O. (2008). Selectivity of a 50-mm diamond mesh knotless polyethylene codend for commercially important fish species in the Aegean Sea. *Journal of Applied Ichthyology*, 24: 311-315.

Tsagarakis, K.,Vassilopoulou, V., Kallianiotis, A., Machias, A. (2012). Discards of the purse seine fishery targeting small pelagic fish in the Eastern Mediterranean Sea. *Scientia Marina,* 76: 561–572.

Tschernij, V., R. Holst. 1999. *Evidence of factors at vessel-level affecting codend selectivity in Baltic cod demersal fishery*. ICES CM 1999/R02: 12 pp.

Tuck, I. D., Hall, S. J., Roberston, M. R., Amstrong, E., Basford, D.J. (1998). Effects of physical trawling disturbance in a previously unfished sheltered Scottish sea loch. *Mar. Ecol. Prog. Ser.,* 162: 227–242.

Tudela, S. (2004). Ecosystem effects of fishing in the Mediterranean: An analysis of the major threats of fishing gear and practices to biodiversity and marine habitats. General Fisheries Commission for the Mediterranean (FAO). *Studies and Reviews*, 74. 58 pp.

Tupper, E. (1981). *Introduction to naval architecture.* Butterworth Heinemann. ISBN 0750625295

Valdemarsen, J.W., Jørgensen, T. Engås, A. (2007). Options to mitigate bottom habitat impact of dragged gears. *FAO Fisheries Technical Paper.* No. 506. Rome, FAO. 2007. 29p.

Valdemarsen, J. W., Suuronen, P. (2003). Modifying fishing gear to achieve ecosystem objectives. In M. Van Marlen B. (2000) Technical modifications to reduce the by-catches and impacts of bottom-fishing gears. In M.J. Kaiser and S.J. de Groot, eds. *The effects of fishing on nontarget species and habitats: biological conservation and socio-economic issues,* pp. 253 - 268. Oxford, UK, Blackwell Science. 399 pp.

Vitale, S., Cannizzaro, L., De Stefano, G., Milazzo, A., Barraco, B. (2011). *The role of the artisanal fishery in Mediterranean coastal zone management: casework on Sicilian coast, SI 64* (Proceedings of the 11th International Coastal Symposium),. Szczecin, Poland, 1946-1949. ISSN 0749-0208.

Walsh, S. J., Engås, A., Ferro, R., Fonteyne, R., Van Marlen, R. (2002). Improving fishing technology to catch (or conserve) more fish: the evolution of the ICES Fishing Technology and Fish Behaviour Working Group during the past century. *ICES Mar. Sci. Symp.,* 215, 493-503.

Watling. L., Norse, E. A. (1998). Disturbance of the seabed by mobile fishing gear: a comparison to forest clear cutting. *Conservation Biology,* 12(6): 1180-1197.

Wileman, D. A., Ferro, R. S. T., Fonteyne, R., Millar, R.B. (1996). Manual of methods of measuring the selectivity of towed fishing gears. *ICES Coop. Res. Rep.*, 215: 126 pp.

References on Turtle Bycatch

Aguilar R, Más J, Pastor X (1993) *Las tortugas marinas y la pesca con palangre de superficie en el Mediterráneo.* Greenpeace. Proyecto internacional.

Aguilar, R., Mas, J., Pastor, X. (1995). Impact of Spanish swordfish longline fisheries on the loggerhead sea turtle Caretta caretta population in the western Mediterranean. In: Richardson JI, Richardson TH (eds) Proc. 12th Annual Workshop on Sea Turtles Biology and Conservation, pp. 1-6. NOAA Technical Memorandum NMFS-SEFSC- 361.

Alós, J., Palmer, M., Grau, A. M., Deudero, S. (2008). Effects of hook size and barbless hooks on hooking injury, catch per unit effort, and fish size in a mixed-species recreational fishery in the western Mediterranean Sea. *ICES J Mar Sci:* 65(6):899-905.

Argano, R., Baldari, F. (1983). Status of Western Mediterranean Sea Turtles. *Rapp Comm Int Mer Médit* 28(5): 233-235.

Argano, R., Basso, R., Cocco, M., Gerosa, G. (1992). New data on loggerhead (Caretta caretta) movements within Mediterranean. *Boll Mus Ist Biol Univ Genova* 56-57:137-163.

Atabey, S., Taskavak, E. (2001). A preliminary study on the prawn trawls excluding sea turtles. *Urun Derg J Fish Aquat Sci* Vol. 18, no. 1-2, pp. 71-79. 2001.

Báez, J. C., Camiñas, J. A., Rueda, L. (2006). Accidental fishing capture of marine turtles in South Spain. *Mar Turtle Newsl* 111:11–12.

Báez, J.C., Real, R., García-Soto, C., De la Serna, J. M., Macías, D., Camiñas, J. A. (2007). Loggerhead sea turtle bycatch depends on distance to the coast, independent of fishing effort: implications for conservation and fisheries management. *Mar Ecol Prog Ser* 338: 249–256.

Baran, I., Kasparek, M. (1989). *Marine Turtles Turkey: Status Survey 1988 and Recommendations for Conservation and Management.* WWF, Heidelberg, 123 pp.

Bentivegna, F. (2002). Intra-Mediterranean migrations of loggerhead sea turtles (Caretta caretta) monitored by satellite telemetry. *Mar Biol* 141: 795-800.

Bjorndal, K. A., Bolten, A. B., Lagueux, C. J. (1994). Ingestion of marine debris by juvenile sea turtles in coastal Florida habitats. *Mar Poll Bull* 28(3): 154–158.

Bolten, A. B., Bjorndal, K. A., Martins HR, Dellinger T, Biscoito MJ, Encalada SE, Bowen BW (1998) Transatlantic developmental migrations of loggerhead sea turtles demonstrated by mtDNA sequence analysis. *Ecol Applic* 8:1–7.

Bolten, A. B., Bjorndal, K. A., Martins, H. R. (1994). Life history model for the loggerhead sea turtle (Caretta caretta) population in the Atlantic: potential impacts of a longline fishery. In: Balazs, G.H. and Pooley, S.G. (Eds.), Research plan to assess marine turtle hooking mortality: results of an expert workshop held in Honolulu, Hawaii, Nov. 16-18, 1993. NOAA, National Marine Fisheries Service, Southwest Fisheries Science Center, Honolulu Laboratory, *Administrative Report* H-93-18. 48-54.

Bradai, M. N. (1992). Les captures accidentelles de Caretta caretta au chalut benthique dans le Golfe de Gabés. *Rapp Comm int Mer Médit* 33: 285.

Bradai, M. N. (1993). *La tortue marine Caretta caretta dans le sud-est de la Tunisie* (Peche accidentelle - Utilisation - Législation). MAP/UNEP. 27 pp.

Bradai, M. N. (1994). Observations sur la tortue marine Caretta caretta en Tunisie. Actes des Premières Journées Tunisiennes des Sciences de la Mer, Kerkennah, 18-20 Décembre 1994. *Bull Inst Nat Sci Techn Mer* 32-34.

Bradai, M. N. (1995). Impact de la pêche sur la tortue marine Caretta caretta sur les côtes sud-est de la Tunisie. *Rapp Comm int Mer Médit* 34: 238.

Broderick, A. C., Glen, F., Godley, B. J., Hays, G. C. (2002). Estimating the number of green and loggerhead sea turtles nesting annually in the Mediterranean. *Oryx* 36(3): 227-236.

Broderick, A. C., Coyne, M. S., Fuller, W. J., Glen, F. G., Godley, B. J. (2007). Fidelity and over-wintering of sea turtles. *Proc Biol Sci* 274(1617): 1533–1539.

Caillouet, C. W. Jr., Duronslet, M. J., Landry, A.M. Jr., Revera, D. B., Shaver, D. J., Stanley, K. M., Heinly, R. W., Stabenau, E. K. (1991). Sea turtle strandings and shrimp fishing effort in the northwestern Gulf of Mexico, 1986-1989. *Fish Bull* 89(4): 712-718.

Cambiè, G., Camiñas, J. A., Franquesa, R., Mingozzi, T. (2008). Impact of surface longline on loggerhead sea turtle Caretta caretta: a case study along the southern Ionian coast of Calabria (Italy). *VII Congresso Nazionale della Società Herpetologica Italica.* Oristano (Italy) 1-5 October 2008.

Camiñas, J. A., De la Serna, J. M. (1995). The loggerhead distribution in the western Mediterranean Sea as deduced from captures by the Spanish longline fishery. *Sci Herpet:* 316-323.

Camiñas, J. A., (1997a). Relacion entre las poblaciones de la tortuga boba (Caretta caretta Linnaeus, 1758) procedentes del Atlantico y del Mediterraneo en la Region del Estrecho de Gibraltar y areas adyacentes. *Revista espagnola de Herpetologia* 11: 91-98.

Camiñas, J. A. (1997b). Capturas accidentales de tortuga boba (Caretta caretta, L. 1758) en el Mediterráneo Occidental en la pesquería de palangre de superfície de pez espada (Xiphias gladius L.). *ICCAT Collective Volume of Scientific Papers XLVI* (4): 446–455.

Camiñas, J. A. (1988). Incidental Captures of Caretta caretta (L.) with surface longlines in the Western Mediterranean. *Rapports et process-Verbaux des Re´ unions du la Commission Internationale Pour l'Exploration Scientifique de la Mer Méditerranée* 31 (2): 285. CIESM, Montecarlo.

Camiñas, J. A., Valeiras, J. (2001). Marine turtles, mammals and sea birds captured incidentally by the Spanish surface longline fisheries in the Mediterranean Sea. *Rapp Comm Int Mer Medit* 36: 248.

Camiñas, J. A., Valeiras J., De La Serna, J. M. (2001). Spanish surface longline gear types and effects on marine turtles in the western Mediterranean Sea. Pages 88-93. In: *Proceedings, First Mediterranean Conference on Marine Turtles.* Rome, 2001.

Camiñas, J. A. (2004). Sea turtles of the Mediterranean Sea: population dynamics, sources of mortality and relative importance of fisheries impacts. *FAO Fish Rep* 738: 27-84.

Camiñas, J. A., Báez, J. C., Real, R., Sagarminaga, R., Valeiras, X. (2006a). Analysis of loggerhead (Caretta caretta, linnaeus, 1758) and swordfish (Xiphias gladius, linnaeus, 1758) capture distribution within sets in the Spanish Mediterranean surface drifting longline fishery. In: Frick M, Panagopoulou A, Rees A, Williams K (eds) *Book of abstracts of the 26th Annual Symposium on Sea Turtle Biology and Conservation.* Island of Crete, Greece, 3-8 April 2006.

Camiñas, J. A., Báez, J. C., Valeiras, J., Real, R. (2006b). Differential loggerhead by-catch and direct mortality due to surface longlines according to boat strata and gear type. *Sci mar* 70 (4): 661-665.

Carreras, C., Cardona, L., Aguilar, A. (2004). Incidental catch of the loggerhead sea turtle Caretta caretta off the Balearic Islands (western Mediterranean). *Biol Cons* 117: 321–329

Casale, P., Laurent, L., De Metrio, G. (2004). Incidental capture of marine turtles by the Italian trawl fishery in the north Adriatic Sea. *Biol Cons* 119: 287–295.

Casale, P., Freggi, D., Basso, R., Argano, R. (2005a). Interaction of the static net fishery with loggerhead sea turtles in the Mediterranean: insights from mark-recapture data. *Herp J* 15: 201-203.

Casale P, Freggi D, Basso R, Argano R (2005b) Interaction of the static net fishery with loggerhead sea turtles in the Mediterranean: insights from mark-recapture data. Short note, *Herp J,* 15: 201-203.

Casale P (2005c) Holes in the circle. *A critical review of circle hooks as a measure for reducing the impact of longline fishery on sea turtles.* Report June 2005. (Unpublished report to WWF).

Casale P, Catturino L, Freggi D, Rocco M, Argano R (2007a) Incidental catch of marine turtles by Italian trawlers and longliners in the central Mediterranean. *Aquatic Conserv:*

Mar Freshw Ecosyst Published online in Wiley InterScience. (www.interscience.wiley.com) DOI: 10.1002/aqc.841.

Casale P, Freggi D, Rocco M (2007b) Mortality induced by drifting longline hooks and branchlines in loggerhead sea turtles, estimated through observation in captivity. Aquatic Conserv: Mar Freshw Ecosyst DOI: 10.1002/acq.894.

Casale P (2008) *Incidental catch of marine turtles in the Mediterranean Sea: captures, mortality, priorities*. WWF Italy, Rome. 64 pp.

Deflorio M, Aprea A, Corriero A, Santamaria N, De Metrio G (2005) Incidental captures of sea turtles by swordfish and albacore longlines in the Ionian Sea. *Fish sci* 71: 1010–1018.

De La Serna JM, Ortiz De Urbina JM, García Barcelona S (2008). Factores estratégicos y tecnológicos que influyen en la captura de especies asociadas en la pesquería de pez espada con palangre de superficie en el Mediterráneo. *Col Vol Sci Pap ICCAT,* 62(6): 1039-1051.

Delaugerre M (1987) Statut des tortues marines de la Corse (et de la Mediterranee). *Vie Milieu 37*(3-4): 243-264.

Dellinger T (2000) *Conservation support project for North Atlantic Caretta caretta sea turtles - Life Nature Project contract no. B4-3200/96/541 (Life96Nat/P/3019).* Final Technical Activity Report, pp. 56, CITMA, Funchal.

Dellinger T, Ferreira T (2005) Diving behaviour of juvenile loggerhead sea turtles (Caretta caretta) and its relation to deep-sea longline fishing in Madeiran Waters. *Final Technical Report to the Portuguese Science Foundation FCT for project* PDCTM-POCTI/P/MAR/15248/1999, pp. 46, Universidade da Madeira, Funchal.

De Metrio G, Megalofonou P (1988) Mortality of marine turtles (Caretta caretta L. and Dermochelys coriacea L.) consequent to accidental capture in the Gulf of Taranto. Rapp Comm int Mer Médit 31(2): 285.

Demetropoulos, A. (1998) Rehabilitation of habitats and management of resources. Euro Turtle web page: http://www.ex.ac.uk/telematics/EuroTurtle/.

Di Natale A. (1995) Drift net impact on protected species: observer data from the Italian fleet and a proposal for a model to assess the number of cetaceans in the bycatch. *ICCAT,* 44 (1): 255-263.

Echwikhi K, Jribi I, Bradai MN, Bouain A (2006) Interaction of marine turtles with longline fisheries in the region of Zarzis (gulf of Gabes, Tunisia). In: Frick M, Panagopoulou A, Rees A, Williams K (eds) *Book of abstracts of the 26th Annual Symposium on Sea Turtle Biology and Conservation.* Island of Crete, Greece, 3-8 April 2006.

Epperly SP, Braun J, Chester AJ, Cross FA, Merriner JV, Tester PA (1995) Winter distribution of sea turtles in the vicinity of Cape Hatteras and their interaction with the summer flounder trawl fishery. *Bull Mar Sci* 56(2): 547-568.

Epperly SP., (2003) Fisheries-related mortality and Turtle Excluder Devices (TEDs). In: Lutz PL, Musick JA (eds), *The Biology of Sea Turtles.* CRC Marine Science Series, CRC Press, Inc., Boca Raton, Florida: 339-353.

FISH/2005/28A - Service Contract SI2.439703. (2008). *Field study to assess some mitigation measures to reduce bycatch of marine turtles in surface longline fisheries.* Final report. 215 pp.

Gerosa G, Casale P (1999) *Interaction of Marine Turtles with Fisheries in the Mediterranean.* Mediterranean Action Plan – UNEP Regional Activity Centre For Specially Protected Areas. 59 pp.

Gerosa G, Aureggi M (2001) Sea turtle handling guidebook for fishermen. United Nations Environment Programme Mediterranean Action Plan - UNEP -MAP, Regional Activity Centre For Specially Protected Areas. 31 pp.

GFCM-SAC (2008) *Report of the Transversal Working group on bycatch/incidental catches.* Rome, Italy. 24 pp.

Gilman E, Zollet E, Beverly S, Nakano H, Davis K, Shiode D, Dalzell P, Kinan I (2006). Reducing sea turtle bycatch in pelagic longline fisheries. *Fish Fish* 7: 2–23.

Gilman E, Kobayashi D, Swenarton T, Brothers N, Dalzell P, Kinan-Kelly I (2007) Reducing sea turtle interactions in the Hawaii-based longline swordfish fishery. *Biol Cons* 139: 19-28.

Godley BJ, Gucu AC, Broderick AC, Furness RW, Solomon SE (1998) Interaction between marine turtles and artisanal fisheries in the eastern Mediterranean: a probable cause for concern? *Zool Middle East* 16: 49-64.

Gramentz D (1989) Marine turtles in the central Mediterranean Sea. Centro 1(4): 41-56.

Groombridge B (1990) Marine turtles in the Mediterranean, distribution, population status, conservation. Council of Europe, *Nature and Environment Service.* 48: 98 pp.

Guglielmi P., Di Natale, A., Pelusi, P. (2000). *Effetti della pesca col palangaro derivante sui grandi pelagici e sulle specie accessorie nel Mediterraneo centrale.* Rapporto al Ministero per le Politiche Agricole e Forestali. DGPA Roma.

Hare, S. (1991). Turtles caught incidental to demersal finfish fishery in Oman. *Mar Turtle Newsl* 53: 14-16.

Henwood, T. A., Stuntz, W. E. (1987). Analysis of sea turtle captures and mortalities during commercial shrimp trawling. *Fish Bull* 85: 813–817.

Hilton-Taylor, C. (2000). *IUCN Red List of Threatened Species.* IUCN, Gland, Switzerland and Cambridge, UK.

Jribi, I., Bradai, M. N., Bouain, A. (2004). Étude de l'Interaction Tortue Marine Caretta Caretta. Chalut Bentique dans le Golfe de Gabès (Tunisie). *Rapp Comm Int Mer Médit* 37: 528.

Jribi, I., Bradai, M. (2008). A bibliographic overview on marine turtles bycatch in the Mediterranean Sea. In: *Report of the GFCM-SCMEE/SCSA Transversal Working Group on bycatch/incidental catches.* FAO, Rome, Italy, 15-16 September 2008.

Jribi, I., Echwikhi, K., Bradai, M. N., Bouain, A. (2008). Incidental capture of sea turtles by longlines in the Gulf of Gabès (South Tunisia): a comparative study between bottom and surface longlines. *Sci Mar* 72(2): 337-342.

Kapantagakis, A., Lioudakis, L. (2006). Sea turtle bycatch in the Greek drifting long line fishery. In: Frick M, Panagopoulou A, Rees A, Williams K (eds) *Book of abstracts of the 26th Annual Symposium on Sea Turtle Biology and Conservation.* Island of Crete, Greece, 3-8 April 2006.

Laurent, L. (1990). Les tortues marines en Algerie et au Maroc (Mediterranee). *Bulletin Societ´e Herpetologique de France* 55:1–23.

Laurent, L. (1991). Les tortues marines des côtes francaises méditerranéennes continentales. *Faune de Provence* (CEEP), 12: 76-90.

Laurent, L., Lescure, J. (1994). L'hivernage des tortues caouannes Caretta caretta (L.) dans le sud Tunisien. *Rev Ecol* (Terre Vie) 49: 63-86.

Laurent, L. (1996a). *Synthese historique de la presence de tortues marines sur les côtes de France (côtes méditerranéennes).* Paris, Ministère Francais de l'Environnement.

Laurent, L., Abd El-Mawla, E. M., Bradai, M. N, Demirayak, F., Oruc, A. (1996b). *Reducing sea turtle mortality induced by Mediterranean fisheries: trawling activity in Egypt, Tunisia and Turkey.* Report for the WWF International Mediterranean Programme. WWF Project 9E0103. 32 pp.

Laurent, L., Camiñas, J. A., Casale, P., Deflorio, M., De Metrio, G., Kapantagakis, A., Margaritoulis, D., Politou, C. Y., Valeiras, J. (2001). Assessing marine turtle bycatch in European drifting longline and trawl fisheries for identifying fishing regulations. Project-EC-DG Fisheries 98-008. Joint project of BIOINSIGHT, IEO, IMBC, STPS and University of Bari. Villeurbanne, France. Available at http://www.seaturtle.org/documents/EMTP-FINAL-REPORT.pdf.

Lazar, B., Tvrtkovic, N. (1995). Marine turtles in the eastern part of the Adriatic Sea: preliminary research. *Natura Croatica* 4: 59-74.

Lazar, B., Margaritoulis, D., Tvrtkovic, N. (1998). Migrations of loggerhead sea turtle (Caretta caretta) into the Adriatic Sea. In : *Memorieas de 18 Simposium International de Biologa y Conservacion de Tortugas Marinas, Mazaltan, Sinaoa (Mexico).* 100-101.

Lazar, B., Margaritoulis, D., Tvrtkovic, N. (2004). Tag recoveries of the loggerhead sea turtle Caretta caretta in the eastern Adriatic Sea: implications for conservation. *J Mar Biol Ass* UK 84: 475-80.

Lazar, B., Ziza, V., Tvrtkovic, N. (2006). Interactions of gillnet fishery with loggerhead sea turtles Caretta caretta in the northern Adriatic Sea. In: Frick M, Panagopoulou A, Rees A, Williams K (eds) *Book of abstracts of the 26th Annual Symposium on Sea Turtle Biology and Conservation. Island of Crete*, Greece, 3-8 April 2006.

Lee, H. A., Poland, G. C. R. (1998). Threats by fishing. Euro turtle (available at www.Ex.ac.uk/telematics/Euroturtle).

Lescure, J. (1987). *Tortues marines de l'Atlantique ouest. National Report for Martinique Western Atlantic Symposium II, Mayagüez,* Puerto Rico, September 1987. Unpublished, 27 pp.

Lewison, R. L., Freeman, S. A., Crowder, L. B. (2004). Quantifying the effects of fisheries on threatened species: the impact of pelagic longlines on loggerhead and leatherback sea turtles. Ecology Letters 7 (3): 221-231.

Life Nature – NAT/IT/00163 (2003). Progetto Delta Life Natura - Riduzione impatto attivita' umane su Caretta e tursiope e loro conservazione in Sicilia.

Løkkeborg, S. (2004). A review of existing and potential longline gear modifications to reduce sea turtle mortality. Pages 165-169. *Papers presented at the Expert Consultation on Interactions between Sea Turtles and Fisheries within an Ecosystem Context.* Rome, 9–12 March 2004. FAO Fish Rep 738, Suppl. Rome, FAO. 238 pp.

Lucchetti, A., Palumbo, V., Antolini, B., Affronte, M., Clò, S., Sala, A. (2008). Reduction of loggerheah turtle (Caretta caretta) bycatch in Mediterran bottom trawl fisheries. *Biol mar medit* 15(1): 336-337.

Lucchetti, A., Sala A., Affronte M. (2008). Turtle Excluder Devices Experiments in the Central Adriatic Sea. Report of the ICES-FAO Working Group on Fish Technology and Fish Behaviour (WGFTFB). 21-25 April 2008, *Tórshavn, Faroe Islands.* ICESCM 2008/FTC:02. BRUXELLES

Lucchetti, A., Sala, A. (2008*). Loggerhead Turtle (Caretta caretta) bycatch, case study: Mediterranean Sea.* 21-25 April 2008, Tórshavn, Faroe Islands. ICESCM 2008/FTC:02. 171-179.

Lucchetti, A., Sala, A. (2010). *An overview of Loggerhead sea turtle (Caretta caretta) bycatch and technical mitigation measures in the Mediterranean Sea.* Reviews in Fish Biology and Fisheries. 20:141–161.

Lutcavage, M. E., Lutz, P. L. (1997a). Diving physiology. In: Lutz PL, Musick JA (Eds) *The Biology of Sea Turtles.* CRC Marine Science Series, CRC Press, Inc., Boca Raton, Florida: 277-296.

Lutcavage, M. E., Plotkin, P., Witherington, B. E., Lutz, P. L. (1997b). Human impacts on sea turtle survival. In: Lutz PL, Musick, JA (Eds) *The Biology of Sea Turtles. CRC Marine Science Series*, CRC Press, Inc., Boca Raton, Florida: 387-409.

Lutz, P. L., Dunbar-Cooper, A. (1987). Variations in the blood chemistry of the loggerhead sea turtle, Caretta caretta. *Fish Bull* 85(1): 37-44.

Maffucci, F., Kooistra, W. H. C. F., Bentivegna, F. (2006). Natal origin of loggerhead turtles, Caretta caretta, in the neritic habitat off the italian coasts, Central Mediterranean. *Biol Cons vol.* 127(2): 183-189.

Margaritoulis, D. (1988). Post-nesting movements of loggerhead sea turtles tagged in Greece. *Rapp Comm Int Mer Médit* 31(2): 284.

Margaritoulis, D., Politou, C. Y., Laurent, L. (2001). Assessing marine turtle bycatch in the trawl fisheries of Greece. Pages 176-180. In *Proceedings of the First Mediterranean Conference on Marine Turtles. Rome*, 2001.

Margaritoulis, D., Argano, R., Baran, I., Bentivegna, F., Bradai, M. N., Camiñas, J. A., Casale, P., De Metrio, G., Demetropoulos, A., Gerosa, G., Godley, B., Houghton, J., Laurent, L., Lazar, Y. B. (2003). Loggerhead sea turtles in the Mediterranean: Present knowledge and conservation perspectives. In: Bolten AB, Witherington BE (eds) *Ecology and conservation of loggerhead sea turtles.* Washington, DC, Smithsonian Institution Press.

Mayol, J., Muntaner, J., Aguilar, R. (1988). Incidencia de la Pesca Accidental sobre las Tortugas Marinas en el Mediterraneo Espanol. *Butlletı′ de la Societat d'Història Natural de les Balears* 32 : 19–31.

Moein, S. E., Musick, J. A., Keinath, J. A., Barnard, D. E., Lenhardt, M., George, R. (1994). Evaluation of seismic sources for repelling sea turtles from hopper dredges. *Report from Virginia Institute of Marine Science, Gloucester* Point, VA, to US Army Corps of Engineers.

Nada, M., Casale, P. *Sea turtle killing and consumption in the Mediterranean coast of Egypt.* International Sea turtle Society. Available in the web-site: http://www.seaturtle.org/ists/PDF/final/2548.pdf.

National Marine Fisheries Service Southeast Fisheries Science Center (2001) Stock assessments of loggerhead and leatherback sea turtles and an assessment of the impact of the pelagic longline fishery on the loggerhead and leatherback sea turtles of the Western North Atlantic. *NOAA Tech. Mem.* NMFS-SEFSC-455.

National Research Council (1990*) Decline of the Sea Turtles: Causes and Prevention.* National Academy Press, Washington, DC, 355 pp.

Oros, J., Calabuig, P., Deniz, S. (2004). Digestive pathology of sea turtles stranded in the Canary islands between 1993 and 2001. *Veterinary Record* 155(6): 169–174.

Oruç, A., Demirayak, F., Sat, G. (1996). *Trawl Fisheries in the Eastern Mediterranean and its Impact on Marine Turtles.* The Conclusive Report. World Wildlife International and Dogal Hayati Koruma Dernegi. Istanbul, Turkey.

Oruç, A. (2001). Trawl fisheries in the eastern Mediterranean and its impact on marine turtles. *Zool Middle East* 24: 119-125.

Panou, A., Antypas, G., Giannopoulos, Y., Moschonas, S., Mourelatos, D. G., Mourelatos, C., Toumazatos, P., Tselentis, L., Voutsinas, N., Voutsinas, V. (1992). Incidental catches of loggerhead sea turtles, Caretta caretta, in swordfish longlines in the Ionian Sea, Greece. *Testudo* 3(4): 46-57.

Panou, A., Tselentis, L., Voutsinas, N., Mourelatus, C., Kaloupi, S., Voutsinas, V., Moschonas, S. (1999). Incidental catches of marine turtles in surface longline fishery in the Ionian Sea (Greece). *Contributions to the Zoogeography and Ecology of the Eastern Mediterranean Region* 1: 435-445.

Parga, M. L. (2008). Oral presentation of preliminary results of Proyecto TECNO funded by the Spanish Ministry of Environment and Fisheries, Secretaría General del Mar (2006 — 2008) and Bycatch Mitigation Projects I and II funded by the NOAA — NMFS (2005 — 2009). Report of the Transversal Working group on bycatch/incidental catches Rome, Italy, 15-16 September 2008.

Pinedo, M. C., Polacheck, T. (2004). Sea turtle bycatch in pelagic longline sets off southern 427 Brazil. *Biol Cons* 119: 335–339.

Piovano, S., Affronte, M., Balletto, E., Barone Dell'Anna, L., Di Marco, S., Dominaci, A., Giacoma, C., Mari, F., Maglietta, F., Zannetti, A. (2002). Experimental evaluation of the attractiveness of swordfish-fishing devices on loggerhead sea turtles: testing the floats' effect. In: Seminoff JA (ed) *Proceedings of the twenty-second annual symposium on sea turtle biology and conservation.* 4-7 April 2002 Miami, Florida, USA

Piovano, S., Balletto, E., Di Marco, S., Dominici, A., Giacoma, C., Zannetti, A. (2004). loggerhead sea turtle (Caretta caretta) bycatches on longlines: the importance of olfactory stimuli. *Ital J Zool* (Suppl. 2): 213–216.

Piovano, S., Celona A, Basciano A, Di Marco S, Giacoma C (2005) At-sea evaluation of artificial bait impact on longline target and bycatch species. In: *Proceedings of the Second Mediterranean Conference on Marine Turtles.* 4-7 May 2005 Kemer, Antalya, Turkey. 48 pp.

Piovano, S., Swimmer, Y., Giacoma, C. (2009). Are circle hooks effective in reducing incidental captures of loggerhead sea turtles in a Mediterranean longline fishery? *Aquatic Conservation: Marine and Freshwater Ecosystems.* Published online in Wiley InterScience (www.interscience.wiley.com). DOI: 10.1002/aqc.1021.

Polovina, J. J., Howell, E. A., Parker, DM, Balazs GH (2003) Dive depth distribution of loggerhead (Caretta caretta) and olive ridley (Lepidochelys olivacea) turtles in the central North Pacific: Might deep longline sets catch fewer turtles? *Fish Bull* (Wash DC) 101:189–193.

Rueda, L., Sagarminaga, R. J., Baez, J., C., Camiñas, J. A., Eckert, S. A., Boggs, C. (2006). *Testing mackerel bait as a possible bycatch mitigation measure for the Spanish Mediterranean swordfish longlining fleet.* In: Frick M, Panagopoulou A, Rees A, Williams, K. (eds) Book of abstracts of the 26th Annual Symposium on Sea Turtle Biology and Conservation. Island of Crete, Greece, 3-8 April 2006.

Rueda, L., Sagarminaga, R. (2008). Reducing bycatch of loggerhead sea turtles in the southwest Mediterranean via collaborative research with fishermen. *Poster presented to the 28th International Sea Turtle Symposium Loreto,* Baja California Sur, Mexico, 19-26 January 2008.

Sasso CR, Epperly SP (2006) Seasonal sea turtle mortality risk from forced submergence in bottom trawls. *Fish Res* 81: 86–88.

Silvani L., Gazo, M., Aguilar, A. (1999). Spanish driftnet fishing and incidental catches in the western Mediterranean. *Biol Conserv* 90(1): 79-85.

Southwood, A., Fritsches, K., Brill, R., Swimmer, Y. (2008). Sound, chemical, and light detection in sea turtles and pelagic fishes: sensory-based approaches to bycatch reduction in longline fisheries. *Endang Species Res:* 1-14.

Stabenau, E.K., Heming, T. A., Mitchell, J. F. (1991). Respiratory, acid-base and ionic status of Kemp's ridley sea turtles (Lepidochelys kempi) subjected to trawling. *Comp Bioch Phys* 99: 107–111.

Stabenau, E. K., Vietti, K. R. N. (2003). The physiological effects of multiple forced submergences in loggerhead sea turtles (Caretta caretta). *Fish Bull* 101: 889-899.

Suggett, D. J., Houghton, J. D. R. (1998). Possible link between sea turtle bycatch and flipper tagging in Greece. Mar Turtle Newsl 81: 10-11.Swimmer J, Brill R, Laurs M (2002) *Behavior and Physiology Experiments Aimed at Reducing Pelagic Longline Interactions with Marine Turtles.* Presented at the Meeting of the American Society of Limnology and Oceanography, Victoria, British Colombia, Canada, 10–14 June 2002.

Swimmer, Y., Arauz, R., Higgins, B., McNaughton, L., McCracken, M., Ballestero, J., Brill, R. (2005). Food color and marine turtle feeding behaviour: Can blue bait reduce turtle bycatch in commercial fisheries? *Mar Ecol Prog Ser* 295: 273–278.

Swimmer, Y., Brill, R. (2006). *Sea Turtle and pelagic fish sensory biology: developing techniques to reduce sea turtle bycatch in longline fisheries.* NOAA Technical Memorandum NMFS-PIFSC-7. 107 pp.

Tomas, J., Aznar, F. J., Raga, J., A. (2001). Feeding ecology of the loggerhead sea turtle Caretta caretta in the western Mediterranean. *Journal of Zoology* 255 : 525-532.

Tudela, S. (2000). Ecosystem effects of fishing in the Mediterranean: an analysis of the major threats of fishing gear and practices to biodiversity and marine habitats. *FAO project for the preparation of a Strategic Action Plan for the conservation of biological diversity (SAP BIO) in the Mediterranean Region.* Rome. 45 pp.

Tudela, S. (2004). Ecosystem effects of fishing in the Mediterranian: an analysis of the major threats of fishing gear and practices to biodiversity and marine habitats. *Studies and Reviews. General Fisheries Commission for the Mediterranean.* No. 74. Rome, FAO: 44p.

UE-FISH/2005/28-A. *Field study to assess some mitigation measures to reduce bycatch of marine turtles in surface longline fisheries.* EU project, Ref. No. FISH/2005/28A - Service Contract SI2.439703. 217 pp.

UNEP MAP RAC/SPA, (2007). *Action Plan for the conservation of Mediterranean marine turtles.* Ed. RAC/SPA, Tunis, 40pp.

Wang, J. H., McAlister, J., Fuxjager, M., Higgins, B., Lohmann, K. J. (2006). Light sticks used in longline fisheries attract juvenile loggerhead sea turtles: potential mitigation strategies. In: Frick M.., Panagopoulou A, Rees A, Williams K (eds) *Book of abstracts of the 26th Annual Symposium on Sea Turtle Biology and Conservation Island of Crete,* Greece, 3-8 April 2006.

Watson J., Foster D., Epperly S., Shah A. (2002). *Experiments in the Western Atlantic Northeast Distant Waters to Evaluate Sea Turtle Mitigation Measures in the Pelagic*

Longline Fishery. Report on Experiments Conducted in 2001. US National Marine Fisheries Service, Pascagoula, MS, USA.

Watson J. W., Hataway, B. D., Bergmann, C. E. (2003). *Effect of hook size on ingestion of hooks by loggerhead sea turtles.* Report of NOAA National Maritime Fisheries Service, Pascagoula, Miss., USA.

Watson, J. W., Foster DG, Epperly S., Shah, A. (2004). *Experiments in the western Atlantic Northeast Distant Waters to evaluate sea turtle mitigation measures in the pelagic longline fishery.* Report on experiments conducted in 2001. 135 pp.

Watson, J. W., Epperly, S. P., Shah, A. K., Foster, D. G. (2005). Fishing methods to reduce sea turtle mortality associated with pelagic longlines. *Can J Fish Aquat Sci* 62: 965-981.

INDEX

D

E

N

O

P

Q

R

S

T

U

V

W

Y

Z